Introduction to Bayesian Estimation and Copula Models of Dependence

Introduction to Bayesian Estimation and Copula Models of Dependence

Arkady Shemyakin
University of St. Thomas

Alexander Kniazev
Astrakhan State University

The right of Arkady Shemyakin and Alexander Kniazev to be identified as the author(s) of this work has been asserted in accordance with law.

Registered Offices
John Wiley & Sons, Inc., 111 River Street, Hoboken, NJ 07030, USA

Editorial Office
111 River Street, Hoboken, NJ 07030, USA

For details of our global editorial offices, customer services, and more information about Wiley products visit us at www.wiley.com

Wiley also publishes its books in a variety of electronic formats and by print-on-demand. Some content that appears in standard print versions of this book may not be available in other formats.

Library of Congress Cataloging-in-Publication Data:

Names: Shemyakin, Arkady. | Kniazev, Alexander (Mathematician)
Title: Introduction to Bayesian estimation and copula models of dependence /
    Arkady Shemyakin, Alexander Kniazev.
Description: Hoboken, New Jersey : John Wiley & Sons, Inc., [2017] | Includes index.
Identifiers: LCCN 2016042826 | ISBN 9781118959015 (cloth) | ISBN 9781118959022 (epub)
Subjects: LCSH: Bayesian statistical decision theory. | Copulas (Mathematical statistics)
Classification: LCC QA279.5 .S435 2017 | DDC 519.5/42—dc23
LC record available at https://lccn.loc.gov/2016042826

Cover image: hakkiarslan/Gettyimages
Cover design: Wiley

Set in 10/12pt WarnockPro by Aptara Inc., New Delhi, India

10 9 8 7 6 5 4 3 2 1

To all our families:
special thanks to
Evguenia Savinkina and
Ninel Kostenko

Contents

List of Figures

List of Tables

List of Tables

Acknowledgments

We express our sincere appreciation to all of our friends and colleagues who contributed to this book in many different ways. We are grateful for many discussions we had and many comments we received from Paul Alper, Susan Callaway, Oleg Lepekhin, Alex McNeil, Olga Demidova, and many others.

We deeply appreciate the assistance we have received with collecting supplementary materials and preparing files for the companion website from Laura Hanson, Kathryn McKee, Cheryl Heskin, Matthew Galloway, Shannon Currier, Natalie Vandeweghe, and Stephanie Fritz.

We are grateful to our collaborators on the research projects which became the foundation for the book: Heekyung Youn, Natalia Kangina, Alicia Johnson, Vadim Gordeev, Matthew Galloway, Ellen Klingner, Nicole Lenz, Nicole Lopez, Kelsie Sargent, and Katheryn Wifvat.

We want to thank all of those to whom the material in the book was presented over the years and who contributed with their thoughtful comments regarding both the form and the content of the book. Special thanks are due to Sarah Millholland, Valentin Artemiev, Yuri Chepasov, Anastasia Rozhkova, Ekaterina Savinkina, Alexander Zyrianov, Doug Swisher, Laura Fink, Eric Schlicht, and many other students and seminar participants.

We would like to acknowledge the staff of our schools, Astrakhan State University and the University of St. Thomas in Minnesota, for their support of our work and making this project possible.

We also would like to thank all the editorial staff of Wiley: Susanne Steitz-Filler, Sari Friedman, Amy Hendrickson, Divya Narayanan, and others who at various stages helped with the completion of the book.

We are also grateful to Ekaterina Kniazeva and Alexandra Savinkina for proof-reading the text and making suggestions which helped immensely in improving the style of the book and hopefully making it readable.

Finally, our deepest gratitude goes to all our families, without whose support and companionship the project would not be possible.

A.S. and A.K.

Acronyms

ABS	asset-backed security
AR	autoregressive
ARCH	autoregressive conditional heteroskedasticity
ARIMA	autoregressive integrated moving averages
BB1	Joe's BB1
BG	beta-geometric mixture model
BUGS	Bayes using Gibbs sampler (software)
CDO	collateralized debt obligation
CDS	credit default swap
CFTP	coupling from the past
CLT	the Central Limit Theorem
CMLE	canonical maximum likelihood estimation (estimates)
c.d.f.	cumulative distribution function
ENIAC	electronic numerical integrator and computer
FGM	Farlie–Gumbel–Morgenstern copulas
FTD	first-to-default (swaps)
GARCH	generalized autoregressive conditional heteroskedasticity
HAC	hierarchical Archimedean copula
HKC	hierarchical Kendall copula
IFM	inference from margins
i.i.d.	independent identically distributed
IMA	independent Metropolis algorithm
ISBA	International Society for Bayesian Analysis
JAGS	just another Gibbs sampler (software)
LLN	the Law of Large Numbers
MANIAC	mathematical analyzer, numerator, integrator, and computer
MCMC	Markov chain Monte Carlo
MHA	Metropolis–Hastings algorithm
MLE	maximum likelihood estimation (estimates)
MPLE	maximum pseudo-likelihood estimation (estimates)
PCC	pair copula construction

PD	probability of default
p.d.f	probability density function
RLUF	Rodriguez Lallena and Ubeda Flores copulas
RTO	regression through the origin
RWMHA	random walk Metropolis–Hastings algorithm
SACF	sample autocorrelation function
TTD	time-to-default
TTF	time-to-failure
WG	Weibull-gamma model
WGM	Weibull-gamma mixture model
WS	Weibull segmentation model

Glossary

$E(X)$	expected value (mean)
$Var(X)$	variance
$Bin(n,p)$	binomial distribution with n trials and success probability p
$Poiss(\lambda)$	Poisson distribution
$NB(p,r)$	negative binomial distribution
$N(\mu, \sigma^2)$	normal (Gaussian) distribution with mean μ and variance σ^2
$T(\eta)$	(Student) t-distribution with η degrees of freedom
$Gamma(z)$	gamma function
$Gamma(\alpha, \lambda)$	gamma distribution with shape α and rate λ
$Beta(\alpha, \beta)$	beta distribution
$Cov(X, Y)$	covariance
$\rho(X, Y)$	correlation
$MN(\mu, \Sigma)$	multivariate normal distribution
$L(x; \theta)$	likelihood function with data x and parameter θ
$\Lambda(x)$	likelihood ratio
$PG(1, r)$	polygamma function order r
$X \sim F$	random variable X has distribution F
$\propto$	proportional
$\preceq$	less or equal in sense of a special natural ordering

About the Companion Website

This book is accompanied by a companion website:

http://www.wiley.com/go/shemyakin/bayesian_estimation

The website includes:

- Solutions to selected exercises
- Excel dataset
- Excel simulation templates
- Appendices for Chapter 8
  - Datasets and results
  - Code in R

Introduction

Why does anyone need another book in Bayesian statistics? It seems that there already exist a lot of resources for those interested in the topic. There are many excellent books covering specific aspects of Bayesian analysis or providing a wide and comprehensive background of the entire field: Berger, Bernardo and Smith, Gamerman and Freitas Lopes, Gelman et al., Robert and Cassella, and many others. Most of these books, though, will assume a certain mathematical and statistical background and would rather fit a reader's profile of a graduate or advanced graduate level. Out of those aimed at a less sophisticated audiences, we would certainly recommend excellent books of William Bolstad, John Kruschke, and Peter Lee. There also exist some very good books on copulas: comprehensive coverage by Nelsen and Joe, and also more application-related Cherubini et al., Emrechts et al., and some others. However, instead of just referring to these works and returning to our extensive to-do lists, we decided to spend considerable amount of time and effort putting together another book—the book we presently offer to the reader.

The main reason for our endeavor is: we target a very specific audience, which as we believe is not sufficiently serviced yet with Bayesian literature. We communicate with members of this audience routinely in our day-to-day work, and we have not failed to register that just providing them with reading recommendations does not seem to satisfy their needs. Our perceived audience could be loosely divided into two groups. The first includes advanced undergraduate students of Statistics, who in all likelihood have already had some exposure to main probabilistic and statistical principles and concepts (most likely, in classical or "frequentist" setup), and may (as we probably all do) exhibit some traces of Bayesian philosophy as applicable to their everyday lives. But for them these two: statistical methods on one hand and Bayesian thinking on the other, belong to very different spheres and do not easily combine in their decision-making process.

The second group consists of practitioners of statistical methods, working in their everyday lives on specific problems requiring the use of advanced quantitative analysis. They may be aware of a Bayesian alternative to classical methods and find it vaguely attractive, but are not familiar enough with formal Bayesian analysis in order to put it to work. These practitioners populate analytical departments of banks, insurance companies, and other major businesses. In short, they might be involved in predictive modeling, quantitative forecasting, and statistical reporting which often directly call for Bayesian approach.

In the recent years, we have frequently encountered representatives of both groups described above as our collaborators, be it in undergraduate research or in applied consulting projects, or both at once (such things do happen). We have discovered a frequent need to provide to them a crash course in Bayesian methods: prior and posterior, Bayes estimation, prediction, MCMC, Bayesian regression and time series, Bayesian analysis of statistical dependence. From this environment we get the idea to concisely summarize the methodology we normally share with our collaborators in order to provide the framework for successful joint projects. Later on this idea transformed itself into a phantasy to write this book and hand it to these two segments of the audience as a potentially useful resource. This intention determines the content of the book and dictates the necessity to cover specific topics in specific order, trying also to avoid any unnecessary detail. That is why we do not include a serious introduction to probability and classical statistics (we believe that our audience has at least some formal knowledge of the main principles and facts in these fields). Instead, in Chapter 1 we just offer a review of the concepts we will eventually use. If this review happens to be insufficient to some readers, it will hopefully at least inspire them to hit the introductory books which will provide a more comprehensive coverage.

Chapter 2 deals with the basics of Bayesian statistics: prior information and experimental data, prior and posterior distributions, with emphasis on Bayesian parametric estimation, just barely touching Bayesian hypothesis testing. Some time is spend on addressing subjective versus objective Bayesian paradigms and brief discussion of noninformative priors. We spend just so much time with conjugate priors and analytical derivation of Bayes estimators that will give an idea of the scope and limitations of the analytical approach. It seems likely that most readers in their practical applications will require the use of MCMC—Markov chain Monte Carlo method—the most efficient tool in the hands of modern Bayesians. Therefore, Chapter 3 contains the basic mathematical background on both Markov chains and Monte Carlo integration and simulation. In our opinion, successful use of Markov chain Monte Carlo methods is heavily based on good understanding on these two components. Speaking of Monte Carlo methods, the central idea of variance reduction nicely transitions us to MCMC and its diagnostics. Equally important, both Markov chains and Monte Carlo methods have numerous important applications outside of Bayesian setting, and these applications will be discussed as examples.

Chapter 4 covers MCMC *per se*. It may look suspicious from traditional point of view that we do not particularly emphasize Gibbs sampling, rather deciding to dwell on Metropolis–Hastings algorithm in its two basic versions: independent Metropolis and random walk Metropolis-Hastings. In our opinion, this approach allows us to minimize the theoretical exposure and get close to the point using simple examples. Also, in more advanced examples at the end of the book, Gibbs sampling will rarely work without Metropolis. Another disclosure we have to make: there exists a huge library of MCMC computer programs, including those available online as freeware. All necessary references are given in the text, including OpenBUGs and several R packages. However, we also introduce some very rudimentary computer code which allows the readers to "get inside" the algorithms. This expresses the authors' firm belief that do-it-yourself is often the best way if not to actually apply statistical computing, but at least to learn how to use it.

This might be a good time to explain the authors' attitude to the use of computer software while reading the book. Clearly, working on the exercises, many readers would find it handy to use some computing tools, and many readers would like to use the software of their choice (be it SPSS, Matlab, Stata, or any other packages). We try to structure our exercises and text examples in a way that makes it as easy as possible. What we offer from our perspective, in addition to this possibility, is a number of illustrations containing code and/or outputs in Microsoft Excel, Mathematica, and R. Our choice of Mathematica for providing graphics is simply explained by the background of the audience of the short courses where the material of the book has been approbated. We found it hard to refuse to treat ourselves and the readers to nice Mathematica graphical tools. R might be the software of choice for modern statisticians. Therefore we find its use along with this book both handy and inevitable. We can take only limited responsibility for the readers' introduction to R, restricting ourselves to specific uses of this language to accompany the book. There exist a number of excellent R tutorials, both in print and online, and all necessary references are provided. We can especially recommend the R book by Crawley, and a book by Robert and Casella.

The first four chapters form the first part of the book which can be used as a general introduction to Bayesian statistics with a clear emphasis on the parametric estimation. Now we need to explain what is included in the remaining chapters, and what the link is between the book's two parts. Our world is a complicated one. Due to the recent progress of communication tools and the globalization of the world economy and information space, we humans are less and less like separate universes leading our independent lives (was it ever entirely true?) Our physical lives, our economical existences, our information fields become more and more interrelated. Many processes successfully modeled in the past by probability and statistical methods assuming independent behavior of the components, become more and more intertwined. Brownian motion is still around, as well as Newtonian mechanics is, but it often fails to serve as a good model for many complicated systems with component interactions. This explains an increased interest to model statistical dependence: be it dependence of physical lives in demography and biology, dependence of financial markets, or dependence between the components of a complex engineering system. Out of the many models of statistical dependence, copulas play a special role. They provide an attractive alternative to such traditional tools as correlation analysis or Cox's proportional hazards. The key factor in the popularity of copulas in applications to risk management is the way they model entire joint distribution function and are not limited to its moments. This allows for the treatment of nonlinear dependence including joint tail dependence and going far beyond the standard analysis of correlation. The limitations of more traditional correlation-based approaches to modeling risks were felt around the world during the last financial crisis.

Chapter 5 is dedicated to the brief survey of pre-copula dependence models, providing necessary background for Chapter 6, where the main definitions and notations of copula models are summarized. Special attention is dedicated to a somewhat controversial problem of model selection. Here, due to a wide variety of points of view expressed in the modern literature, the authors have to narrow down the survey in sync with their (maybe subjective) preferences.

Two types of copulas most popular in applications: Gaussian copulas and copulas from Archimedean family (Clayton, Frank, Gumbel–Hougaard, and some others) are introduced and compared from model selection standpoint in Chapter 7. Suggested principles of model selection have to be illustrated by more than just short examples. This explains the emergence of the last two sections of Chapters 7 and 8, which contain some cases dealing with particular risk management problems. The choice of the cases has to do with the authors' recent research and consulting experience. The purpose of this chapter is to provide the readers with an opportunity to follow the procedures of multivariate data analysis and copula modeling step-by-step enabling them to use these cases as either templates or insights for their own applied research studies. The authors do not take on the ambitious goal to review the state-of-the-art Bayesian statistics or copula models of dependence. The emphasis is clearly made on applications of Bayesian analysis to copula modeling, which are still remarkably rare due to the factors discussed above as well as possibly to some other reasons unknown to the authors. The main focus of the book is on equipping the readers with the tools allowing them to implement the procedures of Bayesian estimation in copula models of dependence. These procedures seem to provide a path (one of many) into the statistics of the near future. The omens which are hard to miss: copulas found their way into Basel Accord II documents regulating the world banking system, Bayesian methods are mentioned in recent FDA recommendations.

The material of the book was presented in various combinations as the content of special topic courses at the both schools where the authors teach: the University of St. Thomas in Minnesota, USA, and Astrakhan State University in Astrakhan, Russia. Additionally, parts of this material have been presented as topic modules in graduate programs at MCFAM—the Minnesota Center for Financial and Actuarial Mathematics at the University of Minnesota and short courses at the Research University High School of Economics in Moscow, Russia, and U.S. Bank in Minneapolis, MN.

We can recommend this book as a text for a full one-semester course for advanced undergraduates with some background in probability and statistics. Part I (Chapters 1–4) can be used separately as an introduction to Bayesian statistics, while Part II (Chapters 5–8) can be used separately as an introduction to copula modeling for students with some prior knowledge of Bayesian statistics. We can also suggest it to accompany topics courses for students in a wide range of graduate programs. Each chapter is equipped with a separate reference list and Chapters 2–8 by their own sets of end-of-the-chapter exercises. The companion website contains Appendices: data files and demo files in Microsoft Excel, some simple code in R, and selected exercise solutions.

Part I

Bayesian Estimation

1

Random Variables and Distributions

Chapter 1 is by no means suggested to replace or replicate a standard course in probability. Its purpose is to provide a reference source and remind the readers what topics they might need to review. For a systematic review of probability and introduction to statistics we can recommend excellent texts by DeGroot and Schervish [4], Miller and Miller [10], and Rice [12]. In-depth coverage of probability distributions in the context of loss models is offered by Klugman et al. in [8]. If the reader is interested in a review with a comprehensive software guide, we can recommend Crawley's handbook in R [3].

Here we will introduce the main concepts and notations used throughout the book. The emphasis is made on the simplicity of explanations, and often in order to avoid technical details we have to sacrifice mathematical rigor and conciseness. We will also introduce a library of distributions for further illus-trations. Without a detailed reference to the main facts of probability theory, we need to however emphasize the role played in the sequel by the concept of conditional probability, which becomes our starting point.

1.1 Conditional Probability

Let A and B be two random events, which could be represented as two sub-sets of the same *sample space* S including all possible outcomes of a chance experiment: $A \subseteq S$ and $B \subseteq S$. *Conditional probability of B given A* measures the chances of B to happen if A is already known to occur. It can be defined for events A and B such that $P(A) > 0$ as

$$P(B \mid A) = \frac{P(A \cap B)}{P(A)}, \tag{1.1}$$

where $P(A \cap B) = P(B \cap A)$ is the probability of intersection of A and B, the event indicating that both A and B occur. This conditional probability should

Introduction to Bayesian Estimation and Copula Models of Dependence, First Edition.
Arkady Shemyakin and Alexander Kniazev.
© 2017 John Wiley & Sons, Inc. Published 2017 by John Wiley & Sons, Inc.
Companion Website: http://www.wiley.com/go/shemyakin/bayesian_estimation

not be confused with the conditional probability of A given B defined as

$$P(A \mid B) = \frac{P(A \cap B)}{P(B)}, \tag{1.2}$$

which shares the same numerator with $P(B \mid A)$, but has a different denominator.

The source of possible confusion is a different choice of the "sample space" or "reference population"—whatever language one prefers to use—in (1.1) and (1.2) corresponding to the denominators in these formulas. In case of (1.1) we consider only such cases that A occurs, so that the sample space or reference population is reduced from S to A, while in (1.2) it changes from S to B.

To illustrate this distinction, we will use a simple example. It fits the purpose of the book, using many illustrations from the fields of insurance and risk management, to begin with an example related to insurance.

In a fictitious country of Endolacia, people drive cars and buy insurance against accidents. Accidents do happen on a regular basis, but not too often. During the last year, which provides the reference timeframe, 1000 accidents were recorded in the entire country. In all but one case the driver of car in an accident was a human being. In one particular case it was verified to be a dog. It probably was a specially trained dog as soon as it was trusted the steering wheel. The question we want to ask is: based on our data, are dogs safe drivers? The answer to this question will be central if we consider a possibility of underwriting an insurance policy to a dog-driver. Considering events A (an accident happened) and B (the driver was a dog), we can estimate the probability of a dog being the driver in case of an accident as

$$P(B \mid A) = \frac{P(A \cap B)}{P(A)} = \frac{1}{1000}.$$

But what exactly does this probability measure? It can be used to properly measure the share of responsibility dog-drivers carry for car accidents in Endolacia, which is indeed rather small, because 1 is a relatively small fraction of a 1000. However, the key question: "are dogs safe drivers?" is not addressed by this calculation. We can even suggest that based on the above information we do not have sufficient data to address this question. What piece of data is missing?

In order to evaluate the safety of dog-drivers, we need to estimate a different conditional probability: $P(A \mid B)$, which determines the probability of an accident for a dog-driver. In order to do it, we need to estimate $P(B)$, which is the probability that a random car in Endolacia at a random time (in accident or not) happens to be driven by a dog. It requires some knowledge of the size of the population of drivers in Endolacia $n(S)$ and the size of the population of dog-drivers $n(B)$ so that $P(B)$ can be estimated as $n(B)/n(S)$.

Let us say that there were 1,000,000 drivers on the roads of Endolacia last year, and only one dog-driver (the one who happened to get into an accident). Then we can estimate

$$P(A \mid B) = \frac{1/1,000,000}{1/1,000,000} = 1,$$

which is much higher than $1/1000$ from the previous calculation and is the probability which should be used as the risk factor or the risk rating of a dog-driver. Looking at this number, we would not want to insure a dog, since it is a very risky business operation.

Correct understanding of conditional probability is a key to understanding in what ways Bayesian statistics is different from classical or frequentist statistics. The details follow in Chapter 2 and further chapters of Part I of the book. It also provides a key to understanding the underlying principles of construction of models of statistical dependence discussed in Chapter 5 and further chapters of Part II.

1.2 Discrete Random Variables

Random variable is a variable that takes on a certain value which becomes known as the result of an experiment, but is not known in advance. For example, an insurance company offers 1000 contracts during a year. Insurance events will happen and claims will be filed, but the number and amount of such claims is unknown before the end of the year. Thus, the number of claims is a random variable, and the total amount of claims is a different though related variable. Another example of random variable is return on investment. If we buy a share of stock and plan to sell it in a month, we can make gain or suffer a loss. The return on such an investment transaction is not known now, though it will be known in 1 month. Let us suppose someone has started a new business. How long will it stay in the market? "The life time" of the business is unknown beforehand, it is a random variable.

A random variable is defined primarily by the set of its possible values: outcomes of a chance experiment. If this set consists of isolated points, the random variable is called ***discrete***. For example, the number of insurance claims in a year is a discrete random variable, while the return on investment is not. We will denote all possible values of a discrete random variable X as $x_i, i = 1, 2, \ldots$. Throughout the text we will try to reserve capital letters to denote random variables and lowercase letters will correspond to their specific numerical values. To define a random variable, along with the set of its values, one needs to define the probabilities of these values. The set of all possible values of a discrete random variable and their respective probabilities is called the ***probability distribution*** of a discrete random variable. The sum of probabilities of all the possible values equals to one. If the number of values is finite, distribution can be described as a finite table:

X	x_1	x_2	...	x_n
P	p_1	p_2	...	p_n

Expected value or **mean** of a discrete random variable X is defined as

$$E(X) = \mu_x = \sum_{k=1}^{n} x_k p_k. \tag{1.3}$$

The upper limit of this sum can be set at infinity. If the number of values is infinite and the corresponding infinite series converges, then expected value exists. Expected value defines the average position of values of a discrete random variable on the numeric line or their central tendency.

Variance of a discrete random variable X is defined as

$$Var(X) = \sigma_x^2 = \sum_{k=1}^{n} (x_k - E(X))^2 p_k. \tag{1.4}$$

Variance describes the spread in values of a random variable around the average. The square root of variance $\sigma(X) = \sqrt{Var(X)}$ is known as **standard deviation**.

The simplest binary random variables are known to have **Bernoulli distribution**, which allows for only two possible outcomes: success with probability p and failure with probability $1 - p$. The table of values for this distribution can be reduced to one formula:

$$P(X = k) = p^k(1 - p)^{1-k}, \quad k = 0, 1.$$

In this formula probability of success p is the only **parameter** of Bernoulli distribution, which can take any value between 0 and 1. **Binomial distribution** describes a random variable Y, the number of successes in n independent experiments, where in each experiment only two results (success and failure) are possible. Such a distribution is described by the following law of probability distribution

$$P(Y = k) = \binom{n}{k} p^k (1 - p)^{n-k}, \quad k = 0, 1, \dots, n, \tag{1.5}$$

where p is the probability of success in any experiment, and $1 - p$ is the probability of failure. Binomial variable can be defined as the sum of n identical Bernoulli variables associated with independent experiments. Here and further on $\binom{n}{k} = \frac{n!}{(n-k)!k!}$. We will say that $Y \sim Bin(n, p)$, if Y is a binomial variable with parameters $p \in [0, 1]$ and integer $n \geq 1$.

Let us assume that a card is drawn out of a deck with 52 cards, and the suit of the card is recorded. After that the card is returned to the deck and the deck is shuffled. Let us suppose, four cards were drawn consecutively using this procedure (drawings with replacement). Let us consider random number X of spades among those four cards. The set of possible values of this random variable consists of numbers: 0; 1; 2; 3; 4. It is obvious that the probability of drawing a spade is the same for every draw and equals $\frac{1}{4}$. The distribution of the random variable X is a binomial distribution. Let us write down the distribution table for this random variable.

X	0	1	2	3	4
P	81/256	27/64	27/128	3/64	1/256

It is easy to calculate that the expected value of this variable equals to 1 and its variance equals to $\frac{3}{4}$. Things get more interesting and more complicated when drawings are performed without replacement, but we will not formally discuss this situation.

A **random flow of events** can be represented as a sequence of events occurring at random moments of time. Let us assume that the probability of k events within a time interval $(s, s + t)$ does not depend on the starting time of the interval s, but only on the length of the interval t (stationary flow). Assume also that the number of events occurring within the interval $(s, s + t)$ does not depend on what had happened before time s, or is going to happen after time $s + t$ (flow with independent increments). We will also assume that two or more events cannot happen at the same time or at an infinitely small interval from each other (condition of ordinary flow). Let us define the counting variable X_t as the number of events within a stationary ordinary flow with independent increments happening within a time interval length t. The law of probability distribution of this random variable is defined for any nonnegative integer k by the formula

$$P(X_t = k) = \frac{(\lambda t)^k e^{-\lambda t}}{k!}, \quad k = 0, 1, 2, \dots . \tag{1.6}$$

This distribution is called **Poisson distribution**, $X \sim Poiss(\lambda)$, where parameter λ known as the *intensity* of the flow is the average number of events which occur in a unit interval $t = 1$.

Let us assume that the average number of cars passing by a certain marker on the highway in 1 minute, equals to 5. If we take observations during any short time interval, then the traffic stream of cars passing by the marker may be considered a Poisson flow. Let us calculate the probability of 8 cars passing by the marker in 2 minutes. Having applied formula (1.6) with $\lambda = 5$, we get 0.1126.

Let the trials with binary outcomes (success with probability p or failure with probability $1 - p$) be performed until the first success (or failure). The distribution of the number of trials X needed to achieve the first success is defined by the formula

$$P(X = k) = (1 - p)^{k-1}p, \quad k = 1, 2, \ldots. \tag{1.7}$$

This distribution is known as the **geometric distribution** with parameter p, $X \sim Geom(p)$.

A natural generalization of the geometric distribution is the **negative binomial distribution**, which considers the number of trials Y needed to achieve r successes. The negative binomial distribution has two parameters: $Y \sim NB(p, r)$ if

$$P(Y = k) = \binom{k-1}{r-1}(1 - p)^{k-r}p^r, k = r, r + 1, \ldots. \tag{1.8}$$

1.3   Continuous Distributions on the Real Line

In this section we shall consider random variables whose set of possible values is the entire real line $(-\infty, \infty)$. It means that such a random variable may take on both positive and negative values. Return on investment or gain/loss in a business transaction could serve as examples.

The **cumulative distribution function** (commonly abbreviated as c.d.f.) of a random variable X is defined as $F(x) = P(X \leq x)$. It is obvious that the function $F(x)$ is nondecreasing and may assume values on the bounded interval $[0, 1]$. It is also evident that $P(a < X \leq b) = F(b) - F(a)$.

If the cumulative distribution function has a derivative $f(x) = F'(x)$, then this derivative is called the **distribution density function**, also known as probability density function, abbreviated in the sequel to p.d.f., or simply "density." If the density exists, the random variable is said to have a **continuous distribution** or just a "continuous random variable." It is evident that any density is a nonnegative function and $P(a \leq X \leq b) = \int_a^b f(x)dx$.

The expected value or **mean** of a random variable X with a continuous distribution is defined as the integral (if it converges)

$$E(X) = \mu = \int_{-\infty}^{\infty} xf(x)dx. \tag{1.9}$$

The **variance** of a continuous random variable X

$$Var(X) = \sigma^2 = \int_{-\infty}^{\infty} (x - E(X))^2 f(x)dx. \tag{1.10}$$

A distribution is called symmetric or **not skewed** if the third central moment equals to zero, that is

$$\int_{-\infty}^{\infty} (x - E(X))^3 f(x) dx = 0. \tag{1.11}$$

The distribution defined by the following density is known as the **normal distribution**

$$f(x) = \frac{1}{\sigma\sqrt{2\pi}} e^{-\frac{(x-\mu)^2}{2\sigma^2}}. \tag{1.12}$$

The normal distribution depends on two parameters allowing for a very easy interpretation: mean μ and variance σ^2. The normal variable $X \sim N(\mu, \sigma^2)$ is not skewed, and its density can be depicted as a bell-shaped Gaussian curve, which is symmetric with respect to the vertical line $x = \mu$, has a peak (mode) at μ, and two inflection points at $\mu - \sigma$ and $\mu + \sigma$.

Figure 1.1 shows the Gaussian curve for $\mu = 3$ and $\sigma = 1$. The dependence of the curve on the value of σ is demonstrated in Figure 1.2, where for all three curves $\mu = 0$. The dashed line corresponds to $\sigma = 0.5$, the solid line to $\sigma = 1$, and the dotted line to $\sigma = 2$. The solid line represents the special case known as *standard normal distribution* $Z \sim N(0, 1)$: $\mu = 0, \sigma = 1$.

Figure 1.3 shows the histogram of daily percentage returns of Australian stock exchange index (ASX). The Gaussian curve is laid over the histogram, with sample mean taken for μ and sample standard deviation for σ. We can observe a certain similarity of shapes of the histogram and the Gaussian curve. Why?

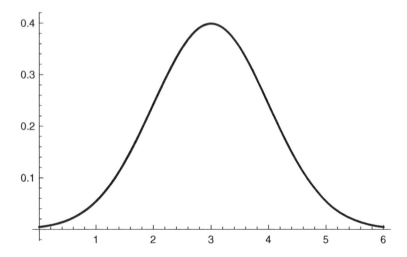

Figure 1.1 Gaussian curve with $\mu = 3$, $\sigma = 1$.

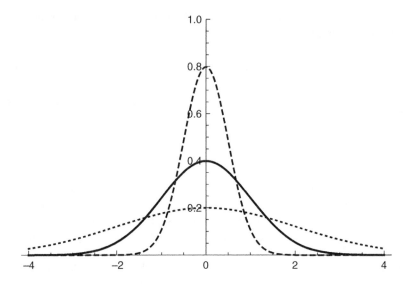

Figure 1.2 Gaussian curves for different values of σ.

The Central Limit Theorem (CLT) is a famous theorem of probability theory stating that for a large number n of independent variables $X_1, \ldots, X_n$ identically distributed with $E(X) = \mu$ and $Var(X) = \sigma^2$, their sum $S_n = \sum_{i=1}^{n} X_i$ is approximately normal, and the distribution of the random variable $Z_n = \frac{S_n - n\mu}{\sqrt{n}\sigma}$

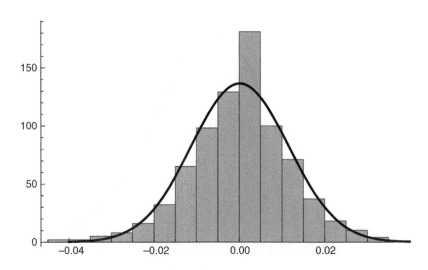

Figure 1.3 Australian stock index ASX.

converges to standard normal when $n \to \infty$. The assumptions of the CLT can be weakened to include weekly dependent and closely but not identically distributed random variables.

This fact explains the observation we made regarding Figure 1.3: when such a variable as daily percentage return of ASX is the result of many relatively small factors, weekly dependent and closely distributed, the distribution of such variable can be successfully approximated by the normal distribution. However, if we look at the graph more attentively, we see that the fit is not ideal, with the central part and the tails of the histogram sticking above the Gaussian curve. This behavior indicates a possible violation of some assumptions of the CLT.

Let $X_0, X_1, \ldots, X_\eta$ denote independent random variables identically normally distributed with zero mean. Consider the random variable

$$t_\eta = \frac{X_0}{\sqrt{\frac{1}{\eta} \sum_{k=1}^{\eta} X_k^2}}. \tag{1.13}$$

This random variable has a **t-distribution** (Student distribution) with η degrees of freedom, $X \sim T(\eta)$. The density of this distribution is defined by the formula

$$f(x) = \frac{1}{\sqrt{\pi \eta}} \cdot \frac{\Gamma\left(\frac{\eta+1}{2}\right)}{\Gamma\left(\frac{\eta}{2}\right)} \cdot \left(1 + \frac{x^2}{\eta}\right)^{-\frac{(\eta+1)}{2}}, \tag{1.14}$$

where $\Gamma(z)$ is the gamma function. The special case of Student distribution with $\eta = 1$ is also known as Cauchy distribution. Let us consider the generalization of this distribution for a random real number $\eta \geq 2$. In this case η is also known as the **tail parameter**. The shape of the density of t-distribution resembles the Gaussian curve, however the t-distribution exhibits heavier tails, allowing for a higher probability of large deviations. When the parameter η increases, t-distribution converges to normal quite quickly so that it could be approximated by normal distribution reasonably well even for $\eta \geq 30$.

Figure 1.4 represents three t-distributions for $\eta = 1$ (the solid graph, Cauchy distribution), for $\eta = 3$ (the dotted graph), and for $\eta = 10$ (the dashed graph). The larger the value of η, the closer the graphs resemble the standard Gaussian curve.

Student t-distribution traditionally plays an important role in statistics, since it is widely used in construction of confidence intervals and testing statistical hypotheses for normal samples. It has recently become more popular in applications for a different reason: It provides a good model for symmetric heavy-tailed variables, when the assumptions of the central limit theorem do not hold.

Distributions that simultaneously demonstrate heavy tails and asymmetry (skewness) have recently been demanded by applications, especially in statistical analysis of financial data. There exist several generalizations of

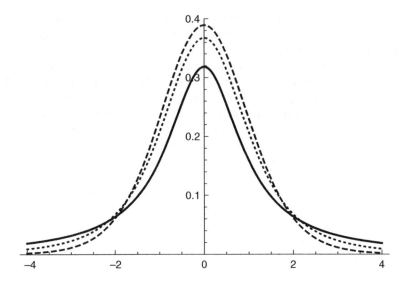

Figure 1.4 Density of *t*-distribution.

t-distribution, which allow us to introduce asymmetry. In the sequel in Chapter 8 we use the modification offered in [7]. The density function of this distribution has the form of

$$d(z; \lambda, \eta) = bc\left(1 + \frac{1}{\eta - 2}\left(\frac{bz + a}{1 + \lambda \, \text{sign}(bz + a)}\right)^2\right)^{-\frac{\eta+1}{2}}, \qquad (1.15)$$

where $a = 4c\lambda\left(\frac{\eta-2}{\eta-1}\right)$, $b = \sqrt{1 + 3\lambda^2 - a^2}$, $c = \frac{\Gamma\left(\frac{\eta+1}{2}\right)}{\sqrt{\pi(\eta-2)}\Gamma\left(\frac{\eta}{2}\right)}$, η is the number of degrees of freedom (the tail parameter), λ is the **skewness parameter**, $\Gamma(x)$ is the gamma function. The tail parameter can take on any real values, such that $\eta > 2$. The skewness parameter can take on values from the interval $(-1, 1)$.

Figure 1.5 presents the graphs of three asymmetric t-distributions for $\eta = 3$, $\lambda = -0.5$ (the solid line), $\eta = 10$, $\lambda = -0.5$ (the dashed line), and $\eta = 3$, $\lambda = 0.5$ (the dotted line).

Figure 1.6 shows the histogram of the daily percentage returns of the Mexican stock exchange controlled for trend and volatility. We can observe that this data exhibits heavy tails and a noticeable asymmetry. The left tail looks heavier than the right one. We assume that this distribution can be modeled by asymmetric *t*-distribution. The parameters of this distribution were estimated from the data.

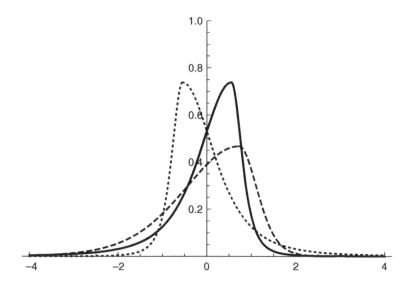

Figure 1.5 Density of asymmetric *t*-distribution.

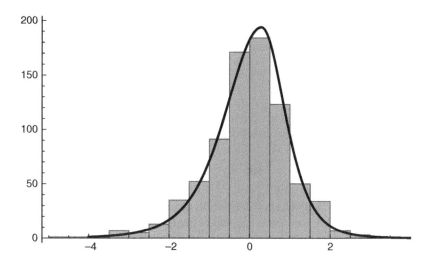

Figure 1.6 Mexican stock index IPC.

1.4 Continuous Distributions with Nonnegative Values

Let us now consider continuous asymmetric distributions with zero density for negative values of x. Such variables often arise in applications to insurance, reliability, or general survival analysis, where X measures the length of life or time to failure of a specific object. Such distributions are conveniently characterized not only by their density $f(x)$ or its c.d.f. $F(x)$, but also by two other functions: ***survival function***

$$S(x) = 1 - F(x) = \int_x^\infty f(t)dt$$

and ***hazard rate***, which is also known in insurance as the force of mortality:

$$h(x) = \frac{f(x)}{S(x)} = -\frac{d}{dx} \ln S(x).$$

While the c.d.f. $F(x)$ describes the chances of our object of interest to fail at or before time x, and density function $f(x)$ measures the unconditional instantaneous failure rate (chances to fail exactly at time x or very close to this time), the survival function $S(x)$ describes the chances to survive to time x, and the hazard rate is the instantaneous failure rate at time x conditioned to survival.

The simplest distribution of this type is ***exponential***, characterized by

$$F(x) = 1 - e^{-\lambda x}, f(x) = \lambda e^{-\lambda x}, \ S(x) = e^{-\lambda x}, h(x) \equiv \lambda, x \ge 0, \lambda > 0. \qquad (1.16)$$

We can suggest a nice relationship between the exponential and Poisson distributions. If we consider a Poisson flow (1.6) with intensity λ, the intervals between adjacent events in the flow are distributed exponentially with parameter λ. There exist simple formulas for the mean and the variance of the exponential distribution:

$$E(x) = \frac{1}{\lambda}, \ Var(X) = \frac{1}{\lambda^2}. \qquad (1.17)$$

There exists a different parameterization of the exponential distribution creating certain confusion in probability textbooks. If we denote $\theta = 1/\lambda$, we can rewrite the formulas (1.16) as

$$F(x) = 1 - e^{-\frac{x}{\theta}}, f(x) = \frac{1}{\theta} e^{-\frac{x}{\theta}}, \ S(x) = e^{-\frac{x}{\theta}}, \ h(x) \equiv \frac{1}{\theta}, \ x \ge 0, \theta > 0.$$

For this parameterization the distribution mean $\theta = E(X)$ becomes the natural ***scale*** parameter. However, (1.16) will be more convenient and we will stick to it unless otherwise specified. Notice that the hazard rate of exponential distribution is determined by the only parameter and does not depend on time x. That would correspond to the failure rate being independent of physical age, history, or experience. It makes it a poor model for biological lives, where chances of a life to fail often increase with age (increasing hazard rate), or models with a positive role of experience, where chances to fail decrease with time (decreasing hazard rate). In order to address the possibilities of the hazard rate to change with time, one can consider two alternative generalizations of the exponential distribution which will play an important role in the further chapters of the book.

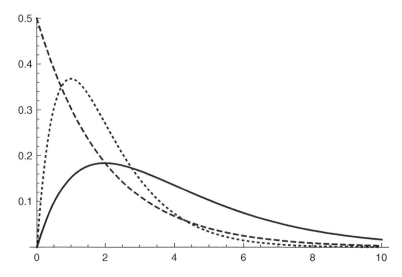

Figure 1.7 Gamma density graphs.

Gamma Distribution

Let us take a look at the distribution defined by such a density function $f(x)$ that when $x < 0, f(x) = 0$ and when $x \geq 0$,

$$f(x) = \frac{\lambda^\alpha}{\Gamma(\alpha)} x^{\alpha-1} e^{-\lambda x}, \alpha > 0, \lambda > 0. \tag{1.18}$$

This distribution is known as the ***gamma distribution***, it depends on two parameters: the ***shape*** α and the ***rate*** or inverse scale λ. Gamma family *Gamma*(α, λ) is known for many nice properties, and describes a very wide class of distributions.

Figure 1.7 represents three gamma density graphs for $\alpha = 2$, $\lambda = 2$ (the solid line), $\alpha = 1$, $\lambda = 2$ (the dashed line), and $\alpha = 2$, $\lambda = 1$ (the dotted line). When $\alpha = 1$, we obtain exponential distribution. Unfortunately, the closed form expressions for c.d.f., survival function, and hazard rate involve untractable integrals containing expressions for the Gamma function. However, there exist convenient expressions for the moments (mean and variance):

$$E(X) = \frac{\alpha}{\lambda}, \; Var(X) = \frac{\alpha}{\lambda^2}. \tag{1.19}$$

Similar to the exponential distribution, many authors introduce an alternative parameterization involving a natural *scale* parameter $\theta = 1/\lambda$, in which case $E(x) = \alpha\theta$ and $Var(X) = \alpha\theta^2$. However, we will prefer (1.18) unless otherwise specified.

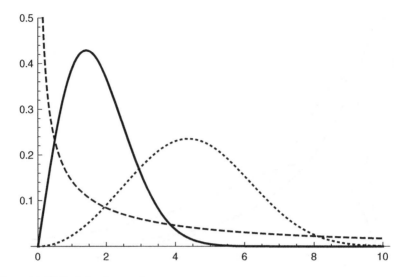

Figure 1.8 Weibull density graphs.

It is also worth noticing that if we define n independent exponential variables X_i with the same parameter λ, then $Y = \sum_{i=1}^{n} X_i$ will have gamma distribution $Y \sim Gamma(n, \lambda)$.

The Weibull Distribution

Here is another distribution often used for lifetime modeling. The density of this distribution for nonnegative values of x is defined by the formula

$$f(x) = \lambda \tau x^{\tau-1} e^{-\lambda x^{\tau}}. \tag{1.20}$$

The density equals to 0 for all negative values of x. Shape parameters τ and λ are both positive. Evidently, the particular case of $\tau = 1$ corresponds to the exponential distribution with parameter λ.

Figure 1.8 represents three Weibull distribution density graphs for $\lambda = 2$ and $\tau = 2$ (the solid line); $\lambda = 5$, $\tau = 0.5$ (the dashed line); $\lambda = 5$, $\tau = 3$ (the dotted line).

It is possible to conveniently represent the c.d.f., survival function, and the hazard rate for the Weibull distribution as:

$$F(x) = 1 - e^{-\lambda x^{\tau}}, S(x) = e^{-\lambda x^{\tau}}, h(x) = \tau \lambda x^{\tau-1},$$

though formulas for the moments are more sophisticated. For instance,

$$E(X) = \lambda^{-1/\tau} \Gamma \left(1 + \frac{1}{\tau}\right).$$

We should also mention an alternative parameterization involving the natural scale parameter $\theta = \lambda^{-1/\tau}$. In the sequel we will often prefer expression (1.20). One of the reasons for that is the special role of parameter λ as a factor in the expression for the hazard function $h(x)$: a linear change in lambda brings about a proportional change in the hazard rate. However, in some instances we will also use the parameterization with shape and natural scale, which is especially convenient for the survival function:

$$S(x) = \exp(-(x/\theta)^{\tau}). \tag{1.21}$$

Other Distributions

By $X_1, \ldots, X_\nu$ denote independent normally distributed continuous random variables with mean 0 and variance 1. Let us consider a random variable $\sum_{k=1}^{\nu} X_k^2$. The distribution of this random variable is called the χ^2 **(chi-square) distribution** with ν degrees of freedom. When $x < 0$, the density of this distribution equals to 0; and when $x \geq 0$ it is defined by the formula

$$f(x) = \frac{x^{\frac{\nu}{2}-1} e^{-\frac{x}{2}}}{2^{\frac{\nu}{2}} \Gamma\left(\frac{\nu}{2}\right)}. \tag{1.22}$$

It is easy to prove that this distribution is a special case of the gamma distribution with $\alpha = \frac{\nu}{2}$, $\lambda = \frac{1}{2}$. As the number of degrees of freedom increases, the χ^2 distribution approaches the normal distribution.

Figure 1.9 represents three χ^2 distributions for $\nu = 1$ (the dashed line), $\nu = 4$ (the solid line), $\nu = 10$ (the dotted line). The χ^2 distribution plays an important role in statistics.

Let $X_1, \ldots, X_{\nu_1}$, $Y_1, \ldots, Y_{\nu_2}$ denote independent identically normally distributed random variables with zero mean. Let us take a look at the random variable

$$X = \frac{\frac{1}{\nu_1} \sum_{k=1}^{\nu_1} X_k^2}{\frac{1}{\nu_2} \sum_{k=1}^{\nu_2} Y_k^2}. \tag{1.23}$$

This random variable has **F-distribution** (or Fisher–Snedecor distribution). The density of this distribution equals to 0 when $x < 0$; and when $x \geq 0$, it is described by the formula

$$f(x) = \frac{\Gamma\left(\frac{\nu_1+\nu_2}{2}\right) \cdot \nu_1^{\frac{\nu_1}{2}} \cdot \nu_2^{\frac{\nu_2}{2}} \cdot x^{\frac{\nu_1}{2}-1}}{\Gamma\left(\frac{\nu_1}{2}\right) \Gamma\left(\frac{\nu_2}{2}\right) \cdot (\nu_1 x + \nu_2)^{\frac{\nu_1+\nu_2}{2}}}. \tag{1.24}$$

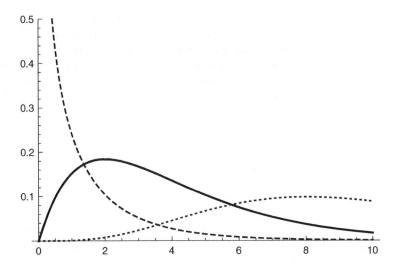

Figure 1.9 Densities of χ^2 distribution.

Figure 1.10 represents three F-distribution graphs for $v_1 = 10$, $v_2 = 10$ (the solid line); $v_1 = 10$, $v_2 = 3$ (the dashed line); $v_1 = 3$, $v_2 = 10$ (the dotted line). The Fisher–Snedecor distribution also plays an important role in statistics.

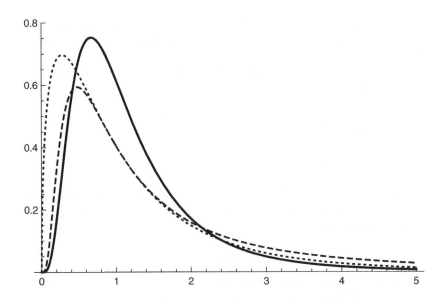

Figure 1.10 Densities of F-distribution.

1.5   Continuous Distributions on a Bounded Interval

In the previous sections we described distributions with density function which takes on nonzero values on an unbounded set. For example, the density of the normal distribution is nonzero on the entire real line and the density of the gamma distribution is nonzero on the positive semiaxis. In applications one also may encounter continuous random variables taking their values on a bounded interval. The following functions are used to model such distributions.

A random variable X is called **uniform** or **uniformly distributed** on a bounded interval $[a, b]$, $X \sim Unif(a, b)$, if the probability of this random variable to fall onto any subinterval inside $[a, b]$ depends only on the length of this subinterval. The density of such a distribution equals to zero outside the interval $[a, b]$ and when $x \in [a, b]$, it is defined by the formula $f(x) = \frac{1}{b-a}$. Let us notice that for any random variable X with a distribution function $F(x)$, the random variable $U = F(X)$ is uniformly distributed on the bounded interval $[0, 1]$.

Beta distribution is a more general type of distribution on a bounded interval. Beta family of distributions contains uniform distribution as a special case. The density of the beta distribution equals to 0 outside the bounded interval $[0,1]$ and when $x \in [0, 1]$, it takes the form of

$$f(x) = \frac{\Gamma(\alpha + \beta)}{\Gamma(\alpha)\Gamma(\beta)} x^{\alpha-1}(1 - x)^{\beta-1}. \tag{1.25}$$

Figure 1.11 shows three density graphs for the beta distribution for $\alpha = 2$, $\beta = 4$ (the solid line); $\alpha = 4$, $\beta = 2$ (the dashed line); $\alpha = 0.5$, $\beta = 0.5$ (the dotted

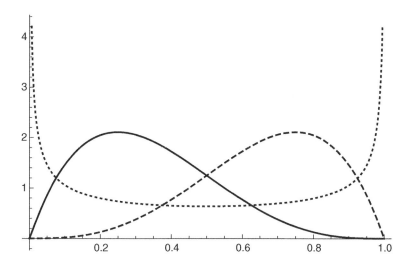

Figure 1.11 Beta distribution densities.

line). Random variables from beta class $X \sim Beta(\alpha, \beta)$ arise in many natural applications, for instance, when we consider random probabilities.

If $X \sim Beta(\alpha, \beta)$, there exist convenient formulas for its moments:

$$E(X) = \frac{\alpha}{\alpha + \beta}, \ Var(X) = \frac{\alpha\beta}{(\alpha + \beta)^2(\alpha + \beta + 1)}. \tag{1.26}$$

There exist some natural relationships between beta and gamma families, especially with the exponential distribution.

If $X \sim Gamma(\alpha, \lambda)$ and $Y \sim Gamma(\beta, \lambda)$ are independent,

$$\frac{X}{X + Y} \sim Beta(\alpha, \beta). \tag{1.27}$$

Also, if for $\alpha < 1$ we obtain independent $X \sim Beta(\alpha, 1 - \alpha)$ and $W \sim Exp(1)$, then

$$XW \sim Gamma(\alpha, 1). \tag{1.28}$$

These properties will be used in Chapter 3 for generating random values from different distributions.

1.6   Joint Distributions

Most nontrivial applications of statistics require analyzing more than one random variable. In this case we have to consider several random variables simultaneously. Let X and Y denote two random variables. Their **joint cumulative distribution function** or joint c.d.f. is defined as

$$F(x, y) = P(X \leq x, Y \leq y). \tag{1.29}$$

If the second mixed partial derivative of this function exists, we will call it **joint density function** or just joint density:

$$f(x, y) = \frac{\partial^2 F(x, y)}{\partial x \partial y}. \tag{1.30}$$

If out of the two variables X and Y only one, say X, is important for our analysis and the values of the other (Y) are irrelevant or we have no information regarding Y, we may "integrate Y out" and consider **marginal density function** of X as "the density of X whatever Y" and define it as the integral of the joint density over y

$$f_X(x) = \int_{-\infty}^{\infty} f(x, y) dy. \tag{1.31}$$

Accordingly, we can define **marginal c.d.f.** as

$$F_X(x) = \lim_{y \to \infty} F(x, y). \tag{1.32}$$

If on the contrary, the specific value y that variable Y takes is not only relevant, but exactly known, we can consider the **conditional distribution of X given Y**:

$$F_{X|Y}(x|y) = P(X \le x | Y = y),\tag{1.33}$$

with the density

$$f_{X|Y}(x|y) = \frac{f(x, y)}{f_Y(y)},\tag{1.34}$$

obtained as the ratio of the joint density and the marginal density of Y.

Random variables are called **independent** if for all possible values of x and y the joint c.d.f. is equal to the product of marginal c.d.f.'s or equivalently, the joint density of the joint distribution of these variables equals to the product of their marginal densities. It is also worth noting that in case when X and Y are independent, the conditional density of X given $Y = y$ for any y coincides with the marginal density of X:

$$f_{X|Y}(x|y) = f_X(x).\tag{1.35}$$

It is natural to define the vector of means as a column vector μ with two components $(E(X), E(Y))^T$. Another important numerical characteristic of a pair of random variables X and Y is their **covariance**

$$Cov(X, Y) = E((X - E(X)) \cdot (Y - E(Y)) = E(XY) - E(X)E(Y),\tag{1.36}$$

measuring the degree of coordination in the way X and Y might change: covariance is an integral with a positive contribution in case both specific values of X and Y lie above or below their respective means, and a negative contribution otherwise. It is easy to see from the definition of the means as integrals over respective densities that $Cov(X, Y) = 0$ if X and Y are independent (the converse is certainly not true, there exist simple examples of dependent variables with zero covariance). It is also easy to see that $Cov(X, Y) = Cov(Y, X)$ and $Cov(X, X) = Var(X)$.

Correlation coefficient is a measure of dependence between two random variables defined by their variances and covariance:

$$\rho(X, Y) = \frac{Cov(X, Y)}{\sqrt{Var(X)Var(Y)}}.\tag{1.37}$$

It takes the values between -1 and 1, these two cases correspond to perfect linear dependence between X and Y: $Y = aX + b$. Along with covariance, it is a convenient measure of linear dependence, but unfortunately it fails to detect some simple nonlinear relationships. For example, $\rho(X, X^2) = 0$ in case of standard normal $X \sim N(0, 1)$.

We can also consider the case of more than two random variables. Let $X = (X_1, \ldots, X_n)^T$ denote a random column vector of dimension n. The joint distribution function of this vector can be defined by the equality

$$F(x_1, \ldots, x_n) = P(X_1 \leq x_1, \ldots, X_n \leq x_n). \tag{1.38}$$

If a mixed partial derivative of n-th order for all the variables of this function exists and is continuous, then it is called the density function or just the density. Thus,

$$f(x_1, \ldots, x_n) = \frac{\partial^n F(x_1, \ldots, x_n)}{\partial x_1 \ldots \partial x_n}. \tag{1.39}$$

We can obtain marginal distribution functions of all vector components from the joint distribution function. For example, the distribution function of the first component $F_1(x_1)$ is calculated as the limit of the joint distribution function $F(x_1, \ldots, x_n)$, when all the arguments except the first one tend to $+\infty$. Accordingly, $f_1(x_1) = \frac{\partial F_1(x_1)}{\partial x_1}$ is the marginal density. One can similarly define the functions of joint distribution of several components and the corresponding density function. For example, let $F(x_1, x_2, x_3, x_4)$ denote the joint distribution of four random variables. Then the distribution function of variables X_1 and X_2 is defined by the equality

$$F_{12}(x_1, x_2) = \lim_{x_3, x_4 \to \infty} F(x_1, x_2, x_3, x_4). \tag{1.40}$$

Accordingly, the two-dimensional density function is defined by the equalities

$$f_{12}(x_1, x_2) = \frac{\partial F_{12}(x_1, x_2)}{\partial x_1 \partial x_2} = \int_{-\infty}^{+\infty} \int_{-\infty}^{+\infty} f(x_1, x_2, x_3, x_4) dx_3 dx_4. \tag{1.41}$$

If the values of some components of multivariate random variable are known, we can consider the *conditional distribution* of the other components. For example, if $x_3 = c_3, x_4 = c_4$, then

$$F_{12|34}(x_1, x_2 | x_3 = c_3, x_4 = c_4) = P(X_1 \leq x_1, X_2 \leq x_2 | x_3 = c_3, x_4 = c_4), \tag{1.42}$$

$$f_{12|34}(x_1, x_2 | x_3 = c_3, x_4 = c_4) = \frac{f(x_1, x_2, c_3, c_4)}{f_{34}(c_3, c_4)}. \tag{1.43}$$

Random variables are called independent if the density of the joint distribution of these variables equals to the product of the marginal densities.

Important characteristic of a random vector $X = (X_1, \ldots, X_n)^T$ is the **covariance matrix** Σ with its elements defined as pairwise covariances

$$\sigma_{ij} = Cov(X_i, X_j) = E((X_i - E(X_i)) \cdot (X_j - E(X_j))). \tag{1.44}$$

For $i = j$ this formula defines the diagonal elements of the matrix as the variances $Var(X_i) = \sigma_{ii}$. That is why the covariance matrix is sometimes called *variance–covariance matrix*.

Correlation matrix R with elements

$$\rho_{ij} = \frac{\sigma_{ij}}{\sqrt{\sigma_{ii}\sigma_{jj}}} \tag{1.45}$$

is often used to describe the structure of dependence of components of a random vector.

A very special role in applications is played by the **multivariate normal distribution**. Let $X = (X_1, \ldots, X_n)^T$ denote a multivariate random vector with the means $\mu_i = E(X_i)$ and the vector of the means defined as the column vector $\mu = (\mu_1, \ldots, \mu_n)^T$. Let Σ denote the covariance matrix of this multivariate random variable. The vector has multivariate normal distribution if its density has the following form:

$$f(\mathbf{x}) = \frac{1}{\sqrt{(2\pi)^n |\Sigma|}} \exp\left(-\frac{1}{2}(\mathbf{x} - \mu)^T \Sigma^{-1}(\mathbf{x} - \mu)\right), \tag{1.46}$$

where $\mathbf{x} = (x_1, \ldots, x_n)^T$ is the column vector corresponding to our set of variables X. We will use notation $X \sim MN(\mu, \Sigma)$ for such random vectors.

The following graphical examples correspond to the multivariate normal case of dimension two (*bivariate normal distribution* is also denoted by $BN(\mu, \Sigma)$). If a vector (X, Y) has bivariate normal distribution, both marginals are normal, $X \sim N(\mu_x, \sigma_x^2)$ and $Y \sim N(\mu_y, \sigma_y^2)$. Also, it can be easily checked that conditional distributions of Y given $X = x$ (and vice versa) are normal for any x. That means that all vertical cross-sections of the following graphs will show similar bell-shaped plots. In the standard case : $n = 2$, $\mu_x = \mu_y = 0$, $\Sigma_{11} = \sigma_x^2 = 1$, $\Sigma_{22} = \sigma_y^2 = 1$, $\Sigma_{12} = \Sigma_{21} = \rho$, and formula (1.46) allows for a somewhat simpler representation:

$$f(x, y) = \frac{1}{2\pi\sqrt{1 - \rho^2}} exp\left(-\frac{x^2 + y^2 - 2\rho xy}{2(1 - \rho^2)}\right). \tag{1.47}$$

Density graphs of the standard bivariate normal distribution (1.47) and contour lines in Figures 1.12, 1.13, and 1.14 correspond to different values of correlation coefficient $\rho = 0$; $\rho = -0.8$; $\rho = 0.8$.

The structure of dependence between the components of a multivariate normal random variable is defined completely by the covariance matrix. This is a huge modeling convenience in applications when dependence between the components is linear. However, in case of considerable nonlinear dependence multivariate normal model clearly does not suffice. The other important feature of the normal multivariate distribution for modeling are light tails, corresponding to very fast decrease of the density away from the center of the distribution.

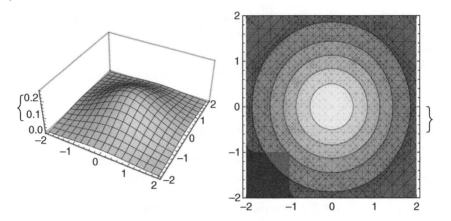

Figure 1.12 Density of bivariate normal distribution, $\rho = 0$.

This feature makes multivariate normal inappropriate for modeling substantial probabilities of the large deviations from the mean known as "heavy tails."

In order to provide a better tool for modeling multidimensional heavy tails, one may use the multivariate t-distribution (Student's distribution). We shall use the same notations for the vector of means and covariance matrix as for the multivariate normal distribution. The density of the multivariate t-distribution with η degrees of freedom is defined by the formula [9]

$$f(\mathbf{x}) = \frac{\Gamma\left(\frac{\eta+n}{2}\right)}{\Gamma\left(\frac{\eta}{2}\right)\sqrt{(2\eta)^n |\Sigma|}} \left(1 + \frac{1}{\eta}(\mathbf{x} - \mu)\Sigma^{-1}(\mathbf{x} - \mu)^T\right)^{-\frac{\eta+n}{2}}, \tag{1.48}$$

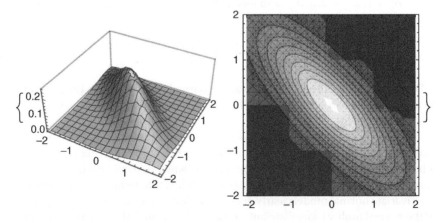

Figure 1.13 Density of bivariate normal distribution, $\rho = -0.8$.

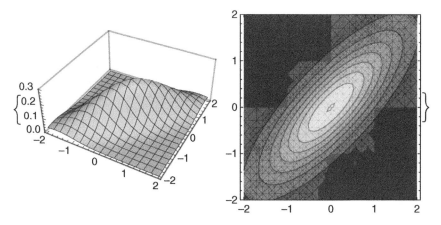

Figure 1.14 Density of bivariate normal distribution, $\rho = 0.8$.

where $\Gamma(z)$ is the gamma function. Smaller values of η correspond to heavier tails.

Figures 1.15, 1.16, and 1.17 represent density graphs of the bivariate t-distribution with the same number of degrees of freedom $\eta = 2.5$ and the same values of means, variances, and correlation as were chosen for the normal distribution ($\mu_1 = \mu_2 = 0; \sigma_1 = \sigma_2 = 1; \rho = 0, \rho = -0.8, \rho = 0.8$). The contour lines in these graphs appear similar to the contour lines of the normal distribution, but one should pay attention to the fact that the values of density outside of the contour lines of the t-distribution are 10 times higher than those for the normal distribution. This is an indication of heavier tails.

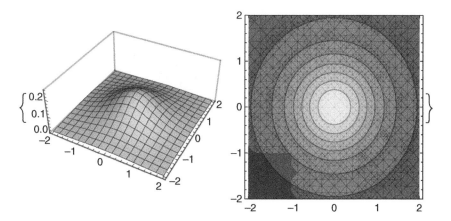

Figure 1.15 Density of bivariate t-distribution, $\rho = 0$.

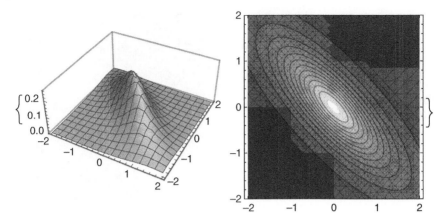

Figure 1.16 Density of bivariate t-distribution, $\rho = -0.8$.

Figure 1.18 represents samples of daily residual percentage returns for two pairs of market indexes after controlling for trend and volatility (see the time series description below). The sample estimates of correlation are given for each pair of indexes. The first image represents the joint distribution of Mexican (IPC) and Japanese (NIKKEI) indexes. Looking at this image it is possible to assume that the two-dimensional normal distribution can be a good model. The second image represents French (CAC) and German (DAX) indexes, which are quite strongly related, especially in the tails (lower left and upper right). The two-dimensional t-distribution seems to be a more appropriate model for this joint distribution.

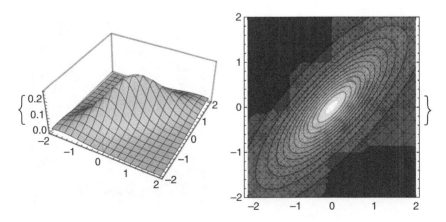

Figure 1.17 Density of bivariate t-distribution, $\rho = 0.8$.

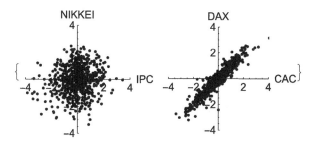

Figure 1.18 Scatterplots of daily index returns: NIKKEI versus IPC and DAX versus CAC.

Let us point out that the calculation of specific values of the multivariate distribution functions turns out to be a calculation of multidimensional integrals, which is often a matter of considerable technical difficulty. Most modern software packages for statistical analysis contain built-in tools for multivariate integration. Typical implementation of such integration requires some form of Monte Carlo methods further discussed in Chapter 3.

1.7 Time-Dependent Random Variables

In practice we often come across sequences of random variables, observed on regular basis (yearly, monthly, daily, etc.). As an example, let us consider a series of daily values of Standard & Poors 500 index (SPX) S_t spanning the 3-year period from January 1, 2009 to December 31, 2011.

Such ordered sequences of random variables are known as ***time series*** or stochastic processes with discrete time. For more systematic information on time series we can recommend a textbook by Bowerman et al. [2] and a more advanced monograph by Wei [13]. A specific observed trajectory s_t of the stochastic process S_t is shown in Figure 1.19. In some cases one may notice certain patterns in the time series values, such as cycles, seasonal changes, and overall tendency to grow or go down. Such a tendency is called ***trend***. It may often be modeled with a polynomial, logarithmic, or exponential deterministic function $Tr(t)$. For example, the series of SPX values for the given 3-year period clearly demonstrates an up-trend. Allowing irregularities in the tendency to grow, one can suggest a standard representation for a time series $X_t, t = 1, \ldots, n$:

$$X_t = Tr(t) + Y_t, \tag{1.49}$$

where Y_t corresponds to the irregular component of the time series, its random part. While estimation of the trend as the major feature of the data is of primary importance, the further analysis of the irregular component is often required

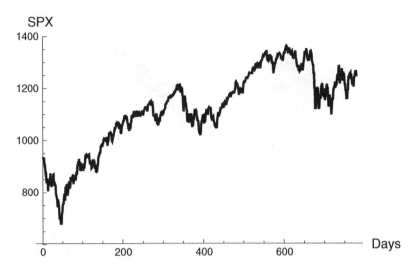

Figure 1.19 Daily values of SPX.

for effective modeling and prediction of the time series. If we believe that this irregular component has certain distribution structure, using observations x_t one may estimate and take out the trend, and then concentrate on the study of the "detrended" time series Y_t.

Ideally, we would like to reduce Y_t to the *white noise*: a sequence of independent normal variables with zero mean and constant variance, $Y_t \sim N(0, \sigma^2)$. However, it is not always possible. For instance, distribution of Y_t may have heavier tails than the normal. Sometimes random terms Y_t can also expose some kind of fine dependence known as ***autocorrelation***, when the adjacent values are correlated to each other, say $\rho(Y_t, Y_{t-1}) \neq 0$. This autocorrelation can be addressed by ***autoregressive models***, the simplest of which, AR(1), suggests the representation

$$Y_t = \alpha_0 + \alpha_1 Y_{t-1} + \varepsilon_t, \quad |\alpha_1| < 1, \tag{1.50}$$

where uncorrelated error terms $\varepsilon_t \sim N(0, \sigma^2)$ correspond to the white noise. In this model errors from the previous steps of the series $t - k, k = 1, 2, \ldots$ are partially inherited (with a discounting factor α_1^k) by its value at time t. The role of coefficient α_1 is important: $\alpha_1 = \rho(Y_t, Y_{t-1})$.

Both detrended model (1.49) with $Y_t \sim N(0, \sigma^2)$ and AR(1) (1.50) can be shown to be ***stationary***, which requires three properties to hold:

- $E(Y_t) = \mu$ for any t (constant mean);
- $Var(Y_t) = \sigma^2$ for any t (constant variance or homoskedasticity);
- $Cov(Y_t, Y_s) = f(|t - s|)$, a function depending only on the difference between s and t.

Stationarity implies that main distribution characteristics of the process Y_t such as central tendency and variation do not change with time. Stationarity will be violated when coefficient $\alpha_1 = 1$. In this case, error terms of the time series ε_t are not partially but completely inherited or accumulated for all future steps. This behavior can be modeled as a *random walk* (also discussed in Chapter 3), with independent random errors $\varepsilon_s, s = 1, \ldots, t$ added together to form the irregular component of the series as a sum

$$Y_t = \alpha_0 t + Y_0 + \sum_{s=1}^{t} \varepsilon_s, \tag{1.51}$$

while the first two terms on the right-hand side represent a linear trend. This formula may be rewritten as a difference equation:

$$Z_t = Y_t - Y_{t-1} = \alpha_0 + \varepsilon_t, \tag{1.52}$$

where ε_t are independent random error terms and Z_t is a stationary time series. This approach, representing an alternative to direct detrending, is known as **differencing**. An indication of nonstationarity of the time series Y_t, which may be transformed into a stationary Z_t by differencing is arriving at the solution $\alpha_1 = 1$ (unit root) when estimating coefficients in AR(1). Possibility of a unit root is checked via **unit root tests**, with its most popular versions being Dickey–Fuller test [5] and Phillips–Perron test [11].

Consecutive differences of a higher order for a time series Y_t can be defined by induction:

$$\Delta Y_t = Y_t - Y_{t-1}, \quad \Delta^p Y_t = \Delta(\Delta^{p-1} Y_t). \tag{1.53}$$

The effect of taking consecutive differences is similar to higher order differentiation. If a time series X_t contains a polynomial trend of the pth degree, then the time series $\Delta^{p+1} X_t$ does not contain a trend. Both detrending for time series with structure (1.49) and differencing for time series with structure (1.51) reduce analysis of a more complex series to a simpler one.

Along with detrending and differencing, we may consider functional transformations $Y_t = g(X_t)$, which pursue the same goal: to isolate the random structure of the time series from its temporal structure, ideally ending up with a sequence of independent random variables. These transformations can be combined with detrending and differencing. An example of such transformation for the SPX values is the transition from sheer index values to logarithmic returns, which behave similar to percentage returns:

$$R_t = \ln \frac{S_t}{S_{t-1}} = \ln S_t - \ln S_{t-1} \approx \frac{S_t - S_{t-1}}{S_{t-1}}. \tag{1.54}$$

This transformation may be treated as combining differencing with logarithmic transform, or more specifically, applying differencing to the logarithms of the

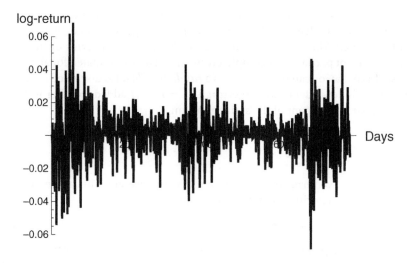

Figure 1.20 Daily logarithmic returns on SPX.

initial time series of index values S_t. Figure 1.20 represents the series of daily logarithmic returns on SPX.

The unit root test can be performed to test whether higher order differencing may be needed. In R, Phillips–Perron test can be implemented under the name of PP.test. Performing it, we get:

	Phillips–Perron Unit Root Test	
	data: SPX	
Dickey–Fuller = −30.8208,	Truncation lag parameter = 6,	p-value = 0.01.

Small p-value indicates the absence of evidence for the unit root, thus further differencing is not required.

However series (1.54) does not look like white noise at all. Autocorrelations are detected in the following R output providing results of parametric estimation in AR(1) model.

Call: ARIMA(x = SPX, order = c(1, 0, 0))		
	ar1	Intercept
Coefficients:	−0.1023	4e−04
s.e.	0.0356	5e−04

This output can be translated to the language of formulas as

$$\hat{R}_t = -0.1023 + 0.0356\hat{R}_{t-1}, \tag{1.55}$$

z-series

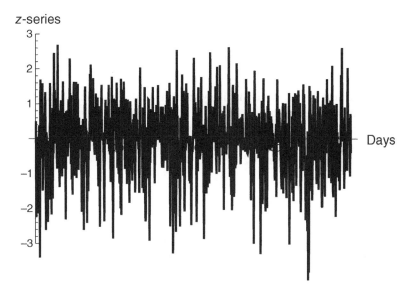

Figure 1.21 Series z_t.

where $\hat{R}_t$ are model fitted values, approximating actual observed values of R_t, and the ***residuals*** $e_t = R_t - \hat{R}_t$ serve as proxy for errors ε_t.

We may also notice the periods of high variation in R_t and the periods of low variation in Figure 1.20. This indicates that $Var(R_t)$, which in finance is known as the **volatility**, is changing with time. The same goes for the variance of the residual series e_t. This contradicts the homoskedasticity, which is one of the requirements of stationarity. In order to model the conditional variance h_t of the residual time series e_t at time t, we assume that it depends on the previous values of conditional variance, as well as on the latest update of the time series. We can choose the simplest form of such dependence and obtain the following model:

$$h_t = \omega + \beta_1 e_{t-1}^2 + \beta_2 h_{t-1}. \tag{1.56}$$

The first model addressing heteroskedasticity in financial series was introduced by Engle [6] as ARCH. It was later generalized to (1.56) by Bollerslev [1] and is known as the first order *generalized autoregressive conditional heteroskedasticity model* commonly abbreviated as GARCH(1,1). The coefficients β_i of this model can be estimated using available data. For the residual SPX series e_t in AR(1) model (1.55) R gives us

$$h_t = 0.0002 + 0.067 e_{t-1}^2 + 0.0915 h_{t-1}. \tag{1.57}$$

Having evaluated the conditional variance h_t, we can adjust for the variable volatility dividing every observation by its conditional standard deviation

$$z_t = \frac{e_t}{\sqrt{h_t}} \tag{1.58}$$

and obtain a time series, which is likely to represent a sequence of independent random variables, whose distribution may be further analyzed. That completes the process of separating the temporal structure of the time series from its random component. Having performed this procedure for the SPX series, we obtain the series in Figure 1.21.

References

1 Bollerslev, T. (1986). Generalized autoregressive conditional heteroskedasticity. *Journal of Economics*, 31, 307–327.

2 Bowerman, B. L., O'Connell, R. T., and Koehler, A. B. (2005). *Forecasting, Time Series, and Regression. An Applied Approach*, 4th ed. Belmont, CA: Thomson Brooks/Cole.

3 Crawley, M. J. (2007). *The R Book*. Chichester: John Wiley & Sons, Ltd.

4 DeGroot, M. H., and Schervish, M. J. (2002). *Probability and Statistics*. Addison Wesley.

5 Dickey, D. A., and Fuller, W. A. (1979). Distribution of the estimators for autoregressive time series with a unit root. *Journal of the American Statistical Association*, 74 (366), 427–431.

6 Engle, R.F. (1983). Estimates of the variance of U.S. inflation based on the ARCH model, *Journal of Money Credit and Banking*, 15, 286–301.

7 Hansen, B. E. (1994). Autoregressive conditional density estimation. *International Economic Review*, 35, 705–730.

8 Klugman, S. A., Panjer, H. H., and Wilmott, G. E. (2012). *Loss Models. From Data to Decisions*, 4th ed. Hoboken, NJ: John Wiley & Sons, Inc.

9 Kotz, S., and Nadarajah, S. (2004). *Multivariate t-Distributions and Their Applications*. Cambridge: Cambridge University Press.

10 Miller, I., and Miller, M. (2014). *John E. Freund's Mathematical Statistics with Applications*, 8th ed. Pearson.

11 Phillips, P. C. B., and Perron, P. (1988). Testing for a unit root in time series regression. *Biometrika*, 75 (2), 335–346.

12 Rice, J. A. (2007). *Mathematical Statistics and Data Analysis*, 3rd ed. Belmont, CA: Thomson Brooks/Cole.

13 Wei, W. W. S. (2006). *Time Series Analysis. Univariate and Multivariate Methods*, 2nd ed. Pearson.

2

Foundations of Bayesian Analysis

2.1 Education and Wages

Common wisdom states that one's salary or wages should have something to do with one's education level. Otherwise, why is higher education so expensive and time consuming? Is it not reasonable that more educated people should be paid more?

Let us define two variables, representing (a) education level of a random person (denote it by E) defined as the number of years of formal education and (b) hourly wage level of the same person (denote it by W) in USD per hour. These variables can be measured for the population of US wage earners. As usual, we will reserve the capital letters E and W for the random variables, while by lowercase e and w we denote particular data: numeric values of these variables. We collect such data for 534 respondents, which constitutes a sample from 1985 census according to Berndt [5].

As we have suggested in the beginning, these two variables must be somewhat related. And if they are, we can use one of these variables measured for a new subject (outside of our sample) to predict the other. For instance, it is logical to believe that if the education level of a person is known, we can make certain predictions regarding their wage. On the other hand, if a person's wage is known, we can make some suggestions regarding their education level. Let us make the latter our primary objective. We will make statistical inference on an unknown parameter (education level) based on available data (wages).

The relationship between these two variables is anything but deterministic. This may be due to many factors which we will not consider for a simple reason: our data contains no information on these factors. There exists a taunt often used by less educated people toward the more educated: If you are so smart, why are you so poor? No doubt, there is a grain of sad truth in this taunt. The authors of the book, both professors of statistics, have to agree.

However, even if this relationship is not deterministic, it still deserves a rigorous probabilistic or statistical treatment. For that, we will consider a function of these two variables: E and W. With both variables being nonnegative,

Introduction to Bayesian Estimation and Copula Models of Dependence, First Edition.
Arkady Shemyakin and Alexander Kniazev.
© 2017 John Wiley & Sons, Inc. Published 2017 by John Wiley & Sons, Inc.
Companion Website: http://www.wiley.com/go/shemyakin/bayesian_estimation

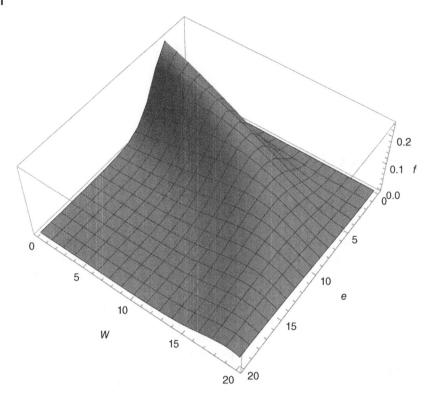

Figure 2.1 Model function of *E* and *W*.

analysis of our data for 534 respondents suggests the following functional representation

$$f(W, E) = \frac{(1.16 - 0.05E)^{4.37} \cdot W^{3.37}}{\Gamma(4.37) \cdot \exp\left((1.16 - 0.05E)W\right)}. \tag{2.1}$$

Model function (2.1) is graphically represented in three dimensions in Figure 2.1. This representation can go two ways: first, we will take the ***probabilistic*** approach and consider the value of education level $E = e$ known and fixed. Then we look at the distribution of hourly wages W given $E = e$. This is the direct view (logically and chronologically), since education comes first, and wages are paid later on, as a consequence. This way E appears to be the parameter of the distribution of random variable W, which can be represented for our data by the probability density function

$$p(w \mid E = e) = f(w, e) = \frac{(1.16 - 0.05e)^{4.37} \cdot w^{3.37}}{\Gamma(4.37) \cdot \exp\left((1.16 - 0.05e)w\right)} \tag{2.2}$$

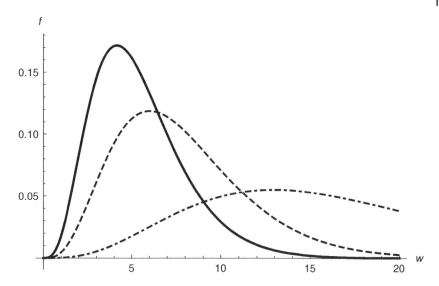

Figure 2.2 Graphs of p.d.f.'s.

which for any given value of parameter E could be recognized as a member of gamma distribution family (1.18). Figure 2.2 shows the graphs of such density functions for the parameter values: 7, 12, and 18 years. We see that for larger values of the parameter the distribution mode shifts to the right and the shape of the graph becomes less steep.

This is a familiar view of several density functions from one family differing only by their parametric values. However, let us take the second view, which can be characterized as ***statistical***. We will consider the hourly wages W (our data) fixed and known. The function of unknown parameter E

$$p(e \mid W = w) = f(w, e) = \frac{(1.16 - 0.05e)^{4.37} \cdot w^{3.37}}{\Gamma(4.37) \cdot \exp\left((1.16 - 0.05e)w\right)} \tag{2.3}$$

is known as the ***likelihood function***. Figure 2.3 demonstrates the graphs of this function for different data values of W equal to 8, 10, and 15 USD per hour.

Formally speaking, these are not the graphs of probability density functions. The area under each graph does not have to equal to one. Notice that the modes shift right with the increase of W, and for our chosen wage levels the most likely education levels are approximately 12 (high school), 14 (two-year college), and 17 (four-year college plus).

This is an example of statistical inference based on likelihood. Starting with the matched sample of 534 pairs (education, wages), we develop model (2.1). This model is used to construct the three-dimensional graph in Figure 2.1, and also its cross-sections: likelihood curves in Figure 2.3. Then, we pick the

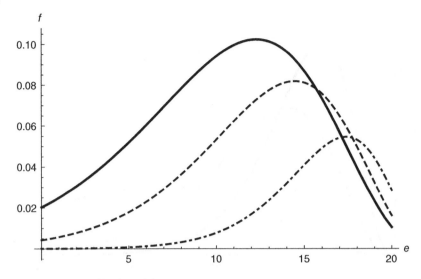

Figure 2.3 Graphs of likelihood functions.

likelihood curve corresponding to a new data point W (from the sample or outside of the sample). Finally, we determine the most likely value of the parameter E corresponding to this W. The objective of the inference is met. This example illustrates a typical application of statistical inference based on the likelihood function. We will need another example to demonstrate some problems related to this approach.

2.2 Two Envelopes

This problem has been discussed extensively in a number of papers and online communications, see a comprehensive review by Bliss [6] or Tsikogiannopoulos [22]. We will consider a specific version of the problem which will help us to illustrate a very important point for further discussion. Suppose you are looking for a job. You are at the end of a tedious interview process, but finally you are hired by a company, and you are ready to take the offer. The only issue remaining to be discussed is your salary. The boss, who is genuinely nice but rather eccentric, suggests that you play a game. He puts two envelopes on the table and informs you that each envelope contains your contract with the exact salary figure, but one number is two times greater than the other. You are allowed to choose one of the envelopes, open it and read the number. Then you have to make the decision: will you take the open envelope, or opt for the closed one (without being able to read the number it contains). What shall you do? Let us take a formal standpoint. The open envelope contains a number, say W_1. The closed one contains an unknown number W_2, which is either $2W_1$ or $\frac{W_1}{2}$. Your choice of an envelope was random and you were equally likely to pick a smaller or a larger number. So we have to deduce that

$$P(W_2 = 2W_1 \mid W_1) = P(W_2 = 0.5W_1 \mid W_1) = 0.5. \tag{2.4}$$

Then, we have to opt for the closed envelope! As we can see from the definition of expected value for discrete variables,

$$E(W_2) = 0.5 \cdot 2W_1 + 0.5 \cdot 0.5W_1 = 1.25W_1 > W_1. \tag{2.5}$$

If we think a little harder, we will have to agree that this solution is absurd. Why so? Let us suppose this line of thought is correct and **whatever** is the number in the open envelope, you should take the other one. Then, why do you have to read this number? You can just open one envelope and (without reading the salary figure) choose the one in the closed envelope. Do you not see the problem yet? Taking it a step further, why do you need to open an envelope at all? You just barely touch it, and immediately opt for the other. But then: what do you do with the other? Open? As soon as you open it, you would be better with a different choice! So you have to walk around a vicious circle: You can never open any envelope.

Why is this solution wrong? It can be explained on two levels; intuitive and mathematical. The intuitive point is that in the real world any solution ignoring the salary figure in the open envelope is very suspicious. You must have an idea of what you are worth on the labor market, coming from you prior experience, which we would call **prior information**. If you see 100K ($100,000) in the first envelope, would you expect 200K or 50K in the second one? If you see 30K in the first, what is more likely to be in the second: 60K or 15K? It is interesting that the prior information is highly subjective: it depends both on your actual worth and its perception. Two people might never agree on it, but two different people should indeed have different solutions to this problem.

At the mathematical level, we have made a wild assumption to obtain (2.4). In reality,

$$p(w) = P(W_2 = 2W_1 \mid W_1 = w) \neq P(W_2 = W_1/2 \mid W_1 = w) \tag{2.6}$$

at least for some numeric values w between 0 and infinity. It is not difficult to check that the correct expression for conditional expected value of the number in the second envelope is given not by (2.5), but by

$$E(W_2 \mid W_1 = w) = p(w) \cdot 2w + (1 - p(w)) \cdot 0.5w = 0.5w(3p(w) + 1). \tag{2.7}$$

Mathematically, the most convenient way to quantify prior information would be to define a continuous distribution by a distribution density function $\pi(w)$ defined for all w from 0 to infinity. We can call it *prior distribution*. In terms of prior distribution, we can rewrite (2.6) as

$$p(w) = P(W_2 = 2W_1 \mid W_1 = w) = \frac{\pi(2w)}{\pi(2w) + \pi(0.5w)}, \tag{2.8}$$

and

$$q(w) = 1 - p(w) = P(W_2 = 0.5W_1 \mid W_1 = w) = \frac{\pi(0.5w)}{\pi(2w) + \pi(0.5w)}. \tag{2.9}$$

These two probabilities, taking into account both prior information and the data of the experiment (the number in the first envelope), will be called *posterior probabilities*. They correspond to the only possible outcomes, describing entire probability distribution of the number in the second envelope W_2 conditional on the value in the first one, w (we will also call it *posterior distribution*).

We can also rewrite (2.7) as conditional expected value

$$E(W_2 \mid W_1 = w) = w\frac{2\pi(2w) + 0.5\pi(0.5w)}{\pi(2w) + \pi(0.5w)} \tag{2.10}$$

which, as we can easily see, could be either greater or less than w, depending on the prior distribution. In fact, the only case when (2.6) gives the same answer as (2.4) corresponds to $\pi(w) \equiv c$ for all w from 0 to infinity (uniform on the positive semiaxis), which is not a proper probability density function.

Let us consider a numeric example. Suppose, random variable W (our worth) can be described by the uniform distribution from 25K to 100K. It corresponds to

$$\pi(w) = \begin{cases} \dfrac{1}{75}, & \text{if } 25 \leq w \leq 100, \\ 0, & \text{otherwise.} \end{cases} \tag{2.11}$$

Therefore,

$$p(w) = \frac{\pi(2w)}{\pi(2w) + \pi(0.5w)} = \begin{cases} 1, & 25 \leq w < 50 \\ 0.5, & w = 50 \\ 0, & 50 < w \leq 100. \end{cases} \tag{2.12}$$

Using (2.7), you can easily finish the problem.

We have to conclude that for this simple example the approach allowing for the incorporation of prior information brings about a pretty elegant solution. This sequence of steps:

- elicitation of prior distribution,
- formal development of posterior distribution (based both on prior and data),
- making conclusions based exclusively on the posterior,

informally constitutes the ***Bayesian inference***.

Notice that if we wanted now to get back to the first example (education and wages), we could also think about incorporating some prior information. If you know somebody in person, have a chance to talk and get a personal impression, and then you are informed of her exact wages, will the wage be the only source for your conclusions regarding her education level? It would be a little harder to quantify your visual and verbal impressions of a person as a neat prior distribution for her education level (unless you ask directly and trust the answer completely), but in principle it could be done. This process is known as prior elicitation, and we will discuss it more than once in the sequel.

At this time we conclude the informal introduction. The following two sections discuss the main formal concepts of Bayesian inference related to hypothesis testing and parametric estimation. Classical procedures will be reviewed first, and then Bayesian modifications will be suggested allowing us to properly incorporate the prior information.

2.3 Hypothesis Testing

2.3.1 The Likelihood Principle

Let us consider an i.i.d. sample $x = (x_1, \ldots, x_n)$ from an infinite population, characterized by a distribution density function $f(x, \theta)$, where x is the vector of data and θ is a (maybe, vector) parameter. Recall from basic statistics course that the ***likelihood function*** is defined as

$$L(x; \theta) = L(x_1, \ldots, x_n, \theta) = \prod_{i=1}^{n} f(x_i, \theta). \tag{2.13}$$

In the context of incomplete data, which is typical for many risk management and reliability applications, including some of those which will be used in the following chapters, we may observe exact values $x = (x_1, \ldots, x_m), m < n$ and just know the lower limits for the other cases: $x_j > c_j, j = m + 1, \ldots, n$. This corresponds to the so-called right censoring of the dataset. In this case we can modify the likelihood formula:

$$L(x; \theta) = L(x_1, \ldots, x_m; c_{m+1}, \ldots, c_n; \theta) = \prod_{i=1}^{m} f(x_i, \theta) \prod_{j=m+1}^{n} S(c_j, \theta), \tag{2.14}$$

where $S(c, \theta) = P(X > c; \theta)$ is the survival function corresponding to the distribution with density $f(x, \theta)$.

The *likelihood principle* is a practically undisputed basic principle of mathematical statistics. According to the likelihood principle, all information regarding the unknown parameter, which could be found in data, is contained in the likelihood function.

2.3.2 Review of Classical Procedures

Let us review the main definitions and procedures of hypothesis testing for the simplest case of two parametric hypotheses. We believe that there exist two exhaustive and mutually exclusive possibilities regarding the parameter θ of the distribution $f(x, \theta)$. One is designated as the *null hypothesis* and is denoted by $H_0 : \theta = \theta_0$. The other is described as the *alternative* $H_1 : \theta = \theta_1$. A *test statistic* (a function of the sample) $T(x)$ and *critical region* $R_T \subset Range(T)$ are used to define a *test*, which is a basic binary decision rule:

"*Reject H_0*" if $T(x) \in R_T$ or "*Do not reject H_0*" if $T(x) \notin R_T$.

Notice that although in the formal context of two simple hypotheses "*Do not reject H_0*" seems to be equivalent to "*Accept H_0*" or "*Reject H_1*," in common applications we find the first of the three statements the most appropriate. Null hypothesis is also known as status-quo or zero-effect hypothesis requiring no immediate action if not rejected. Typical examples of a null hypothesis would be: "new drug is no good," "surgery is not required," "new technology is exactly as effective as the old one." Rejecting null is something which requires an action: "approve a new drug," "proceed with the surgery," "implement the new technology" and is associated with additional cost or risk. The burden of proof is higher for rejecting the null. Not rejecting the null might call for more testing to be performed and more proofs to be provided before we take the action. It does not necessarily mean accepting the null or rejecting the alternative. In the context of classical hypotheses testing, one usually speaks of *statistical significance* of the results, if the null hypothesis can be rejected.

We can judge the quality of the tests using definitions of two errors which we want to be as small as possible:

Type I error (reject true null) with probability $\alpha = P(T(x) \in R_T \mid H_0)$
Type II error (do not reject false null) with probability $\beta = P(T(x) \notin R_T \mid H_1)$

Quantity $1 - \beta$ is also known as the *power* of the test.

In general, type I error is the one which is more important in the context of the study. Therefore, a common approach is to choose the largest admissible probability of type I error, which is called *significance level*, and then find tests which will make type II error as small as possible. Neyman–Pearson theory of hypothesis testing suggests that in a very general setting the most powerful tests (those minimizing type II error for type I error not exceeding the significance level) should be based on the ***likelihood ratio***:

$$\Lambda(x) = \frac{L(x; \theta_1)}{L(x; \theta_0)}. \tag{2.15}$$

If we use the likelihood ratio as the test statistic, the only thing we need to specify the test procedure is the critical region, which will take the form $R_\Lambda = \{x : \Lambda(x) > c_\alpha\}$. Here c_α is the critical value of the test, which can be determined by α, if the distribution of the likelihood ratio under the condition of null hypothesis is known. In practical applications, we usually look for a simpler test statistic $T(x)$ equivalent to the likelihood ratio in the sense that $R_T = \{x : T(x) > k_\alpha\} = R_\Lambda = \{x : \Lambda(x) > c_\alpha\}$.

Ten Coin Tosses

A random variable X has Bernoulli distribution (see Section 1.2) if it may assume two values only: 0 or 1, and is described by a discrete probability function $f(x, \theta) = \theta^x (1 - \theta)^{1-x}$, where the value of the parameter $\theta = P(X = 1)$ is unknown. Consider a random i.i.d. sample $x = (x_1, \ldots, x_n)$ from this distribution. Write down the likelihood function

$$L(x, \theta) = \theta^{\sum x_i}(1 - \theta)^{n - \sum x_i} = \theta^{n\bar{x}}(1 - \theta)^{n - n\bar{x}}. \tag{2.16}$$

Let us perform a series of ten trials: tosses of the same coin. Suppose we record "heads" seven times. This series is treated as an i.i.d. sample $x = (x_1, \ldots, x_n), n = 10$, where the underlying variable X has Bernoulli distribution and θ is unknown probability of "heads" (success). We will consider null hypothesis $H_0 : \theta = \theta_0 = 0.5$ (fair coin) versus a simple alternative $H_1 : \theta = \theta_1 = 0.7$ (coin biased in a special way). How can we construct the likelihood ratio test?

In our case,

$$\Lambda(x) = \frac{L(x; 0.7)}{L(x; 0.5)} = \frac{0.7^{n\bar{x}} \cdot (1 - 0.7)^{n - n\bar{x}}}{0.5^{n\bar{x}} \cdot (1 - 0.5)^{n - n\bar{x}}} = (3/5)^n \cdot (7/3)^{n\bar{x}} \tag{2.17}$$

so

$$\Lambda(x) > c \iff n\bar{x} > k = \frac{\ln c + n \ln 5 - n \ln 3}{\ln 7 - \ln 3}. \tag{2.18}$$

The exact formula with logarithms on the right-hand side of the last expression is irrelevant as soon as we can see that $R_{\bar{x}} = \{x : n\bar{x} > k_\alpha\} = R_\Lambda = \{x : \Lambda(x) > c_\alpha\}$ for some k_α, and the equivalent form of the likelihood ratio test is:

"***Reject*** H_0" if $n\bar{x} > k_\alpha$ or "***Do not reject*** H_0" if $n\bar{x} \le k_\alpha$.

Picking up an "industry standard" $\alpha = 0.05$, from the table of binomial distribution we can see that, $P(\sum X_i > 7 \mid H_0) = 0.055, P(\sum X_i > 8 \mid H_0) = 0.011$, and the rejection region consists of $n\bar{x} = \sum x_i \in \{9, 10\}$ so in our example with seven heads we do not reject the null hypothesis that the coin is fair. Notice also that our final verdict does not depend on the alternative if instead of $\theta_1 = 0.7$ we choose another value larger than 0.5. The test might have less power, but its significance level stays the same.

2.3.3 Bayesian Hypotheses Testing

Consider the same general setting of two simple hypotheses. Incorporating prior information in this case means assigning prior probabilities to the null hypothesis $P(H_0)$ and the alternative $P(H_1) = 1 - P(H_0)$.

Let us also consider the observed data x and the likelihoods $L(x, \theta_0)$ and $L(x, \theta_1)$. Since the two hypotheses are mutually exclusive and exhaustive, the Law of Total Probability suggests to "integrate out" the parameter and obtain a parameter-free weighted likelihood, where the prior probabilities of the hypotheses play the role of weights:

$$m(x) = L(x, \theta_0)P(H_0) + L(x, \theta_1)P(H_1).$$

As a function of data, $m(x)$ is usually called the **marginal distribution**. For given data it is a constant, and as we will see very soon, it plays very little role in the following analysis. However, we will need it during the derivation of the Bayesian test. According to the definition of conditional probability or a simplified version of the Bayes theorem,

$$P(H_0 \mid x) = \frac{L(x, \theta_0)P(H_0)}{m(x)}, \quad P(H_1 \mid x) = \frac{L(x, \theta_1)P(H_1)}{m(x)},$$

and the ratio of these two expressions, after $m(x)$ cancels out, is

$$\frac{P(H_1 \mid x)}{P(H_0 \mid x)} = \frac{L(x, \theta_1)}{L(x, \theta_0)} \times \frac{P(H_1)}{P(H_0)}. \tag{2.19}$$

The ratio on the left-hand side is known as **posterior odds**. The first ratio on the right-hand side is the likelihood ratio, which we have encountered while developing classical testing procedures. The second is the ratio of prior probabilities or **prior odds**. When the prior odds equal one, the posterior odds has the Bayesian name of the **Bayes factor**, see [14]. In this case of two simple hypotheses, the Bayes factor coincides with the likelihood ratio. In general, both the Bayes factor and the prior odds will be needed to evaluate the posterior odds.

Finally, in order to decide whether to reject the null hypothesis or not, without any prior information (that means, setting prior odds to one, or in other words, assuming $P(H_0) = P(H_1) = 0.5$), we have to establish a reasonable benchmark for the posterior odds. We can consider the posterior odds 3:1 to provide "substantial" evidence against the null and 10:1 to provide "strong" evidence. Jeffreys [13] used these benchmarks for the Bayes factors, see also

[14]. So for substantial evidence against the null, the Bayesian test takes the form:

"***Reject*** H_0" if $\frac{P(H_1|x)}{P(H_0|x)} > 3$ or "***Do not reject*** H_0" if $\frac{P(H_1|x)}{P(H_0|x)} \leq 3$.

For strong evidence,

"***Reject*** H_0" if $\frac{P(H_1|x)}{P(H_0|x)} > 10$ or "***Do not reject*** H_0" if $\frac{P(H_1|x)}{P(H_0|x)} \leq 10$.

Ten Coin Tosses

Returning to the ten coin toss experiment in Bayesian setting, we need to specify the priors. Let us say that our prior belief in fairness of the coin can be expressed by $P(H_0) = 0.5$ or in the prior odds of 1:1. From (2.17) we may calculate the Bayes factor as

$$\Lambda(x) = \frac{3^3 7^7}{5^{10}} \approx 2.28,$$

bringing about the posterior odds of 2.28:1 which gives less than substantial (and much less than strong) evidence against the null hypothesis. Notice that the classical procedure introduced in Subsection 2.3.2 did not require the computation of the likelihood ratio, but it called for the computation of binomial probabilities.

To extend the analogy with the classical procedure, assume that for the same null and alternative and choice of priors we observe a different result: instead of seven, we have nine "heads" in a series of ten coin tosses. Then the Bayes factor is

$$\Lambda(x) = \frac{3^1 7^9}{5^{10}} \approx 12.40,$$

which provides a strong evidence against the null. Noticing that $n\bar{x} = 9$ lies in the critical region for $\alpha = 0.05$, we may conclude that in this case both classical and Bayesian approaches to hypotheses testing lead us to similar results.

2.4 Parametric Estimation

2.4.1 Review of Classical Procedures

Let us recall some concepts of classical theory of statistical estimation. Consider a random sample $x = (x_1, \ldots, x_n)$ from a certain population with p.d.f. $f(x, \theta)$. A statistic is defined as any function of the random sample. A statistic used as an approximate value of unknown parameter of the distribution (population) can be considered an ***estimator*** of the parameter. All estimators are random variables. We have to distinguish an ***estimate*** as a specific numeric value used to approximate the unknown parameter from an estimator which is a rule leading to different numerical estimates for different samples from the same distribution (population). We will try not to mix up these two terms, though it might be confusing at times.

From the classical standpoint, desirable properties of estimators are: consistency, unbiasedness, and efficiency. An estimator is consistent if it converges in a certain sense to the true parameter value when the sample size grows to infinity. An estimator is unbiased if its expected value (mean) coincides with the true value of the estimated parameter. An estimator is called the most efficient within a certain class if for the given class its mean square error is the smallest.

2.4.2 Maximum Likelihood Estimation

There exist several popular methods to obtain parametric estimators in a general setting: the method of moments, the method of least squares, etc. However the method of maximum likelihood is probably the most important for classical statistics. A ***maximum likelihood estimator (MLE)*** $\hat{\theta}$ for unknown k-dimensional vector parameter θ is a statistic, which maximizes the likelihood function defined in (2.13). When the sample sizes are large, or ideally, are allowed to grow to infinity, MLEs have all nice properties: they are consistent, asymptotically unbiased, asymptotically efficient, and even asymptotically normal (their distribution is close to normal). However for small sample sizes MLEs do not always behave that well, and may be substantially inferior to the estimators obtained by other methods.

Let us recall the algorithm used to obtain MLE in regular cases.

- Write down the likelihood function (2.13).
- Calculate logarithmic likelihood or log-likelihood $l(x_1, \ldots, x_n, \theta) = \ln L$ (a technical step simplifying further derivation).
- Calculate partial derivatives of the log-likelihood with respect to parameters.
- Find the critical points of the logarithmic likelihood, that is, the solutions of the system of equations

$$\frac{\partial}{\partial \theta^{(j)}} l(x_1, \ldots, x_n, \theta) = 0, j = 1, 2, \ldots, k. \tag{2.20}$$

- Check if the critical point found is indeed the absolute maximum point (not a local maximum, nor a minimum or a saddle point). The value of absolute maximum obtained herewith $\hat{\theta}$ is the MLE.

This algorithm cannot be used in nonregular cases (e.g., if the partial derivatives are not defined). Sometimes it does not lead to a unique solution. Notice also that the exact solution of system (2.20) is available only in some simple textbook cases. In more complex situations one has to use numerical methods of optimization to find an approximate solution. Let us recall two simple examples illustrating both analytical and numerical methods.

Ten Coin Tosses

Returning to the ten coin toss experiment, we will once again use the likelihood function defined in (2.16). Then we will calculate its logarithm, the log-likelihood

$$l(x_1, \ldots, x_n, \theta) = n\bar{x} \ln \theta + (n - n\bar{x}) \ln (1 - \theta). \tag{2.21}$$

It is easy to find the derivative w.r.t θ and solve the following equation

$$\frac{\partial l}{\partial \theta} = \frac{n\bar{x}}{\theta} - \frac{n - n\bar{x}}{1 - \theta} = 0. \tag{2.22}$$

The solution is not unexpected: $\theta = \bar{x}$. To check if this critical point is a maximum, find the second derivative

$$\frac{\partial^2 l}{\partial \theta^2} = -\frac{n\bar{x}}{\theta^2} - \frac{n(1 - \bar{x})}{(1 - \theta)^2}. \tag{2.23}$$

Since for Bernoulli distributed samples $0 \leq \bar{x} \leq 1$, this second derivative is never positive. Therefore, the critical point we have found is a local maximum point. As soon as it is the only critical point, it is also the absolute maximum. Thus $\hat{\theta} = \bar{x}$ is the MLE for the parameter of Bernoulli distribution.

We perform a series of 10 trials (coin tosses): $n = 10$. Suppose we record "heads" seven times ($x = 7$). So $\hat{\theta} = \bar{x} = 0.7$.

Life Length

Many random variables in insurance or reliability applications representing "life length" or "time to failure" can be conveniently described by gamma distribution with density function

$$g(y; \alpha, \lambda) = \frac{\lambda^\alpha}{\Gamma(\alpha)} y^{\alpha-1} e^{-\lambda y}, \ y \geq 0, \tag{2.24}$$

where the values of two parameters α (shape) and λ (rate or reciprocal scale) are unknown, see also Section 1.4. Gamma class of distribution was introduced in Chapter 1. The variety of forms of gamma densities for different combinations of parameter values (see Figure 1.17) makes this class very attractive in modeling survival.

Suppose we obtain an i.i.d. sample from gamma distribution $(x_1, \ldots, x_n)$. Let us write down the log-likelihood function

$$l(x_1, \ldots, x_n, \alpha, \lambda) = n\alpha \ln \lambda - n \cdot \ln \Gamma(\alpha) + (\alpha - 1) \sum \ln x_i - \lambda \sum x_i. \tag{2.25}$$

As we can see, the approach of the previous example will not work that easily: differentiation in both α and λ is way too difficult to allow for a simple analytical solution. However, numerical optimization procedures might help (see Exercises at the end of the chapter).

An alternative to MLE is known as the ***method of moments***, which is a particular case of a more general ***plug-in principle***. It uses the fact that if there exists a convenient formula expressing the unknown parameters of a distribution through its moments (say, the mean and the variance), then plugging sample moments in instead of theoretical moments often brings about consistent (not necessarily unbiased or efficient) estimators.

From (1.19) we know that $E(X) = \alpha/\lambda$ and $Var(X) = \alpha/\lambda^2$, therefore

$$\alpha = \frac{E(X)^2}{Var(X)}, \ \lambda = \frac{E(X)}{Var(X)},$$

after plug-in leading to the method of moment estimators:

$$\hat{\alpha} = \frac{\bar{x}^2}{s^2}, \ \hat{\lambda} = \frac{\bar{x}}{s^2}.$$

Method of moments estimators sometimes are attractively simple, but are lacking certain large sample optimality properties of MLE. Returning to 10 coins example, we can easily see that due to $E(X) = \theta$, the method of moments estimator of θ coincides with the MLE.

2.4.3   Bayesian Approach to Parametric Estimation

Let us assume that all relevant prior information regarding parameter θ is summarized in $\pi(\theta)$, which is the p.d.f. of the prior distribution. In order to obtain the posterior distribution, we can use continuous version of the Bayes Theorem, ***the most important formula in this chapter***!

$$\pi(\theta \mid x) = \frac{\pi(\theta)L(x,\theta)}{\int \pi(\theta)L(x,\theta)d\theta} \propto \pi(\theta)L(x,\theta). \tag{2.26}$$

The left-hand term of (2.26) is the p.d.f. of the posterior distribution. The right-hand term is just a product of prior and likelihood. Symbol $\propto$ denotes proportionality of its left and right hand sides or equivalence of those to within a constant multiple. The constant multiple in question is the integral in the denominator of the central term of (2.26). This integral depends on x only and does not depend on the parameter which is integrated out. It represents the ***marginal distribution of the data***, and is an absolute constant when data are provided. The bad news is: In most applications this constant is hard to determine. The good news is: In most cases it is not necessary. There exist some ways around.

For now let us suppose that this integral is not a problem, the posterior distribution is defined completely, and it has a nice analytical form. This is possible in a limited number of simple examples. The posterior distribution is what we need to provide Bayesian inference. What information can we get out of the posterior distribution?

Point Estimation

The mean, median, or mode of the posterior distribution (in the sequel: the **posterior mean**, the posterior median, and the posterior mode) can play the role of point estimators of the parameter:

$$\theta^* = E(\theta \mid x) = \int \theta \pi(\theta \mid x) d\theta, \tag{2.27}$$

$$\theta^{**} = med(\theta \mid x), \int_{-\infty}^{\theta^{**}} \pi(\theta \mid x) d\theta = 0.5, \tag{2.28}$$

$$\theta^{***} = argmax(\pi(\theta \mid x)). \tag{2.29}$$

The first one (by far, the most popular choice) minimizes posterior Bayes risk which is defined as the mean square deviation of the estimator from the parameter averaged over all parameter values, while the second minimizes the absolute deviation of the estimator. Sometimes, the posterior mode is the easiest to use.

Interval Estimation

Having chosen the level $1 - \alpha$ in advance, we are looking for an interval (θ_l, θ_u) such that

$$1 - \alpha = P(\theta_l < \theta < \theta_u) = \int_{\theta_l}^{\theta_u} \pi(\theta \mid x) d\theta. \tag{2.30}$$

Such an interval is called a **credible interval**. Unlike classical confidence intervals, Bayesian credible intervals allow for a simple probabilistic interpretation as intervals of given posterior probability. We can always choose a symmetric interval, though it is not the only possible choice:

$$\int_{-\infty}^{\theta_l} \pi(\theta \mid x) d\theta = \frac{\alpha}{2} = \int_{\theta_u}^{\infty} \pi(\theta \mid x) d\theta. \tag{2.31}$$

2.5 Bayesian and Classical Approaches to Statistics

As we have noticed already, the likelihood principle is universally accepted in modern statistics. Neither classical nor Bayesian approaches are disputing it. The main difference between the two approaches is in the interpretation of parameters. In classical statistics parameters are unknown but fixed. There exists the "true" value of the parameter, which Nature stores for us somewhere, though we might never hope to find it. According to Bayesian interpretation, parameters are random variables, which can be described by distribution laws. There is no such thing as the "true" value of the parameter.

This principal difference between two approaches has many dimensions: from simple procedural considerations to the fundamental worldview issues.

On the most general philosophical level, classical statistics is deterministic: It considers populations and processes along with their characteristics such as population mean and variance (true values of parameters) to belong to the objective reality. It also suggests at least a distant possibility to learn or guess the true values of parameters. Moreover, if our methods are good and consistent, we believe that we can approximate these true values very well. All events around us are actually pre-determined. We perceive them as random due to the limitations of human knowledge.

The Bayesian approach, however, considers characteristics of populations and processes being random in principle. All our knowledge regarding random events contains some level of uncertainty and is subjective. Any statement regarding the population parameters expresses just the ***degree of rational belief*** in the characteristics of the population.

Such philosophical differences arise in many fields of knowledge: for instance, in physics they are epitomized by an elegant question: "Does God play dice?" As the history of philosophy testifies, this question cannot be successfully answered, this way or another. As the history of science testifies, both approaches have proven to be fruitful. In particular, representatives of both points of view made substantial contributions in the development of Statistics as a discipline [3].

Going down to a more practical standpoint, we can notice two important differences between Bayesian and classical approaches. The first difference is in the interpretation of data. Classical approach assumes a probability model in which data are allowed to vary for a given parameter value. In education/wages example (education level being the parameter, wages being the data) the reference group for our inference is formed by all respondents with the same education level. A good statistical procedure in the classical sense is a procedure which is good enough for any member of the reference group.

The data we obtain is just one random sample from a potentially huge reference group, a variety including other random samples, which we will possibly never see. We can always consider a possibility (at least, theoretically) of repeating our experiments and increasing the amount of available data. We judge the statistical procedures by their behavior in this unlimited data world, therefore we often tend to integrate over data samples, including those obtained and those only theoretically possible, while keeping our parameters fixed. Probability of an event is the limit of relative frequency of its occurrence in an infinitely long series of identical experiments. That is why classical statistics is also frequently dubbed (especially among the Bayesians) "frequentist."

The Bayesian approach does not assume a possibility of obtaining unlimited data. All inference has to be based solely on the data obtained in the experiment and possibly on some prior information. We do not assume a possibility of our data being different and do not integrate over the reference group. We have just one person whose wages are reported, and we have to make inferences regarding this person's education level. Our reference group is the set of possible parameter values. Parameters of the distribution of data are considered random, and we judge our procedures integrating over possible values of parameters, and keeping the data fixed. Probability rules are used directly to obtain the rules of statistical inference [1, 4, 17].

The choice between these two approaches in applied statistics is often determined by specifics of the application. In some classes of problems we can easily afford repetitive experiments and increasing the sample size is just a technical difficulty. In other classes we are "doomed" to work with small samples, seeing no principal possibility of getting additional data.

The other practical difference between two approaches related to the first one is the use of prior information at our disposal before the statistical data are obtained. Such prior information is very likely to be nontrivial in many problem settings. Let us once again recall the education/wage example, or even better, the problem of two envelopes. The amount and role of prior information widely varies for different applications. However, the classical approach allows the use of prior information in a very narrow sense: at the stage of formalizing the model setting the likelihood. After the likelihood is written down, there is no conceivable way to incorporate any additional information. Bayesian approach allows for the use of prior information in the form of a prior distribution.

2.5.1 Classical (Frequentist) Approach

Classical approach, in our view, found its most logical representation in the ideas and works of a prominent British statistician Sir Ronald Aylmer Fisher (1890–1962) (Figure 2.4). As the portrait reveals, Fisher was an avid smoker and did some research on the effect of smoking tobacco on human health. What is more important, he made a great contribution to the development of Statistics in the twentieth century. He worked in multiple applications and obtained many deep theoretical results. He also contributed to the formulation of basic principles of what we call here the classical approach. He also happened to be very anti-Bayesian.

Let us recall the main principles of classical statistical analysis:

• Parameters (numerical characteristics of populations and processes) are fixed unknown constants.

Figure 2.4 Sir Ronald Aylmer Fisher (1890–1962). Reproduced with permission of R. A. Fisher Digital Archive, The Barr Smith Library, The University of Adelaide.

- Probabilities are always interpreted as the limits of relative frequencies in unlimited series of repetitive experiments.
- Statistical methods are judged by their large sample behavior.

A typical method of statistical estimation is the maximum likelihood method. Hypotheses testing is based on the likelihood ratio. The likelihood function is treated as a function of data and is not considered to describe the distribution of the parameter.

However, in order to cast some doubt in practical infallibility of this approach, we may use a classical experiment performed by Fisher and discussed in his book [10] to illustrate Fisher's exact test. In our description of this experiment we will follow [20], though this story being told by many statisticians has many different versions.

2.5.2 Lady Tasting Tea

At a summer tea party in Cambridge, England, a lady (some sources identify her as Dr. Muriel Bristol) states that tea poured into milk tastes differently than that of milk poured into tea. The lady claims to be able to tell whether the tea or the milk was added first to a cup. Her notion is taken sceptically by the scientific minds of the group. But one guest, happening to be Ronald Aylmer Fisher, proposes to scientifically test the lady's hypothesis. Fisher proposes to give her eight cups, four of each variety, in random order. One could then ask what the probability was for her getting the number she got correct, but just by chance (the null hypothesis). The experiment provides the lady with eight randomly ordered cups of tea, four prepared by first adding milk, four prepared by first adding the tea. She is to select the four cups prepared by one method. She was reported [20] to prove correct all eight times. Using permutation formula, one can suggest that under the null hypothesis her chances to guess correctly, or randomly select one out of $8!/(4!(8 − 4)!) = 70$ possible permutations of eight cups are about 0.014, which is the p-value of Fisher's exact test.

Putting aside interesting connotations of this problem for experimental design and randomization (the latter to be discussed in the next chapter), we can consider a slightly different setup, where each of the eight cups is randomly and independently prepared in one of the two possible fashions (the milk first or the tea first) according to the results of a fair coin toss. Then the chances of the lady to guess correctly all eight times become $1/2^8 = 1/256$, which is approximately 0.004. If we increase the number of cups from 8 to 10, then in the case of all 10 correct guesses the p-value or the significance level of the test goes down to 0.001. However, what exactly does this p-value mean?

Let us cite a letter from an outstanding American Bayesian statistician L. J. Savage (Figure 2.5) to a well-known colleague, biostatistician J. Cornfield, where this experiment (discussing a version with 10 cups) is used to cast some doubt on the practicality of using classical procedures based solely on likelihood and suggesting some role of the context of the experiment and thus the prior information.

"There is a sense in which a given likelihood ratio does have the same sense from one problem to another, at least from the Bayesian point of view. It is always the factor by which the posterior odds differ from the prior odds. However, all schools of thought are properly agreed that the critical likelihood ratio will vary from one application to another. The Neyman-Pearson school expresses this by saying that the choice of the likelihood ratio is subjective and should be made by the user of the data for his particular purpose. As a Bayesian, I would say that a critical likelihood ratio for a dichotomous decision about a simple dichotomy depends (in an evident way) on the loss associated with the decision and on the prior probabilities associated with the dichotomy.

The fact that the reaction to an experiment depends on the content of the experiment as well as on its mathematical structure and whatever economic issues might be involved seems to me to be brought out by the following triplet of examples that occurred to me the first time I taught statistics, when I still had a completely orthodox orientation. Imagine the following three experiments: 1.

Figure 2.5 Leonard Jimmie Savage (1917–1971). Courtesy of Gonalo L. Fonseca.

Fisher's lady has correctly dealt with ten pairs of cups of tea. Many readers will recognize this example to be from R. A. Fisher's famous lady tasting tea experiment. 2. The professor of 18-th century musicology at the University of Vienna has correctly decided for each of 10 pairs of pages of music which was written by Mozart and which by Haydn. 3. A drunk in a parlor has succeeded 10 times in correctly calling a coin secretly tossed by you. These three experiments all have the same mathematical structure and the same high significance level. Can there, however, be any question that your reaction to them is justifiably different? My own would be: 1. I am still skeptical of the lady's claim, but her success in her experiment has definitely opened my mind. 2. I would originally have expected the musicologist to make this discrimination; I would even expect some success in making it myself; he, an expert in the matter, felt sure that he could make it. His success in 10 correct trials confirms my original judgment and leaves no practical doubt that he would be correct in substantially more than half of future trials, though I would not be surprised if he made occasional errors. 3. My original belief in clairvoyance was academic, if not utterly nonexistent. I do not even believe that the trial was conducted in such a way that trickery is a plausible hypothesis, and feel sure that the drunk simply had an unusual run of luck. Of course, these tests are not simple dichotomies, but I think you will find them germane to your question" [21].

Figure 2.6 Rev. Thomas Bayes (circa 1702–1761). https://commons.wikimedia.org/wiki/File:Thomas_Bayes.gif. CC0-1.0 public domain https://en.wikipedia.org/wiki/Public_domain

2.5.3 Bayes Theorem

History of the Bayesian approach in statistics traditionally goes back to the times of Reverend Thomas Bayes (Figure 2.6). One can argue though that Laplace and other prominent contemporaries did a lot of what we now call Bayesian reasoning. Not much is known for sure about Bayes including his date of birth. Even his portrait included in this book, according to modern studies, is no longer believed to be his.

Thomas Bayes in his posthumous treatise formulated the statement known now as the Bayes Theorem. This theorem indicates how one should calculate inverse conditional probabilities of hypotheses after the experiment related to these hypotheses. This theorem can be rigorously formulated for the discrete case

$$P(B_i \mid A) = \frac{P(A \mid B_i)P(B_i)}{\sum_j P(A \mid B_j)P(B_j)}, A \subseteq \cup B_i, B_i \cap B_j = \emptyset. \tag{2.32}$$

Here B_i are hypotheses, and A is some event, which is influenced by the hypotheses, thus the post-event probabilities of the hypotheses are also

Figure 2.7 Sir Harold Jeffreys (1891–1989). Reproduced with permission of the Master and Fellows of St. John's College, Cambridge.

influenced by the event. This formula in a simple form was used in the Bayesian treatment of two simple hypotheses in Section 2.3. In estimation problems we are more likely to use the continuous version of the theorem which was introduced in (2.26).

Bayes' work and ideas were never completely forgotten, but they also were out of the mainstream for a very long time. The revival of Bayesian approach is probably due to two factors. The first factor is the recent expansion of new applications of statistics (insurance, biology, medicine) characterized by relatively small sample sizes (limited data) and ample prior information. The second is the development of computational techniques which substantially extends the practical applicability of Bayesian methods.

It would probably be most accurate to name another outstanding British scientist Sir Harold Jeffreys (1891–1989) (Figure 2.7) as the founder of the modern Bayesian statistics. One of his main contributions was his seminal book [13] containing the foundations of Bayesian statistics. However many other great statisticians, including Bruno De Finetti, the famous proponent of subjective probability; David Lindley; already cited Jimmie Savage; and Arnold Zellner (Figure 2.8) should be honorably mentioned among those who brought attention back to this eighteenth century approach.

Figure 2.8 Arnold Zellner (1927–2010), the first president of ISBA. Reproduced with permission of International Society for Bayesian Analysis.

2.5.4 Main Principles of the Bayesian Approach

Let us summarize the main principles of Bayesian statistical analysis keeping close to the treatment of Bolstad [7].

- True values of parameters are unknown, are not likely to become known, and therefore will be considered random.
- Prior information regarding the parameters before the experiment is allowed and is mathematically expressed in the form of the prior distribution of the random parameter.
- All probabilistic statements expressed by the researcher regarding the parameters express the "degree of rational belief" and are subjective.
- Likelihood function is the only source of information regarding the parameter which is contained in data (weak likelihood principle).
- We do not consider the feasibility of obtaining new data and do not consider the data which could theoretically be obtained.
- Probability rules and laws are applied directly to the statistical inference.
- Our information regarding the parameters is updated using the data.
- Posterior distribution incorporates the prior information and the information from data in the only mathematically consistent way: using the Bayes formula.

The main advantages of Bayesian approach are well known and are discussed in different terms and at different levels in such books as [1, 4, 7, 11], and [16]. From decision theoretic standpoint, we can prove that all admissible estimation procedures can be obtained using Bayesian approach. Sometimes, Bayesian procedures outperform classical procedures even using classical criteria of performance. Bayesian estimation can be substantially more reliable for small samples, while it will lead to the same results as MLE when the sample size grows to infinity. Bayesian procedures behave better than classical ones when there exists substantial amount of prior information which should not be neglected. However, even with very weak prior information, Bayesian approach sometimes performs better than the classical approach. One explanation to this phenomenon is a possible difference in numerical stability between mathematical procedures of optimization (MLE) and integration (Bayesian estimators).

The development of Bayesian procedures in the past faced a very important obstacle. The problem may sound technical, but it was grave. The continuous version of the Bayes theorem (2.26) allows one to determine the posterior density to within a multiplicative constant. Combining (2.26) and (2.27) we can write down the following formula for the most popular **Bayes estimator**: the posterior mean.

$$\theta^* = E\left(\theta \mid x\right) = \frac{\int \theta \pi\left(\theta\right) L\left(x, \theta\right) d\theta}{\int \pi\left(\theta\right) L\left(x, \theta\right) d\theta}. \tag{2.33}$$

Unfortunately, exact analytical solution for these integrals is rarely available, although we will consider some nice exclusions in Section 2.7. Numerical integration is possible but is far from trivial especially for higher dimensions of parametric space (multiple parameters). That is why Bayesian statistics was stuck in a rut for a long time, being restricted to handling very special cases and not being able to spread the wings.

A "Bayesian revolution" took place in the middle of the twentieth century. What Bayes could not dream of, and what Jeffreys could not carry out, was made possible by the development of the toolkit widely known as MCMC— Markov chain Monte Carlo methods. International Society for Bayesian Analysis (ISBA) was founded in 1992 as the result of increased interest to Bayesian approach, and since then Bayes methods have been growing in popularity even more. Modern Bayesian statistics provides ample tools for such integration developed since the 1950s and still under active development. Chapter 4 of the book is dedicated to some of these methods.

However, the first question we have to ask before we face the problems with evaluating the posterior in (2.33) is: How do we specify the prior?

2.6   The Choice of the Prior

Contemporary Bayesian statisticians do not form a homogeneous group. There are serious differences in the ways people understand and apply the Bayesian principles. One of the most discussed issues is the ***prior elicitation*** a.k.a. the choice of the prior in particular problems [15]. In the following three subsections we will illustrate three popular approaches using one simple example of ten coin tosses.

2.6.1   Subjective Priors

Consider a particular problem with a substantial amount of available prior information: Before the experiment we already know something important about the parameter of interest. This knowledge can be expressed in terms of the parameter's probability distribution. If probabilities are understood as the degrees of rational belief, no wonder that different people may have different subjective beliefs which will lead to different subjective distributions. Two medical doctors read a patient's chart and make different observations. Two investors interpret long-term market information in two different ways. They will have different priors.

Assume that both doctors perform the same necessary tests and both investors observe the same results of the recent trades, meaning both parties get the same data and agree on the likelihood. Their final conclusions may still differ because of their priors being different.

This difference in the final decisions due to the difference of priors is both the curse and the blessing of Bayesian statistics. If the use of prior information is allowed, subjectivity is inevitably incorporated in the final decision. Therefore the best thing to do is to develop the rules which will translate subjective prior beliefs into a properly specified prior distribution.

Imagine you are a statistician hired to help with an applied problem: You have to determine how likely a given coin is to fall heads up. One should understand that when we are talking "coin tosses" and "heads," we do not necessarily have actual coins in mind. We might model binary random events with probability of success being represented by a "head," and probability of failure represented by a "tail" in a clinical trial setting, as well as while recommending medical treatment or investment decisions.

Statisticians rarely work on applied problems on their own; they become team members along with experts in the object field. An expert on coins joins you on the team and you two have to state the problem statistically and develop an experiment together. Both of you have agreed that available time and resources allow you to perform 10 independent identical tosses of the coin. The parameter of interest is ***probability of heads*** θ and the experimental data

$X = (X_1, \dots, X_n)$ are results of tosses, $n = 10$, $P(X_i = 1) = \theta$, $P(X_i = 0) = 1 - \theta$, X_i being the random number of heads on each toss. This problem was considered above as an illustration of maximum likelihood estimation.

However, now you are allowed to observe the coin prior to the experiment and summarize the prior information. The coin expert can provide the knowledge and you will assist with putting this knowledge into the form of prior distribution on θ. The expert agrees that chances of the coin landing on its edge or disappearing in thin air during any toss are negligible, so that any outcomes other than heads or tails should not be taken into account. He also observes the coin and it looks symmetric. It is likely that the coin is fair, but the past experience tells the expert that some coins are biased either toward the heads or toward the tails. You ask him to quantify these considerations in terms of average expectations. Suppose that the expert concludes: on the average, θ is about 0.5, but this guess is expected to be off on the average by 0.1. You interpret this expert opinion in terms of two numbers: mean value $E(\theta) = 0.5$ and variance $Var(\theta) = [st.dev.(\theta)]^2 = 0.01$.

Now it is your time to specify the prior. The choice of beta family of distributions introduced in Section 1.5 for random θ is logical. It is a wide class including both symmetric and skewed, flat and peaked distributions: $\theta \sim Beta(\alpha, \beta)$, if its p.d.f. according to (1.25) is

$$\pi(\theta) = \frac{\Gamma(\alpha + \beta)}{\Gamma(\alpha)\Gamma(\beta)} \theta^{\alpha-1}(1 - \theta)^{\beta-1}$$

for $0 < \theta < 1$, $\alpha \geq 0$, $\beta \geq 0$. The cluster of gamma functions on the right-hand side does not depend on θ thus it is irrelevant for further analysis so we can write

$$\pi(\theta) \propto \theta^{\alpha-1}(1 - \theta)^{\beta-1}. \tag{2.34}$$

Several Beta p.d.f.s for various parameter values are depicted in Figure 2.9, see also Figure 1.11.

One of the representatives of the beta family will fit the requirements of $E(\theta) = 0.5$ and $Var(\theta) = 0.01$. Using formulas from Chapter 1 (1.26)

$$E(\theta) = \frac{\alpha}{\alpha + \beta}, \quad Var(\theta) = \frac{\alpha\beta}{(\alpha + \beta)^2(\alpha + \beta + 1)},$$

we can solve for α and β:

$$\frac{\alpha}{\alpha + \beta} = 0.5 \Rightarrow \alpha = \beta,$$

and

$$\frac{\alpha\beta}{(\alpha + \beta)^2(\alpha + \beta + 1)} = \frac{\alpha^2}{4\alpha^2(2\alpha + 1)} = 0.01 \Rightarrow \alpha = \beta = 12.$$

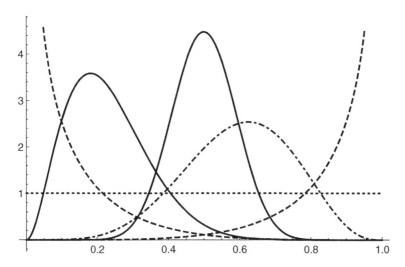

Figure 2.9 Graphs of beta p.d.f.'s.

Then the expert's opinion can be paraphrased as $\theta \sim Beta(12, 12)$. This is our subjective prior. We will use this example in future because it works so well numerically.

Ten Coin Tosses

To complete our analysis and obtain the Bayes estimate for θ, consider an experiment with 7 heads in 10 tosses, $\sum x_i = n\bar{x} = 7$. Using (2.16), we obtain the likelihood

$$L(x, \theta) = \theta^{n\bar{x}}(1 - \theta)^{n - n\bar{x}} = \theta^7(1 - \theta)^3 \tag{2.35}$$

and using (2.26), the posterior

$$\pi(\theta \mid x) \propto \pi(\theta)L(x, \theta) = \theta^{\alpha-1}(1 - \theta)^{\beta-1} \times \theta^7(1 - \theta)^3$$

$$= \theta^{12+7-1}(1 - \theta)^{12+3-1} \sim Beta(19, 15).$$

The posterior is also a beta distribution shown along with the prior in Figure 2.10, so using the first part of formula (2.35) with new values $\alpha = 19$ and $\beta = 15$, we obtain the numerical value of the Bayes estimate for θ

$$\theta^\star = E(\theta \mid x) = \frac{19}{19 + 15} = \frac{19}{34} \approx 0.56, \tag{2.36}$$

while the second part of (1.26) provides an error estimate

$$Var(\theta \mid x) = \frac{19 \times 15}{(19 + 15)^2(19 + 15 + 1)} \approx 0.007. \tag{2.37}$$

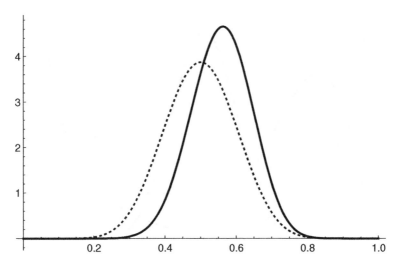

Figure 2.10 Graphs of prior (dotted) and posterior (solid) for subjective prior.

2.6.2 Objective Priors

The main benefit of the Bayesian approach to statistics is the ability to incorporate prior information. Suppose no prior information is available, and even object field experts cannot help. What could be the rationale for applying Bayesian approach instead of classical in such cases anyway? Why should we do it at all? One of possible motivations is the desire to stay within the Bayesian framework, which allows us to use integration instead of optimization (a purely technical reason) or assume randomness of parameters (a philosophical reason). This motivation is especially important in complex multiparameter studies, when prior information can be elicited for some of the parameters, but not for the others. If we could express the idea of no prior information as a special *noninformative* prior distribution, it would allow us to maintain Bayesian setting and derive joint posteriors for our multivariate parameters. For instance, we will be able to use nifty Bayesian software without the necessity to laboriously develop the prior distribution and to justify our subjective point. Just fill the blanks with objective noninformative priors and let computer do the work.

Objective priors are the priors that every expert should agree to use. They are usually weak, nonintrusive, nonviolent priors, which express as little reliance on the prior information as possible and put the emphasis on the data instead. There exist many approaches to the construction of noninformative priors based on certain mathematical principles: Jeffreys prior defined through Fisher information is justified by the idea of invariance to reparameterization [13]; reference priors introduced by Berger and Bernardo maximize the impact of the data on the posterior [2, 4]; maximum entropy priors introduced by Jaynes [12] and Zellner are derived from information-theoretical considerations; and probability matching priors use the idea that Bayesian credible intervals in the absence of prior information should be closely related to the classical confidence intervals.

Bayesian procedures with objective noninformative priors derived by all of these methods bring about results in most cases consistent with the procedures of classical (frequentist) statistics. However the choice of objective prior is not always obvious and sometimes no agreement is achieved between different constructions. Let us use the illustration from the previous subsection.

Ten Coin Tosses

Let us say the coin expert has no past experience with similar coins, and therefore has no opinion at all regarding the probability of heads θ prior to the experiment. Then the envelope is pushed to the statistician (that is you) to decide what prior to put in. An alternative would be to forget about Bayesian approach and resort to MLE, which for 7 observed heads out of 10 tosses provided us in Section 2.4 with the intuitively plausible estimate $\hat{\theta} = 0.7$. However, if we still want to follow the Bayesian approach and take the posterior mean as the estimate of θ, we need to specify the prior, which in our case should be noninformative. An intuitively attractive uniform prior $Beta(1, 1)$ happens to be off. Jeffreys' prior, reference prior approach, and probability matching prior all point at a different choice of $Beta(0.5, 0.5)$ with a horseshoe-shaped p.d.f

$$\pi(\theta) \propto \frac{1}{\sqrt{\theta(1 - \theta)}}.$$

This choice of the horseshoe distribution, though a bit strange at first sight, is justifiable not just mathematically (three different approaches lead to the same decision!) but also intuitively. Just think outside of the coin framework. In general, we are talking about some random event of which we know nothing in advance. What can we say about the probability of this event? Should it be uniformly spread on the interval from 0 to 1? Is it equally likely to be almost impossible (probability close to 0), 50-50 like a fair coin toss (close to the middle of the interval), or almost inevitable (probability close to 1?) Where do the majority of random events in our lives belong?

One can argue that most of the random events in our everyday lives are either almost impossible or almost inevitable. The authors of this book, as most of the other people, have daily routines they tend to follow. Coffee in the morning? Almost inevitable for one of us and almost impossible for the other. Walking to work and back? Almost impossible for one of us and almost inevitable for the other. Both of us show up for classes according to our schedules. There are always slight chances for the opposite to happen, but most of the daily routines are set up firmly. Sun rises, traffic flows, birds fly south in the Fall. You do not often toss a coin to determine whether to eat a lunch or whether to teach a class. If we look closely, there are many random events in our everyday lives, but very few of them have 50-50 chances to happen. Visualize a horseshoe-shaped p.d.f. with the minimum in the middle and most of the probability mass concentrated at both ends.

If you are satisfied with the justification above, let us adopt $Beta(0.5, 0.5)$ prior and evaluate the posterior as we did for subjective prior:

$$\pi(\theta \mid x) \propto \pi(\theta)L(x, \theta) = \theta^{\alpha-1}(1 - \theta)^{\beta-1} \times \theta^7(1 - \theta)^3$$

$$= \theta^{0.5+7-1}(1 - \theta)^{0.5+3-1} \sim Beta(7.5, 3.5).$$

The numerical value of the objective Bayes estimate is

$$\theta^\star = E(\theta \mid x) = \frac{7.5}{7.5 + 3.5} = \frac{7.5}{11} \approx 0.68. \tag{2.38}$$

If you still insist on the uniform (you could), this is still a possibility, even if it lacks a good mathematical justification. If taken as a prior, it will not lead us to a grave mistake. We obtain the posterior

$$\pi(\theta \mid x) \propto \theta^{1+7-1}(1 - \theta)^{1+3-1} \sim Beta(8, 4)$$

and

$$\theta^\star = E(\theta \mid x) = \frac{8}{8 + 4} = \frac{8}{12} \approx 0.67. \tag{2.39}$$

You can see both priors (uniform and horseshoe) and both corresponding posteriors in Figure 2.11. You can see that the posteriors are relatively close. It can be easily seen that for larger dataset sizes the role of prior becomes less and less visible, and all posteriors look alike. By the way, it is also true for subjective priors.

As we see from Figure 2.11, both $Beta(0.5, 0.5)$ and uniform $Beta(1, 1)$ are not completely uninformative: objective Bayes estimates are still numerically different from MLE. In order to assign a prior for which posterior mean is exactly equal to MLE, we will have to consider an even more radical option $Beta(0, 0)$, also known as Haldane's prior. This prior is ***improper***. It means that it is not a proper probability distribution in the sense that the area under its density does not integrate to 1 and it even admits events of infinite probability. Though it seems a horrible thing to happen, improper priors play their role in Bayesian analysis, and their use does not bring about any big problems unless they lead to improper posteriors, which is a real disaster preventing any meaningful Bayesian inference. Haldane's prior with nonempty dataset $n \geq 1$ leads to proper posteriors, thus it is usable. We will not emphasize the use of improper priors though sometimes they provide valid noninformative choices.

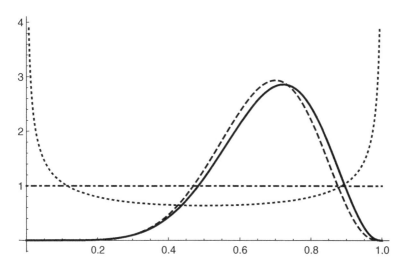

Figure 2.11 Graphs of priors and posteriors for objective priors.

2.6.3 Empirical Bayes

The empirical Bayes approach to the development of priors is all about using empirical information. In general, we should not use the information from our data to choose the prior. That would constitute a clear case of double dipping: One should not use the same data twice—to choose a prior first, and then to develop the posterior. However, if additional information regarding the distribution of parameter is available, it can be used even if it is not directly relevant to our experiment [9, 19].

Ten Coin Tosses

Suppose we have never tossed our coin yet, so the experiment was not yet performed. It is possible though to have a record of five similar experiments with five other coins which have been tossed before. For the sake of simplicity, assume that each of these five coins was tossed 10 times leading to different numbers of heads: 5, 3, 6, 7, 5. These coins are different from ours, so these results are not directly relevant to the value of θ we want to estimate. However, if we believe the five coins in prior experiments to be somewhat similar to ours, they give us an idea of the distribution of θ over the population of all coins.

If we still decide to go for a beta prior, we will use neither expert estimates nor objective priors, but will estimate prior parameters α and β directly from the empirical data of five prior experiments. We can obtain method of moments estimates

$$\widehat{E}(\theta) = 0.52, \ \widehat{Var}(\theta) = 0.022,$$

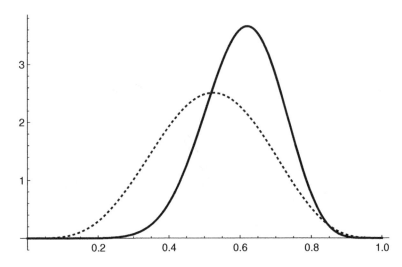

Figure 2.12 Graphs of prior (dotted) and posterior (solid) for empirical prior.

and then using (1.26) solve for $\hat{\alpha}$ and $\hat{\beta}$:

$$\hat{\alpha} = 5.38, \hat{\beta} = 4.97.$$

Then the empirical Bayes prior is $\theta \sim Beta(5.38, 4.97)$ in Figure 2.12.

To obtain the Bayes estimate for θ for the experiment with 7 heads in 10 tosses, $\sum x_i = n\bar{x} = 7$, calculate the posterior

$$\pi(\theta \mid x) \propto \theta^{5.38+7-1}(1 - \theta)^{4.97+3-1} \sim Beta(12.38, 7.97)$$

and the estimate for θ is given by

$$\theta^\star = E(\theta \mid x) = \frac{12.38}{12.38 + 7.97} \approx 0.61. \tag{2.40}$$

The empirical Bayes approach is more powerful than this example may suggest. First of all, in the presence of empirical information it is not always necessary even to define the prior distribution as belonging to a certain parametric family. Instead, one can use nonparametric Bayes approach.

Nonparametric Bayes modeling is a very hot topic in modern statistics. A Bayesian nonparametric model is by definition a Bayesian model on an infinite-dimensional parametric space, which represents the set of all possible solutions for a given problem (such as the set of all densities in nonparametric density estimation). Bayesian nonparametric models have recently been applied to a variety of machine learning problems including regression, classification, and sequential modeling. However this is out of the focus of the book. We will illustrate nonparametric empirical Bayes on the classical compound Poisson example by Robbins [18], see also Carlin and Louis [8].

Compound Poisson Sample

Let us consider an integer-valued sample $x = (x_1, \ldots, x_n)$ from compound Poisson distribution, where each element has Poisson distribution with its own individual parameter value θ_i corresponding to the distribution mean. This sample can represent a heterogeneous population of insurance customers with Poisson claims rates that vary across the population. Our goal is to estimate the entire vector of parameters $\theta = (\theta_1, \ldots, \theta_n)$. Bayesian approach suggests calculating posterior means

$$\theta_i^* = E(\theta_i \mid x_i) = \frac{\int \theta \pi(\theta) L(x_i, \theta) d\theta}{\int \pi(\theta) L(x_i, \theta) d\theta}, \tag{2.41}$$

and we will try to do it without explicitly specifying the prior $\pi(\theta)$. Recall that the integral in the denominator of (2.41) is known as the marginal distribution of the data point x_i:

$$m(x_i) = \int \pi(\theta) L(x_i, \theta) d\theta;$$

and manipulating the formula of Poisson likelihood

$$\theta L(x_i, \theta) = \theta \frac{\theta^{x_i} e^{-\theta}}{x_i!} = \frac{\theta^{x_i+1} e^{-\theta}}{x_i!}$$

$$= \frac{(x_i + 1)\theta^{x_i+1} e^{-\theta}}{(x_i + 1)!} = (x_i + 1)L(x_i + 1, \theta),$$

we can also rewrite the numerator in (2.41) as

$$\int \theta \pi(\theta) L(x_i, \theta) d\theta = (x_i + 1) \int \pi(\theta) L(x_i + 1, \theta) d\theta$$

$$= (x_i + 1)m(x_i + 1).$$

Combining expressions for the numerator and the denominator, obtain

$$E(\theta_i \mid x_i) = (x_i + 1) \frac{m(x_i + 1)}{m(x_i)}.$$

The only way the prior is included in this formula is inside the marginals. But the marginals can be estimated directly from the data. Suppose that the initial sample contains only $k \leq n$ distinct values y_j, so that $x = (y_1, \ldots, y_k)$, $N(y_j)$ is the number of $x_i : x_i = y_j$, and $n = \sum_{j=1}^{k} N(y_j)$. Then it is reasonable to use

$$\hat{m}(y_j) = \frac{N(y_j)}{n}, \quad \theta_j^* = (y_j + 1) \frac{N(y_j + 1)}{N(y_j)}$$

to estimate the Poisson rates of individual observations.

Nonparametric Bayes approach is becoming increasingly popular in the recent years, especially in the context of discrete distributions for relatively small samples with a large number of parameters.

2.7 Conjugate Distributions

The continuous version of the Bayes theorem completely solves the problem of representing the posterior through the prior and the likelihood. However the analytical derivation of the posterior is rarely feasible. In particular, such solution is possible if the data distribution obtains a so-called *conjugate* prior.

Definition A family of parameter distributions is called conjugate with respect to given data distribution (likelihood), if application of the Bayes Theorem to the data with a prior from this family brings about a posterior from the same family.

If we start with a prior, it is not hard to determine whether, after multiplication by the likelihood, the posterior belongs to the same family as the prior. We have done exactly that in the previous section with the ten coin example, where a prior distribution from Beta family seems logical. However, not all likelihoods will have a conjugate family. In some (but not all) interesting examples a sufficient condition for a likelihood to obtain a conjugate prior can be formulated.

2.7.1 Exponential Family

The condition of existence of a conjugate prior (2.42) is satisfied for the exponential family, see [7]. Distribution density function for the *exponential family* can be represented as

$$f(x \mid \theta) = A(\theta)B(x)e^{C(\theta)T(x)}. \tag{2.42}$$

Many well-known distributions belong to the exponential family (do not confuse with the *exponential distribution* defined in Section 1.4: the latter is just a particular case of the exponential family).

The likelihood has a similar form

$$L(x; \theta) = A(\theta)^n \prod_{i=1}^{n} \left(B(x_i)\right)e^{C(\theta)\sum_{i=1}^{n} T(x_i)}, \tag{2.43}$$

but the data are fixed and the parameter is treated as a random variable. Thus the factor $\prod_{i=1}^{n} B(x_i)$ is a constant with no further effect. The conjugate prior has the same form as the likelihood:

$$\pi(\theta) \propto (A(\theta))^k e^{lC(\theta)}. \tag{2.44}$$

Here k and l are some constants determining the shape of the distribution. Since posterior is proportional to the product of the prior and likelihood, we obtain

$$\pi(\theta \mid x_1, \ldots, x_n) \propto (A(\theta))^{k+n} e^{(l+\sum_{i=1}^{n} T(x_i))C(\theta)}. \tag{2.45}$$

Therefore, the posterior belongs to the same conjugate family as the prior with constants $k^* = k + n$; $l^* = l + \sum_{i=1}^{n} T(x_i)$ depending on the likelihood. Let us provide one example of a conjugate family.

2.7.2 Poisson Likelihood

Poisson distribution introduced in Section 1.2 characterizes random number X of events in a stationary flow occurring during a unit of time. Let us denote the intensity parameter (average number of events during the unit of time) by θ. Distribution of this random variable can be presented as

$$f(x \mid \theta) = P(X = x) = \frac{\theta^x}{x!} e^{-\theta}, x = 0, 1, \ldots.$$

Poisson distribution belongs to the exponential family with

$$A(\theta) = e^{-\theta}; B(x) = \frac{1}{x!}; C(\theta) = \ln \theta; T(x) = x.$$

From this expression, the conjugate prior distribution is

$$\pi(\theta) \propto e^{-k\theta} e^{l \ln \theta} \propto \theta^l e^{-k\theta},$$

which corresponds to $Gamma(\alpha, \lambda)$ family with $k = \lambda, l = \alpha - 1$.

Let us arrange a sample of n independent copies of the variable X and denote it by $(x_1, \ldots, x_n)$. Likelihood function then can be expressed as

$$L(x_1, \ldots, x_n; \theta) \propto \theta^{\Sigma x_i} e^{-n\theta}.$$

Therefore the posterior can be represented as

$$\pi(\theta \mid x_1, \ldots, x_n) \propto \theta^{l^*} e^{-k^*\theta} = \theta^{\alpha+\Sigma x_i - 1} e^{-(\lambda+n)\theta},$$

which is also a gamma distribution with parameters $\alpha + \sum x_i$ and $\lambda + n$. Using formulas for the moments of gamma distribution (1.19), we also obtain the general form of Bayes estimator as the posterior mean:

$$\theta^* = E(\theta \mid x_1, \ldots, x_n) = \frac{\alpha + \Sigma x_i}{\lambda + n}.$$

The smooth transition from prior to posterior in the problem of ten coin tosses discussed in Sections 2.3, 2.4, and 2.6 is explained by conjugate beta prior for binomial likelihood.

2.7.3   Table of Conjugate Distributions

Let us list the basic conjugate families of distributions.

Distribution of data (likelihood)	Conjugate prior/posterior
Normal (known variance)	Normal
Exponential	Gamma
Uniform	Pareto
Poisson	Gamma
Binomial	Beta
Pareto	Gamma
Negative binomial	Beta
Geometric	Beta
Gamma	Gamma

References

1 Berger, J. O. (1993). *Statistical Decision Theory and Bayesian Analysis*. Springer Verlag.

2 Berger, J. O., Bernardo, J. M., and Sun, D. (2009). The formal definition of reference priors. *Annals of Statistics*, 37(2), 905.

3 Berger, J. O., and Berry, D.A. (1988). Statistical analysis and the illusion of objectivity. *American Scientist*, 76(2), 159–165.

4 Bernardo, J., and Smith, A. F. M. (1994). *Bayesian Theory*. New York: John Wiley & Sons, Inc.

5 Berndt, E. R. (1991). *The Practice of Econometrics: Classic and Contemporary*. Addison-Wesley.

6 Bliss, E. (2012). A concise resolution to two envelope paradox, arXiv:102.4669v3.

7 Bolstad, W.M. (2007). *Introduction to Bayesian Statistics*, 2nd ed. John Wiley and Sons, Ltd.

8 Carlin, B. P., and Louis, T. A. (2008). *Bayesian Methods for Data Analysis*, 3rd ed. Oxford: Chapman & Hall.

9 Casella, G. (1985). An introduction to empirical Bayes data analysis. *American Statistician*, 39(2), 83.

10 Fisher, R. A. (1935). *The Design of Experiments*, Macmillan.

11 Gelman, A.,Carlin, J., Stern, H., and Rubin, D. (1995). *Bayesian Data Analysis*. London: Chapman & Hall.

12 Jaynes, E. T. (2003). *Probability Theory: The Logic of Science*, Cambridge University Press.

13 Jeffreys, H. (1961). *Theory of Probability*, 3rd ed. Classic Texts in the Physical Sciences. Oxford: Oxford University Press.

14 Kass, R. E., and Raftery, A. E. (1995). Bayes factors. *Journal of the American Statistical Association*, 90(430), 773–795.

15 Kass, R. E, and Wasserman, L. A. (1996). The selection of prior distributions by formal rules. *Journal of the American Statistical Association*, 91(435), 1343–1370.

16 Kruschke, J. (2014). *Doing Bayesian Data Analysis. A Tutorial with R, JAGS and Stan*, 2nd ed. Elsevier, Academic Press.

17 Lee, P. M. (1989). *Bayesian Statistics*. London: Arnold.

18 Robbins, H. (1956). An empirical Bayes approach to statistics. In: Proceedings of the Third Berkeley Symposium on Mathematical Statistics and Probability, Volume 1: Contributions to the Theory of Statistics, 157–163.

19 Rossi, P. E., Allenby, G. M., and McCulloch, R. (2006). *Bayesian Statistics and Marketing*, John Wiley & Sons, Ltd.

20 Salsburg, D. (2002). *The Lady Tasting Tea: How Statistics Revolutionized Science in the Twentieth Century*, W. H. Freeman/Owl Book.

21 Savage, L. J. (1962). Letter to J. Cornfield, *22 February 1962, Leonard Savage Papers (MS 695)*, New Haven, CT: Yale University Library, 6, 161.

22 Tsikogiannopoulos, P. (2014). Variations on the two envelopes problem. *Hellenic Mathematical Society Journal*, 3, 77.

Exercises

2.1 Suppose that the hourly wage W has a gamma distribution. Suppose that the mean value $E(W)$ of this variable is 20 USD per hour and the standard deviation $\sigma(W)$ is 5 USD.

Recall from (1.19) that gamma distribution $Gamma(\alpha, \lambda)$ has moments $E(W) = \alpha/\lambda$ and $Var(W) = \alpha/\lambda^2$.

Suppose that in the setting of the two envelope problem the job offer is applied not to annual salaries, but to hourly wages. The open envelope has an offer of 18 USD per hour. Calculate the values of the distribution density $f(w)$ for $w = 9$ and $w = 36$ (using R or other software). What is the more likely value? Which of the envelopes should be chosen?

2.2 Suppose that the sample of human life lengths in years provided on the companion website in the file *survival.xlsx* comes from a $Gamma(\alpha, \lambda)$ distribution. Estimate it using MLE (numerical optimization in Excel or R using formula (1.19), see *MLE.xlsx*) and the method of moments. Compare the results.

2.3 Supose that the sample of human life lengths in years provided in the file *survival.xlsx* comes from a $Weibull(\lambda, \tau)$ distribution. Estimate its parameters using MLE (write down the likelihood function and use numerical optimization in Excel or R).

2.4 For the following problem of deciding between two simple hypotheses (normal with unknown mean and known variance) by a sample of size 1 (single observation), $H_0 : N(0, 1)$ versus simple alternative $H_1 : N(2, 1)$ use the test statistic $\bar{X}$ and determine its critical region for

   (a) Classical test with significance level 0.05 (also, what is the power of this test?)

   (b) Bayesian test with prior probability $P(H_0) = 3/4$ using posterior odds 1:1.

2.5 Suppose that a single observation is to be drawn from the following p.d.f.: $f(x|\theta) = (\theta + 1)x^{-(\theta+2)}, \ 1 \le x < \infty$, where the value of θ is unknown. Suppose that the following hypotheses are to be tested: null hypothesis $H_0 : \theta = 0$ versus simple alternative $H_1 : \theta = 1$

   (a) Determine (in terms of the rejection region) the Bayesian test procedure, corresponding to prior probability $P(H_0) = 1/2$ and posterior odds (the Bayes factor) $1 : 1$; $3 : 1$ (substantial evidence) and $10 : 1$ (strong evidence).

   (b) Determine (in terms of the rejection region) the Bayesian test procedure, corresponding to prior probability $P(H_0) = 1/3$ and posterior odds $1 : 1$.

   (c) Determine (in terms of the rejection region) the Bayesian test procedure, corresponding to prior probability $P(H_0) = 2/3$ and posterior odds $1 : 1$.

2.6 Estimate posterior mean θ with Poisson likelihood for the exponential prior with the prior mean $E(\theta) = 2$ and the data vector $x = (3, 1, 4, 3, 2)$.

2.7 Random variable X corresponds to the daily number of accidents in a small town during the first week of January. From the previous experience (prior information), local police Chief Smith tends to believe that the mean daily number of accidents is 2 and the variance is also 2. We also observe for the current year the sample number of accidents for 5 days in a row: 5, 2, 1, 3, 3. Let us assume that X has Poisson distribution with parameter θ. Using the gamma prior (Hint: suggest values for the parameters of prior distribution using Chief Smith's previous experience and formulas for the moments of gamma distribution), determine:

   (a) the posterior distribution of the parameter θ given the observed sample;

   (b) according to Chief Smith, the Bayesian estimate of the parameter θ (posterior mean);

   (c) ignoring the prior information, the maximum likelihood estimate of the parameter θ.

2.8 Prove that using posterior mean as an estimator of θ for normal data with unknown mean and known variance $X_i \sim N(\theta, \sigma^2), i = 1, \ldots, n$ with normal prior $\pi(\theta) \sim N(\mu, \tau^2)$ brings about

$$\theta^* = \frac{\tau^2 \bar{X} + \frac{\sigma^2}{n} \mu}{\tau^2 + \frac{\sigma^2}{n}}.$$

(Hint: Use the definition of normal density, basic algebra, and complete the square in the exponential term of the formula for posterior density). Use this formula to obtain Bayesian estimate for the case $n = 10, \bar{x} = 4.3$, $\sigma^2 = 3, \mu = 0, \tau^2 = 2$.

2.9 Calculate the variances of the estimates obtained in (2.38), (2.39), and (2.40). Compare with each other and with the variance obtained in (2.37). What conclusions can you make?

2.10 Build symmetric 95% credible intervals for parameter θ estimated in (2.36), (2.38), (2.39), and (2.40) based on corresponding posterior distributions. Compare with the classical confidence interval for θ based on the assumption of approximate normal distribution of the sample mean.

2.11 Prove that binomial distribution $Bin(n, \theta)$ for fixed n with unknown probability of success θ belongs to the exponential family. Provide a factorization similar to (2.42).

2.12 Using the result of the previous problem, prove that beta family provides conjugate priors for binomial distribution.

3

Background for Markov Chain Monte Carlo

3.1 Randomization

3.1.1 Rolling Dice

It is a very old idea, even an ancient one: The idea that sometimes we lack information needed to make informed deterministic decisions, and therefore we resort to higher authorities. It happened in ancient Egypt and China, in pre-Columbian South and Central America [12], and was well-documented in Greece (e.g., the Oracle at Delphi) [6]. People used to consult the gods before making important decisions: to go to war or not to go to war, when to start harvesting, how to plan families, friendships, and strategic alliances. This communication with the higher powers was usually established through the institutes of priests equipped with special knowledge and special communication devices for communication with gods.

One of the older and more venerable devices for such communication is a die or a set of dice. In its modern version, a die is a six-sided cube with numbers or symbols on each of its sides. Older versions could be five sided (*pichica* of Incas [12]) or even double sided (a coin). Putting aside the esoteric nature of these objects, we can also consider them to be simple and handy pre-computer era tools of **randomization**.

Imagine a story from the old times of tribal wars. The Chief of a tribe, before raiding the neighbors' lands, seeks the advice of the High Priest. Priests might not talk directly to the gods, but they are capable of collecting and analyzing information. Let us say that from a historical perspective, one of the six such raids normally ends with the death of the raiders' leader. The Chief wants a deterministic forecast of the enterprise's success or failure, and does not want to know anything about probabilities. So the High Priest rolls a die. If the number "six" appears on the uppermost side, he predicts the Chief's death. If he observes a number from "one" to "five," he announces the planned raid safe. His decision is based on a single randomized experiment. A valid prediction takes into account available statistics of the raids of the past. Moreover,

Introduction to Bayesian Estimation and Copula Models of Dependence, First Edition.
Arkady Shemyakin and Alexander Kniazev.
© 2017 John Wiley & Sons, Inc. Published 2017 by John Wiley & Sons, Inc.
Companion Website: http://www.wiley.com/go/shemyakin/bayesian_estimation

additional information may be available to the High Priest: Let us say, the Chief is old and weakly. Chief's age seems to increase the chances for a lethal end to the venture. On the other hand, "old" might also mean "wise and experienced," which increases the chances of success. The degree of risk faced may depend on many other factors, which a good High Priest should take into account. However, in order to incorporate additional information, the Priest might need not a single die, but a more sophisticated randomization tool.

Think of your own experience. Have you ever tossed a coin and made a 50-50 decision based on the results of the toss? This is how we often deal with uncertainty in the situations when we need to make a definitive deterministic decision.

Computer-based randomization tools are glorified dice of the modern times. With the high speeds of operation and the extensive memory of modern computer devices, we routinely apply randomized procedures to generate multiple scenarios of the future. Stochastic simulation utilizing such scenarios assists us in decision-making in such fields as business, finance, medicine, and even such a rigorous discipline as engineering.

3.1.2   Two Envelopes Revisited

Let us review the problem of two envelopes introduced in the previous chapter to demonstrate the role of prior information. To use it for a different purpose, we have to rid this problem of its applied flavor and consider a rather abstract theoretical version [4]. Suppose you are given two sealed envelopes, one contains a number A, and the other contains a number B, which is different from A. You know nothing about A and B other than that these two are real numbers (including negative, positive, and zero), and no limits are imposed on their values. Both numbers are generated by our Opponent, say, Nature, using an undisclosed procedure, and then put into the envelopes. You have no idea how these two numbers are selected.

You are allowed to choose and open one of the envelopes, and read the number inside. Then you have to decide which of the numbers is larger: the known one you see in the open envelope or the unknown one contained in the envelope which stays sealed. The trick task is: you have to guarantee that *for any numbers A and B that are chosen by the Opponent*, you can guess the larger one with a probability strictly exceeding 0.5.

Let us formulate this task mathematically:

$$\forall A, B\ P(Guess \mid A, B) > 0.5, \tag{3.1}$$

and then consider some simple strategies.

I. *Fully Random Strategy*: Always say that the number in the opened envelope is larger. Or always say that it is smaller. Or just toss a coin and do what it says. This way you will always achieve the probability of 0.5, but will never exceed it.

II. ***Intelligent Benchmark Strategy***: Pick a number. Say, zero. Justification: Zero is "exactly in the middle of the real axis." Use zero as your benchmark in the following decision.

If $A > 0$, then choose A;
if $A < 0$, then choose B.

Notice that for all you know (or rather do not know) about the procedure with which Nature selects A and B, for the sake of simplicity we can neglect the chance that any of the two numbers is exactly equal to 0.

Strategy II is not bad! As we can easily see, it gives no advantage if A and B are both negative or both positive.

$$P(Guess \mid A < 0, B < 0) = P(Guess \mid A > 0, B > 0) = 0.5. \tag{3.2}$$

However, if two numbers lie on different sides of zero,

$$P(Guess \mid A < 0 < B) = P(Guess \mid B < 0 < A) = 1. \tag{3.3}$$

If we introduce an additional (though unjustified) assumption that both A and B are independently equally likely to be positive or negative, than $P(Guess) = 0.75$, which would be really good. However, you have not achieved the goal: You still have not found a winning strategy which would work ***for any numbers A and B that are chosen by the Opponent***. That means that if a smart opponent knows that you are going to use zero as the benchmark, he or she will always choose an A and a B of the same sign.

Notice that choosing any other fixed benchmark (positive or negative) instead of zero will bring about the same result if a smart opponent steals the value of your benchmark. That corresponds to the idea that any fixed number could be considered to be positioned "exactly in the middle" of the real axis.

III. ***Randomized Benchmark Strategy***: Pick any random real number R.

If $A > R$, then choose A;
if $A < R$, then choose B.

As we can see, similarly to Strategy II,

$$P(Guess \mid A < R, B < R) = P(Guess \mid A > R, B > R) = 0.5. \tag{3.4}$$

However,

$$P(Guess \mid A < R < B) = P(Guess \mid B < R < A) = 1. \tag{3.5}$$

If the probability of R falling between A and B is positive $P(R \in (A, B)) = p(A, B) > 0$ (this should be achieved by an intelligent random choice of R),

$$P(Guess \mid A, B) = p(A, B) + 0.5(1 - p(A, B)) = 0.5 + 0.5p(A, B) > 0.5, \tag{3.6}$$

for any numbers A and B that are chosen by the Opponent. The goal is attained!

We have to confess that there is still a question we tried to avoid so far. This question is: what exactly do we need to do to pick a random real number R so that it is guaranteed to fall between any real numbers A and B with a positive probability? Construction or generation of such random numbers is a problem we will consider now. The example above gives some motivation for the idea of randomization, which requires more thorough analysis of random number generation. Simulation of multiple future scenarios will also require procedures of generation of multiple random numbers with very specific properties.

The following section will contain basic references to the modern tools of random number generation. We will not forget about such classical randomization tools as a roll of a die or even a coin toss. However, the computer era opens up new opportunities.

3.2 Random Number Generation

3.2.1 Pseudo-random Numbers

Virtually any modern software involved with statistics is expected to be able to generate random numbers. To be exact, ***pseudo-random numbers.*** There exist various techniques of such procedures which can be reviewed in [15] or [13]. Microsoft Excel has a function $= RAND()$ and R has a command "runif" which, when applied, activate a uniform pseudo-random number generator. In general, such a generator is an algorithm which starts from an initial value $u_0 : 0 < u_0 < 1$ and uses a transformation D producing a sequence $\{u_i\} = \{D^i(u_0)\}$ of values in the interval $[0, 1]$. The transformation D, which we consider a black box (a secret of software developers) guarantees that for any n the sequence of values $(u_1, u_2, \ldots, u_n)$ reproduces the behavior of an i.i.d. sample $(V_1, V_2, \ldots, V_n)$ of uniform random variables $V_i \sim Unif[0, 1]$. By construction, sequence $\{u_i\}$ is anything but independent: each term is obtained directly from the same "seed value" u_0. However, for our purposes it is irrelevant if the resulting sequence supplies us with a uniform cover of the unit interval emulating the behavior of an independent sample. Special testing tools provide computational and statistical means to verify the performance of pseudo-random number generators, such as suggested in [16], [18], or [28].

Uniform pseudo-random number generator solves a simple problem: simulating an i.i.d. random sample with elements $V_i \sim Unif[0, 1]$. What if we need a uniform sample on an arbitrary interval $[a, b]$? We will transform the sequence $\{u_i\}$ into a sequence $\{a + (b - a)u_i\}$. It is easy to see that this new sequence will emulate the behavior of i.i.d. variables $W_i \sim Unif[a, b]$. This follows from the fact that $W_i = a + (b-a)V_i \sim Unif[a, b]$.

Similar to this simple linear transformation from one uniform sample to another uniform sample defined on a different interval, we can consider more general transformations. They will allow us to transform initial uniform sample $\{u_i\}$ into samples from many other distributions. We will review some basic techniques. For more details one can review classical sources [5, 10, 27]. For more recent development see [2], [7], [21], and [24], and also [8] and [14] with special accent on applications to finance and insurance.

3.2.2 Inverse Transform Method

Let us consider a distribution with continuous c.d.f. $F(x)$. Our goal is to use a uniform sample $\{u_i\}$ to build an i.i.d. sample from this distribution. For this purpose we will define for any $u \in [0, 1]$ the generalized inverse for $F(x)$ as $F^{-1}(u) = \inf\{x : F(x) \geq u\}$.

Lemma 3.1 *If $U \sim Unif[0, 1]$, then $V = F^{-1}(U) \sim F$.*

Proof: For any $u \in [0, 1]$ and $x \in F^{-1}([0, 1])$ it is true that $F(F^{-1}(u)) \geq u$ and $F^{-1}(F(x)) \leq x$. Therefore,

$$\{(x, u) : F^{-1}(u) \leq x\} = \{(x, u) : F(x) \geq u\} \tag{3.7}$$

and

$$P(V \leq x) = P(F^{-1}(U) \leq x) = P(U \leq F(x)) = F(x). \tag{3.8}$$

∎

Thus the sequence $\{F^{-1}(u_i)\}$ provides us with an i.i.d. sample from $F(x)$.

Here are four steps for the implementation of this procedure:

1. Write down the c.d.f. $u = F(x)$.
2. Derive the inverse $F^{-1}(u)$.
3. Using standard "black box" procedure, generate a uniform sample $\{u_i\}$.
4. Using the inverse from step 2, transform the sample from step 3 into

$$(F^{-1}(u_1), F^{-1}(u_2), \ldots, F^{-1}(u_n)) = (x_1, x_2, \ldots, x_n) \sim F(x). \tag{3.9}$$

Let us use a popular (and important) example of exponential distribution.

Exponential Distribution
To construct an i.i.d. sample from exponential distribution with distribution density function $f(x) = \lambda e^{-\lambda x}, 0 \leq x < \infty$ with parameter $\lambda > 0$, we can

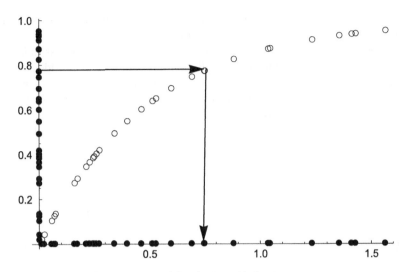

Figure 3.1 Sampling from exponential distribution with $\lambda = 2$.

follow the steps as demonstrated in Figure 3.1:

1. $F(x) = \int_0^x \lambda e^{-\lambda t}\, dt = 1 - e^{-\lambda x}$.
2. $u = 1 - e^{-\lambda x} \Rightarrow e^{-\lambda x} = 1 - u \Rightarrow -\lambda x = \ln(1 - u) \Rightarrow x = F^{-1}(u) = -\lambda^{-1}\ln(1 - u)$.
3. Using standard procedure, generate a uniform sample $(u_1, u_2, \ldots, u_n)$.
4. Using the inverse from step 2, transform the sample from step 3 into

$$(-\lambda^{-1}\ln(1 - u_1), \ldots, -\lambda^{-1}\ln(1 - u_n)) = (x_1, \ldots, x_n) \sim Exp(\lambda).$$

3.2.3 General Transformation Methods

The inverse transform method is not easy to carry out for all distributions of interest. The usual problem is with step 2: obtaining an analytical form of the inverse c.d.f. However, from Chapter 1 we know that there exist some relationships between distribution families facilitating our task. For instance, gamma distribution is related to the exponential (see Section 1.4), and a similar relationship exists between beta and gamma distributions. More details on relationships between different distribution functions may be found in [1]. Being able to efficiently generate samples from exponential distribution using inverse transform is not the end of the road. It also helps to deal with some other distributions. Simple examples of random number generation using transformation methods are summarized on the companion website in **Appendices** an Excel file ***pseudorandom.xlsx***. For examples in R see [25].

Gamma and Beta Distributions for Integer Parameters

Suppose we can generate samples from exponential distribution $\{x_i\} \sim Exp(\lambda)$. Using the fact that if for integer α and independent X_j

$$X_j \sim Exp(\lambda),\ j = 1, \ldots, \alpha,\ \text{then}\ Y = \sum_{j=1}^{\alpha} X_j \sim Gamma(\alpha, \lambda)$$

(see Section 1.4), we can:

1. Generate α different samples from the exponential: $(x_{ij}) \sim Exp(\lambda)$, $j = 1, \ldots, \alpha$, $i = 1, \ldots, n$.
2. Set $y_i = \sum_{j=1}^{\alpha} x_{ij} \sim Gamma(\alpha, \lambda)$ for all $i = 1, 2, \ldots, n$.

Similarly, for integer α and β we can also use (1.27) from Section 1.5 and do the following:

1. Generate $\alpha + \beta$ samples $\{x_{ij}\} \sim Exp(1)$.
2. Set $y_i = \dfrac{\sum_{j=1}^{\alpha} x_{ij}}{\sum_{j=1}^{\alpha+\beta} x_{ij}} \sim Beta(\alpha, \beta)$ for all $i = 1, 2, \ldots, n$.

Normal Distribution (Box–Mueller Algorithm)

The following algorithm is one of the most popular methods to generate samples from standard normal distribution. This or similar methods are built in many standard software packages. We make use of the following representation for bivariate standard normal distribution (zero mean and unit covariance matrix) with two independent components

$$(X_1, X_2) \sim MN(0, I) \Rightarrow X_1^2 + X_2^2 \sim Exp(0.5),$$

$$\arctan \frac{X_2}{X_1} \sim Unif[-\pi/2, \pi/2],$$

and then apply a two-step procedure:

1. Generate $(u_{i1}, u_{i2}) \sim Unif[0, 1]$, $i = 1, \ldots, n$.
2. Define for all $i = 1, 2, \ldots, n$

$$\begin{cases} x_{i1} = \sqrt{-2\ln u_{i1}}\ \cos(2\pi u_{i2}), \\ x_{i2} = \sqrt{-2\ln u_{i1}}\ \sin(2\pi u_{i2}). \end{cases} \tag{3.10}$$

A little extra work, but as a result we obtain two independent samples from standard normal distribution (or one larger sample of size $2n$). This is our first illustration of using a higher-dimensional technique to solve an essentially one-dimensional problem. Such examples are common for the implementation of Monte Carlo methods.

Discrete Distributions on Finite Sample Spaces

Suppose we need to generate data from a discrete distribution. For the sake of simplicity, assume that all possible values are integers and there is a finite number of values $p_k = P(X = k), k = 1, 2, \ldots, m$.

1. Generate a sample $(u_1, u_2, \ldots, u_n) \sim Unif[0, 1]$.
2. Assign for all $i = 1, 2, \ldots, n$

$$x_i = 1, \text{ if } 0 \leq u_i < p_1,$$
$$x_i = 2, \text{ if } p_1 \leq u_i < p_1 + p_2,$$
$$x_i = 3, \text{ if } p_1 + p_2 \leq u_i < p_1 + p_2 + p_3,$$
$$\ldots$$
$$x_i = m \text{ if } 1 - p_m \leq u_i < 1.$$

The sample $(x_1, \ldots, x_n)$ has the desired discrete distribution. This generator will also work for any noninteger values, but for this construction it is important that the sample space is finite.

Two final examples in this subsection are less straightforward. They require more work and are not always very efficient. Moreover, the performance of these two capricious random number generators depends on circumstances and cannot be fully predicted in advance.

Poisson Distribution

In a Poisson process, times between two events are distributed exponentially as mentioned in Section 1.4. It can be expressed as the following statement:

$$X_j \sim Exp(\lambda), Y \sim Poisson(\lambda) \Rightarrow P(Y = k) = P\left(\sum_{i=1}^{k} X_i \leq 1 < \sum_{i=1}^{k+1} X_i \right),$$

indicating that one needs to add up a random Poisson number of exponential variables to get to 1. Then we can

1. Generate exponential variables $x_{i1}, x_{i2}, \ldots$ and add them up until $\sum_{j=1}^{k+1} x_{ij} > 1$.
2. Then the count k which takes you to 1 ($k + 1$ takes you over) is the value of $y_i \sim Poisson(\lambda)$.
3. Repeat for $i = 1, 2, ..n$.

This procedure may request unlimited number of summands on step 1. It becomes prohibitively complicated for large values of parameter λ. Therefore this sampling procedure, though it is elementary, is rarely used for practical purposes. Methods similar to those discussed in the following subsection, are more common.

Gamma and Beta for Arbitrary Parameters

If parameters of beta distribution are not necessarily integer, one can use a general representation

$$\textit{Independent } U, V \sim \textit{Unif}[0,1] \Rightarrow X = U^{\frac{1}{\alpha}} \left(U^{\frac{1}{\alpha}} + V^{\frac{1}{\beta}} \right)^{-1} \sim \textit{Beta}(\alpha, \beta),$$

which holds if $U^{\frac{1}{\alpha}} + V^{\frac{1}{\beta}} \leq 1$. Then the procedure is clear, but the additional condition has to be checked.

1. Generate independent uniform variables $(u_i, v_i) \sim \textit{Unif}[0,1]$.
2. If $u_i^{\frac{1}{\alpha}} + v_i^{\frac{1}{\beta}} \leq 1$ assign $x_i = u_i^{\frac{1}{\alpha}} \left(u_i^{\frac{1}{\alpha}} + v_i^{\frac{1}{\beta}} \right)^{-1} \sim \textit{Beta}(\alpha, \beta)$. Otherwise, skip the ith element.
3. Repeat for $i = 1, 2, \ldots, n$.

The qualifying condition on the first step has to be checked for each i, and if it does not hold, the ith values from the uniform samples are wasted. Effectively, it may bring about much smaller samples from beta distribution. If we can sample from beta, we also can sample from gamma with $\alpha < 1$ using the following property (see Section 1.5, (1.28), also [1]):

$$X \sim \textit{Beta}(\alpha, 1 - \alpha), \alpha < 1; W \sim \textit{Exp}(1) \Rightarrow Y = XW \sim \textit{Gamma}(\alpha, 1).$$

The size of the beta sample will also determine the size of the gamma sample.

Wasting our time and effort on the first step to generate values, which are then rejected on the second step seems undesirable. However the idea of algorithms rejecting some of the generated values will soon be proven to be very productive.

3.2.4 Accept–Reject Methods

If simulation from a distribution is difficult or impossible to organize using the method of inverse transform and generators suggested in the previous subsection, we might use a different trick based on the idea that sometimes a problem which is difficult to solve in one dimension can be reduced to a simpler problem in two dimensions. Let us establish the following theorem which, following Robert and Casella [24], we will characterize as *the fundamental theorem of simulation*.

Suppose that the distribution of interest has p.d.f. $f(x)$ (*target density*). Let us define *uniform coverage* $\{(x_i, y_i)\}, i = 1, \ldots, n$ of a region $A \subset \mathbf{R}^2$ as such a set of independently chosen points in $\mathbf{R}^2$, that for any $B \subseteq A$ the probability $P((x_i, y_i) \in B) = \textit{Area}(B)/\textit{Area}(A)$ and denote $\{(x_i, y_i)\} \sim \textit{Unif}(A)$. Then the following statement holds:

Theorem 3.1 *If a uniform coverage* $\{(x_i, y_i)\} \sim Unif(S)$ *is generated for the region under the graph (subgraph) of the target density* $S = \{(x, u) : 0 < u < f(x)\}$, *the first coordinates of the coverage form an i.i.d. sample from the distribution with target density* $x_i \sim f(x)$, *and for any* x_i *such that* $f(x_i) > 0$ *and is finite, the second coordinates are uniform,* $u_i \sim Unif[0, f(x_i)]$.

Proof: The idea of the proof consists in representing the target density as $f(x) = \int_0^{f(x)} du$. Then it can be interpreted as the marginal density of the joint distribution of $(X, U) \sim Unif(S)$. ∎

Thus the problem of generating a sample from $X \sim f(x)$ is equivalent to the problem of generating a uniform coverage or a sample from two-dimensional uniform distribution. How can adding one dimension simplify the task of sampling from $f(x)$? Direct generation of a uniform coverage over a curvilinear area may not be straightforward. However, in case when the target density has a finite domain $[a, b]$ and is bounded from above $\forall x \in [a, b] f(x) \leq M$, this curvilinear area is fully contained in a rectangle $[a, b] \times [0, M]$ and we can suggest the following two-step ***accept–reject*** procedure:

1. Generate uniform samples $x_i \sim Unif[a, b], u_i \sim Unif[0, M], i = 1, 2, \ldots, n$.
2. For all $i = 1, 2, \ldots, n$ check an additional condition:

if $u_i < f(x)$ then assign $y_i = x_i$ (accept),
if $u_i \geq f(x)$ then skip (reject).

Resulting sample $\{y_i\}$ may be of a smaller size due to possible rejections, but each of its random elements $\{Y_i\}$ will have the target density $f(x)$ since

$$P(Y_i \leq y) = P(X_i \leq y | U_i \leq f(X_i)) = \frac{\int_0^y \int_0^{f(x)} du dx}{\int_0^1 \int_0^{f(x)} du dx} = \int_0^y f(t) dt. \quad (3.11)$$

Beta Distribution
Beta distribution has a finite domain $[0, 1]$ and for $\alpha \geq 1, \beta \geq 1$ is uniformly bounded from above, so it satisfies all the requirements of the above accept–reject construction. We can set $a = 0$, $b = 1$ and determine m as any upper bound for $Beta(\alpha, \beta)$. The initial sample from two-dimensional uniform is shown by circles in Figure 3.2.

Resulting sample $\{y_i\}$ will have a random size K depending on the number of accepted values x_i. It is shown by filled circles in Figure 3.3, while rejected points from the previous picture stay unfilled.

Now we can address two related questions: first, how large or how small K is; and second, how important is the tightness of bound M? Both questions can be addressed by calculating the probability of acceptance and hence the expected

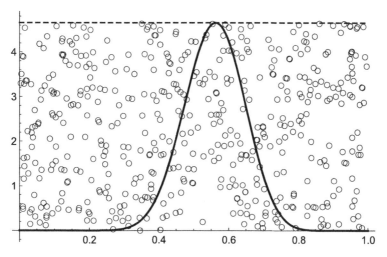

Figure 3.2 Uniform sample and Beta(19,15) p.d.f.

value of a binomial random variable K:

$$E(K) = nP(accept) = nP(U_i \leq f(X_i)) = \frac{n}{M} \int_0^1 \int_0^{f(x)} du\,dx = \frac{n}{M}. \quad (3.12)$$

The tightness of the upper bound determines the relative loss in the resulting sample size with the respect to the initial sample. Therefore, the closer M is to the maximum of the target density, the better. In the special case of

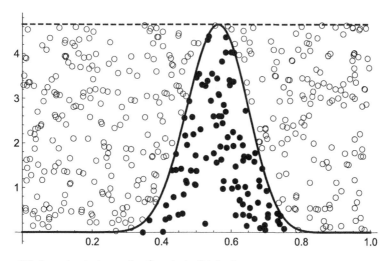

Figure 3.3 Accept–reject sampling from beta distribution.

Table 3.1 Accept–reject method for Beta(19,15)

i	x_i	$f(x_i)$	u_i	y_i
1	0.08	8.8E−11	0.63	Reject
2	0.55	4.63	3.59	0.55
3	0.89	4.17E−5	0.37	Reject

Beta(19, 15) discussed in Chapter 2, we can calculate the maximum obtained at $\theta = (\alpha - 1)/(\alpha + \beta - 2) = 18/32 = 0.5625$ as $f(0.5625) \approx 4.564$, meaning that the sample size drops from n to K on the average by more than four times. In other words, in order to generate a resulting sample of the needed size, one has to take at least four times as many elements of the initial sample. Looks like a bad deal, but this is still a vast improvement relative to the method suggested in the beginning of the subsection, which requires generating $\alpha + \beta = 34$ elements in order to obtain just one element of a beta sample.

Let us provide a short numerical example with $n = 3$. In Table 3.1 we draw three points x_i uniformly from 0 to 1, calculate the values of *Beta*(19, 15) p.d.f. at these points, $f(x_i)$, and compare these values for each $i = 1, 2, 3$ with the draws u_i from the uniform distribution on $[0, M = 4.654]$. If $u_i < f(x_i)$, then accept and record $y_i = x_i$, otherwise reject.

General Case

The requirements of finite domain and uniform upper bound can be released, if there exists such a density $g(x)$ that for all x there exists such a bound M (not necessarily tight) that $f(x) \leq Mg(x)$. In this case the following modification of the accept–reject algorithm is suggested.

1. Generate samples $x_i \sim g(x), u_i \sim Unif[0, 1], i = 1, 2, \ldots, n$.
2. For all $i = 1, 2, \ldots, n$ check an additional condition:

if $u_i < \dfrac{f(x_i)}{Mg(x_i)}$ then assign $y_i = x_i$ (accept),

if $u_i \geq \dfrac{f(x_i)}{Mg(x_i)}$ then skip (reject).

Resulting sample $\{y_i\}$ from the target density may be of a smaller size, which depends on the tightness of the bound M.

Let us consider a familiar example of gamma distribution. Suppose we need to generate a sample from *Gamma*(4, 2), with density

$$f(x) = \frac{2^4}{\Gamma(4)} x^3 e^{-2x} = \frac{8}{3} x^3 e^{-2x}, \tag{3.13}$$

which is not hard to do using the methods from Subsection 3.2.3. However, if we choose accept–reject technique instead, it will require a choice of instrumental

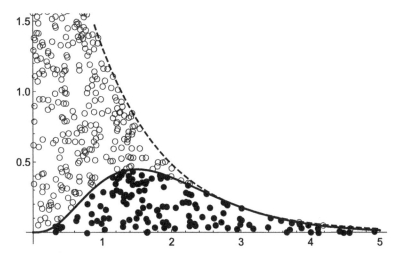

Figure 3.4 Accept–reject sampling from gamma distribution.

density $g(x)$ different from the uniform: we will need instrumental distribution defined on $(0, \infty)$, which will also "dominate" the tail of $f(x)$ in the sense that there exists a constant M such that for all $x \geq 0$

$$f(x) \leq Mg(x). \tag{3.14}$$

Fortunately, such a choice is easy to make: exponential distribution $Exp(1)$ with density $g(x) = e^{-x}$ is easy to generate from, and the ratio $f(x)/g(x)$ is conveniently maximized at $x = 3$ and inequality (3.14) is satisfied with $M = 72e^{-3} \approx 3.585$. The algorithm follows:

1. Generate two independent samples $u_i, v_i \sim Unif[0, 1], i = 1, 2, \ldots, n$.
2. Transform $x_i = -\ln(1 - u_i)$ creating a sample $x_i \sim Exp(1)$.
3. For all $i = 1, 2, \ldots, n$ check an additional condition:

   if $v_i < \dfrac{f(x_i)}{Mg(x_i)} = \dfrac{1}{27}x^3 e^{3-x}$ then assign $z_i = x_i$ (accept), otherwise skip (reject).

Figure 3.4 demonstrates a sample y_i covering the subgraph of $Mg(x)$ with accepted points z_i depicted as filled circles and rejected point staying unfilled. Simple examples of applications of accept–reject technique in Microsoft Excel are offered in ***acceptreject.xlsx***.

3.3 Monte Carlo Integration

Random number generation developed in Section 3.2 is not the ultimate goal. It will be used as a tool assisting with problems of estimating distribution moments, probabilities, or simulating multiple scenarios of the future. All these problems can be better understood in context of numerical integration.

3.3.1 Numerical Integration

Numerical integration is the last resort when an integral cannot be evaluated analytically and closed form solutions are not available. There exists a host of techniques which could be reduced to the following simple scheme. Suppose that $D \subset R$ is a subset of the real axis, $g(x)$ is a real-valued function, and our goal is to evaluate the definite integral

$$I = \int_D g(x)dx.$$

In order to define the procedure of numerical integration we need to define a partition of D into nonoverlapping subintervals Δ_i, $i = 1, 2, \ldots, n$, $\bigcup_{i=1}^{n} \Delta_i = D$ and a choice of points $x_i^* \in \Delta_i$ and weights $w_i : \sum_{i=1}^{n} w_i = 1$. Then integral I can be approximated by

$$I_n = \sum_{i=1}^{n} w_i |\Delta_i| g(x_i^*).$$

The art of numerical integration consists in choosing the partition, points inside the intervals, and weights in such a way that I_n becomes a good approximation of I without requiring too many subintervals. Smart choices bring about good results for relatively small partition size n.

Suppose now that the integral of interest has the form

$$I = \int_D h(x)f(x)dx, \tag{3.15}$$

where $f(x)$ is a p.d.f. of a certain distribution. This representation requires a factorization of the integrand $g(x)$ into two multiples: $h(x)$ and a p.d.f. $f(x)$. This factorization is not always obvious and does not have to be unique. However in order to use the following method of numerical integration, first of all we have to represent the integral of interest in the form (3.15), which requires us to explicitly define both factors $h(x)$ and $f(x)$ forming the integrand $g(x) = h(x)f(x)$.

Monte Carlo is the historical capital of gambling industry having a lot to do with random choices. That is where the name is coming. The idea of the **Monte Carlo method**, instead of trying to find a smart choice of points for numerical integration, is to choose the points randomly. It is interesting how the historical development of Monte Carlo methods is intertwined with the development of modern physics, and in particular with such applications as A-bomb and H-bomb development. As we have mentioned before, the idea of Monte Carlo simulation is not new, but in its modern form its development is closely tied to the emergence of the first computer, ENIAC, in 1946. Monte Carlo approach was discussed and developed for computer implementation by such renowned scholars as Stanislaw Ulam and John von Neumann and christened, probably, by a physicist Nicholas Metropolis. The paper of Metropolis and Ulam [19] was the first publication presenting the method.

Monte Carlo method works in two steps.

1. Generate a random sample i.i.d. $x = (x_1, \dots, x_n) \sim f(x)$.

2. Calculate $\bar{h}_n = \frac{1}{n} \sum_{i=1}^{n} h(x_i)$.

The justification of the method is simple: $\bar{h}_n$ is the sample mean of the function $h(X)$ while integral I is the distribution mean $E_f h(X)$, where the subscript f indicates the distribution of X. Then the Law of Large Numbers (LLN) under very general conditions implies

$$\bar{h}_n \to E_f h(X), \tag{3.16}$$

where the exact form of convergence for $n \to \infty$ (in probability, almost sure, etc.,) requires its own qualifying conditions. Random variable $\bar{h}_n$ by construction has variance

$$Var(\bar{h}_n) = \frac{1}{n} \int_D [h(x) - E_f h(X)]^2 f(x) dx, \tag{3.17}$$

which can be estimated from the sample by

$$v_n = \frac{1}{n^2} \sum_{i=1}^{n} [h(x_i) - \bar{h}_n]^2, \tag{3.18}$$

and the Central Limit Theorem (CLT) will also apply:

$$\frac{\bar{h}_n - E_f h(X)}{\sqrt{v_n}} \sim N(0, 1).$$

This will allow us to estimate the error of approximation in (3.16). Exact conditions for LLN and CLT to apply are discussed, for instance, in [24] and [14], and will be not of our concern for now.

3.3.2 Estimating Moments

Monte Carlo numerical integration can clearly help to estimate distribution moments. Out first example though will be a textbook integral with no analytical solution

$$I = \int_0^1 \sqrt{1 + x^3} dx.$$

We can treat this integral as an integral of form (3.15) with $D = [0, 1]$, $h(x) = \sqrt{1 + x^3}$ and $f(x) \sim Unif[0, 1]$. The algorithm is then clear:

1. Generate a random sample i.i.d. $u = (u_1, \ldots, u_n) \sim Unif[0, 1]$.

2. Calculate $\bar{h}_n = \frac{1}{n} \sum_{i=1}^{n} \sqrt{1 + u_i^3}$.

Let us do it for $n = 100$. We repeat this experiment three times independently, and our answers are $\bar{h}_n = 1.13; 1.10; 1.11$. In this case we cannot obtain the closed form solution, therefore it is difficult to judge the quality of Monte Carlo approximation.

Our second example will involve $\theta \sim Beta(19, 15)$, which has been discussed before. Our goal will be to estimate the mean value $E(\theta)$. Let us pretend for a while that we do not know that the exact solution can be obtained from (2.25) as $E(\theta) = \alpha/(\alpha + \beta) = 19/34 \approx 0.5588$. The benefit of applying a numerical algorithm to a problem with the known exact solution is the ability to obtain the immediate estimate of the error.

If we represent $E(\theta)$ in the form of (3.15) with $X = \theta$, $D = [0, 1]$, $h(\theta) = \theta$, and $f(\theta) \sim Beta(19, 15)$, Monte Carlo method will work as follows:

1. Generate a random sample i.i.d. $\theta = (\theta_1, \ldots, \theta_n) \sim Beta(19, 15)$. Several ways to do it were discussed in Section 3.2.

2. Calculate $\bar{h}_n = \frac{1}{n} \sum_{i=1}^{n} \theta_i$.

If we use, for instance, the accept–reject method from the previous section, which can be easily implemented in Microsoft Excel (see Exercises), we obtain for samples $n = 100$ repeated three times the values $\bar{h}_n = 0.54; 0.55; 0.56$. The discrepancy between the exact and numerical solutions is not huge, but it still reveals an unpleasant truth that the convergence of Monte Carlo method is often painstakingly slow and even in a relatively simple example $n = 100$ is not enough to establish good precision.

3.3.3 Estimating Probabilities

If a probability of interest can be expressed as an integral, Monte Carlo method can help with its estimation. Let us introduce indicator function of set $A \subset D$ as

$$I_A(X) = \begin{cases} 1, & X \in A \\ 0, & X \notin A \end{cases},$$

where X is a random variable with p.d.f. $f(x)$. Then

$$P(A) = \int_A f(x)dx = \int_D I_A(x)f(x)dx = E_f I_A(x), \tag{3.19}$$

and assuming $h(x) = I_A(x)$, probability (3.19) can be estimated by a sample $x = (x_1, \ldots, x_n)$ as

$$\bar{h}_n = \frac{1}{n} \sum_{i=1}^{n} I_A(x_i), \tag{3.20}$$

the proportion of sample points hitting set A.

Number π

Let us inscribe a circle C with center at $(0,0)$ and radius 1 into a square $S = [-1, 1] \times [-1, 1]$. Probability of a point randomly thrown onto S to land inside the circle is

$$P(C) = \frac{Area(C)}{Area(S)} = \frac{\pi}{4}.$$

Then if we generate a random sample uniformly covering S and estimate probability $P(C)$ by $\bar{h}_n = \frac{1}{n} \sum_{i=1}^{n} I_C(x_i)$, the proportion of sample points, we can also obtain an estimate for π: $\hat{\pi} = 4\hat{P}(C) = 4\bar{h}_n$. The exact sequence of actions is:

1. Generate a paired sample $\{(u_i, v_i)\} \sim Unif[0, 1]^2$.
2. Transform it into $\{(x_i, y_i)\} \sim Unif(S)$ setting $x_i = 2u_i - 1$ and $y_i = 2v_i - 1$.
3. Count all points inside the circle by checking condition $x_i^2 + y_i^2 \leq 1$. This count will provide $\bar{h}_n = \frac{1}{n} \sum_{i=1}^{n} I_C(x_i)$.
4. The estimate of number π is $\hat{\pi} = 4\bar{h}_n$.

In three experiments with $n = 100$ in each we obtained $\hat{\pi} = 3.00; 3.24; 3.12$. The precision of estimation is worse than in the previous examples. It reflects the general rule: Higher-dimensional simulation requires larger sample sizes.

Tails of Normal Distribution

One of the biggest inconveniences of the probability theory is that the integral defining the c.d.f. of the standard normal distribution

$$\Phi(y) = \int_{-\infty}^{y} \phi(x)dx = \frac{1}{\sqrt{2\pi}} \int_{-\infty}^{y} e^{-\frac{x^2}{2}} dx$$

cannot be evaluated analytically. We can rely on tables or a black box computer software. But we can also approximate $\Phi(y)$ using Monte Carlo approach with $h(x) = I_{(-\infty, y)}(x)$:

1. Generate a sample $x = (x_1, \ldots, x_n) \sim \phi(x)$.
2. Count all points x_i such that $x_i \leq y$.
3. Estimate $\Phi(y) = E_\phi h(X)$ using $\bar{h}_n = \frac{1}{n} \sum_{i=1}^{n} I_{(-\infty, y)}(x_i)$.

3.3.4   Simulating Multiple Futures

The Black–Scholes formula is a classical Nobel Prize winning result in mathematical finance. It provides a closed form solution to the problem of pricing European options. Let us define European call option as the right to buy (or not to buy) a share of certain stock at a designated future time T (exercise time) for the fixed price K (strike price). An option is "in the money" and should be exercised (the right to buy used) if at time T stock price S_T exceeds K. The option owner's gain is $S_T - K$, the difference between the future market value of stock and its strike price. If however $S_T \leq K$, there is no point in exercising the option, so the option's owner lets the option expire. The value of the call option C_T at time T is determined as $C_T = \max\{S_T - K, 0\}$.

The present value of the option C_0 at time 0 should be equal to the discounted expected value of the call at time T, where r is the risk-free interest rate assumed in the calculation

$$C_0 = e^{-rT} E(C_T) = e^{-rT} E(\max\{S_T - K, 0\}).$$

This expectation is not easy to evaluate. The Black–Scholes formula does exactly that under certain additional assumptions. The most important of these assumptions is the "lognormal" model for the price of underlying stock

$$S_T = S_0 \exp\left\{\left(\mu - \frac{\sigma^2}{2}\right) T + \sigma\sqrt{T}X\right\}, \tag{3.21}$$

where μ is the expected return per year, σ is the measure of spread in prices known as volatility, and X is standard normal variable modeling uncertainty in future stock prices.

An alternative to Black–Scholes formula is a Monte Carlo simulation. For this simulation we will need to:

1. Generate multiple future scenarios $x = (x_1, \ldots, x_n) \sim \phi(x) \sim N(0, 1)$.
2. Calculate corresponding n values $S_T(x_i)$.
3. Evaluate $h(x_i) = \max\{S_T(x_i) - K, 0\}$.
4. Use $\bar{h}_n = \frac{1}{n}\sum_{i=1}^{n} h(x_i)$ to estimate $E_\phi h(X)$.
5. Estimate the present value of the option $\hat{C}_0 = e^{-rT}\bar{h}_n$.

For a numerical example, consider $S_0 = K = 50$, $T = 0.5$ (year), $\mu = r = 0.05$, and $\sigma = 0.3$. Black–Scholes formula yields $C_0 = 4.817$. Running $n = 100$ random future scenarios three times gives us $\hat{C}_0 = 4.57; 4.72; 3.58$. These values are not exactly close. See **MCintegration.xlsx** for a simple Microsoft Excel implementation of this procedure. What is the practical point of using Monte Carlo for option price calculation? First of all, we can use larger sample sizes $n = 10,000$ or even $n = 10^6$ or more. Modern computer will do it in a split second. However, we cannot ever beat the exact solution, can we?

Unfortunately for Black–Scholes model, it has its own limitations. Its rise and fall in modern finance is related to the assumptions of the formula being too stringent and often violated at today's markets. One advantage of the Monte Carlo approach is its additional flexibility w.r.t. the stock price model. If log-normal model (3.21) is modified to allow for variable volatility and nonnormal returns, it is relatively easy to reflect these modifications in Monte Carlo calculations. Black–Scholes formula is much harder to modify.

All the examples from this section were performed and are replicated on the book website in Microsoft Excel format in *MCintegration.xlsx*. All of them share one common feature: The sample sizes we suggest for illustration purposes do not guarantee a high precision of the Monte Carlo method. One way out of this bind would be to use brute force and dramatically increase the sample sizes. Another possible way out is to analyze the precision in terms of variance of Monte Carlo estimates and to recommend some means of variance reduction.

3.4 Precision of Monte Carlo Method

In this section we address the precision of Monte Carlo integration; in other words, the rate of convergence in (3.16). First we discuss the ways to measure precision and then suggest some methods of its improvement.

3.4.1 Monitoring Mean and Variance

As discussed in the previous section, the main idea of Monte Carlo integration is to generate a random sample $x = (x_1, \ldots, x_n)$ and estimate integral (3.15) via the sample mean

$$\bar{h}_n = \frac{1}{n} \sum_{i=1}^{n} h(x_i). \tag{3.22}$$

The simplest way to see how good is the estimate and to observe its improvement with the increase of the sample size is to look at accumulated sample means $\bar{h}_k; k = 1, \ldots, n$. Figure 3.6 demonstrates cumulative sample means $\bar{h}_k$ with $n = 1000$ for the case of $h(x) = (\cos(50x) + \sin(20x))^2$ and $f(x) \sim Unif[0, 1]$, estimating area under the graph in Figure 3.5 as a trigonometric integral

$$I = \int_0^1 (\cos(50x) + \sin(20x))^2 dx. \tag{3.23}$$

We do not need to know the exact value of this integral in order to see that the trajectory of sample means converges somewhere in the vicinity of 1. Variance $Var(\bar{h}_n)$, when it can be easily calculated, clearly measures the precision of

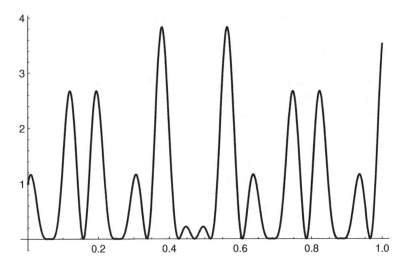

Figure 3.5 Graph of $h(x)$.

estimation and we will use this approach when we can. However, in the absence of convenient formulas for the variance, one can also estimate $Var(\bar{h}_k)$ using accumulated sample squared errors v_k defined for $k = n$ in (3.18) and available directly from the data. Figure 3.7 demonstrates accumulated sample variances $\bar{h}_k$ with $n = 1000$.

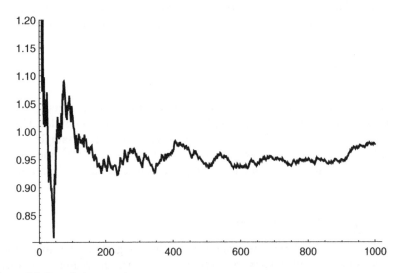

Figure 3.6 Cumulative sample means.

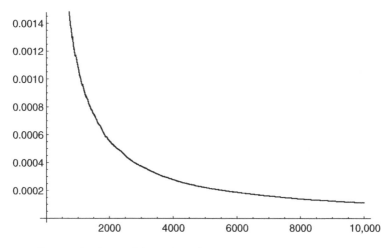

Figure 3.7 Estimated variance of the estimated means.

If the Central Limit Theorem can be applied, we can obtain approximate 95% confidence intervals for $\bar{h}_k$ as can be seen on Figure 3.8, where the dotted lines correspond to the upper and lower bounds of the interval.

An alternative method to monitor variance and precision suggests running parallel multiple samples. The graph of three accumulated means $\bar{h}_k$ for independent samples of the same size 1000 is demonstrated in Figure 3.9.

When all three trajectories visually converge, it is a good indication of precision improving with sample size.

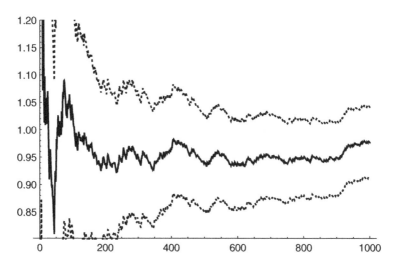

Figure 3.8 Confidence intervals for means.

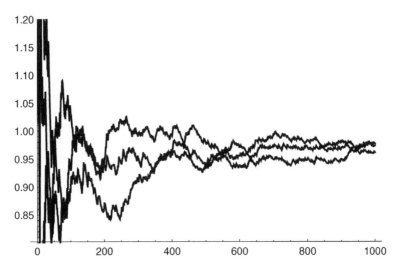

Figure 3.9 Multiple trajectories for estimated means.

3.4.2 Importance Sampling

Let us consider the p.d.f. of Cauchy distribution restricted to positive semiaxis:

$$X \sim f(x) = \frac{2}{\pi} \frac{1}{1+x^2}, \quad 0 < x < \infty. \tag{3.24}$$

We consider the problem of estimating probability $p = P(X > 1)$. Using substitution $u = x^{-1}$ in the integral below, we can easily see that

$$p = \frac{2}{\pi} \int_1^\infty \frac{1}{1+x^2} dx = \frac{2}{\pi} \int_0^1 \frac{1}{1+u^2} du = 1 - p, \tag{3.25}$$

hence 1 is the median of the distribution (3.24) therefore $p = 0.5$. Knowledge that $p = 0.5$ is not required in the further analysis, but it is convenient for assessing the results. As usual, let us pretend that we do not know the exact solution and will try to approximate it by a Monte Carlo method. Denote $h(X) = I_{(1,\infty)}(X)$ and then represent

$$p = P(X > 1) = \int_0^\infty h(x)f(x)dx,$$

which allows us to draw a sample $x = (x_1, \ldots, x_n)$ from the distribution of X and use the estimate

$$\hat{p} = \bar{h}_n = \frac{1}{n} \sum_{i=1}^n I_{(1,\infty)}(x_i).$$

Notice that due to (3.25) we can use a different $h(X) = I_{(0,1)}(X)$ with the same $f(x)$ to estimate $1 - p = P(X \leq 1)$ and then take $\hat{p} = 1 - \bar{h}_n$. The difference is in counting successes versus counting failures in a series of binomial trials. It might be easier in the sense that the points to be counted belong to a finite interval $[0, 1]$, not to the infinite semi-interval $[1, \infty)$. Variance will stay the same due to the symmetry of binomial distribution. Sampling from restricted Cauchy distribution from 0 to ∞ can be carried out by a simple inverse transform taking into account $P(X \leq x) = \frac{2}{\pi} \arctan x$ and creating a sample of $x_i = \tan[\frac{\pi}{2} u_i]$, where $u_i \sim Unif[0, 1]$.

We ran three independent experiments with $n = 100$ and obtained values $\hat{p} = 0.54; 0.47; 0.48$. Notice that in this case experiments are not even needed to judge the precision of estimation. By construction, $n\hat{p} \sim Bin(n, p)$ so $Var(\hat{p}) = \frac{1}{n} p(1 - p)$. For $n = 100$ and $p = 0.5$ we get $Var(\hat{p}) = 0.0025$ corresponding to standard deviation of 0.05. If we do not know the exact value of p, we can use the estimate $\hat{p}$ instead.

However we can significantly reduce the variance of this estimate thus increasing precision of Monte Carlo integration. This improvement can be provided by **importance sampling**. Let us notice that in the formula

$$p = P(X > 1) = 1 - \int_0^{\infty} h(x) f(x) dx = \frac{2}{\pi} \int_0^1 \frac{1}{1 + x^2} dx$$

factorization of the integrand into $h(x)$ and $f(x)$ is not unique. What if we make a different choice:

$$h^*(x) = \frac{2}{\pi} \frac{1}{1 + x^2}, \; f^*(x) \sim Unif[0, 1],$$

and then use the well-established procedure of sampling from uniform distribution to obtain u_i, and then calculate

$$\hat{h}_n^* = \frac{1}{n} \sum_{i=1}^n \frac{2}{\pi} \frac{1}{1 + u_i^2}?$$

The sequence of actions is very transparent. The most important result of this change is the reduced variance. Direct integration yields

$$Var(\hat{p}) = \frac{1}{n} Var\left(\frac{2}{\pi} \frac{1}{1 + x^2}\right) = \frac{4}{n\pi^2} \left(\int_0^1 \frac{dx}{(1 + x^2)^2} - \left(\int_0^1 \frac{dx}{1 + x^2}\right)^2\right)$$

$$= \frac{4 + 2\pi - \pi^2}{4n\pi^2} \approx \frac{0.01}{n},$$

bringing about the standard deviation of 0.01 for $n = 100$. Three experiments we performed with $n = 100$ yielded $\hat{p} = 0.503; 0.488; 0.5$.

General scheme of importance sampling requires designation of **instrumental density** $f^*(x)$ and is based on the equality

$$E_f h(X) = \int h(x) f(x) dx = \int \left[h(x) \frac{f(x)}{f^*(x)} \right] f^*(x) dx = E_{f^*} \left[h(X) \frac{f(X)}{f^*(X)} \right],$$

where $h^*(X) = \left[h(X) \frac{f(X)}{f^*(X)} \right]$ is the importance function. In our restricted Cauchy example $f^*(x) \sim Unif[0, 1]$, but in general we could use any convenient density.

At this point we might notice that accept–reject sampling with uniform instrumental density could be also used instead of importance sampling. However it would end up with a smaller sample size, which could undo the effect of higher precision.

What is the meaning of the term *importance sampling*, and how can it help with variance reduction? In our example, the reduced variance is achieved because the uniform sample is concentrated on the smaller interval $[0, 1]$, where the density $f(x)$ is higher instead of spanning entire semiaxis $(0, \infty)$. All points in the sample x_i are assigned weights $h^*(x_i)$ proportional to the initial density $f(x)$, so the resulting estimate is obtained as a weighted average, where the weights reflect the points' relative importance.

3.4.3 Correlated Samples

As suggested in the beginning of Section 3.4, precision of Monte Carlo estimation can be measured in terms of variance $Var[\bar{h}_n]$. Let us denote for each sample element (not necessarily i.i.d.)

$$\sigma_i^2 = Var[h(X_i)] = \int [h(x_i) - E_f(h(x_i))]^2 f(x_i) dx_i.$$

If all X_i are independent and therefore uncorrelated,

$$Var[\bar{h}_n] = \frac{1}{n^2} \sum_{i=1}^n \sigma_i^2, \tag{3.26}$$

and if in addition all X_i are identically distributed with variance σ^2, $Var[\bar{h}_n] = \frac{1}{n} \sigma^2$. For an i.i.d. sample x statistic

$$v_n(x) = \frac{1}{n^2} \sum_{i=1}^n [h(x_i) - \bar{h}_n]^2]$$

defined in (3.18) is a good estimate for $Var[\bar{h}_n]$. However if X_i can be correlated, (3.26) is no longer true and a more general result follows.

Theorem 3.2 *If $\sigma_i^2 < \infty$ and covariances σ_{ij} are such that $|\sigma_{ij}| = |Cov[h(x_i), h(x_j)]| < \infty$, then*

$$Var[\bar{h}_n] = \frac{1}{n^2}\left(\sum_{i=1}^{n}\sigma_i^2 + 2\sum_{i=1}^{n}\sum_{j=i+1}^{n}\sigma_{ij}\right). \tag{3.27}$$

Proof: Denoting $Y_i = h(X_i)$, we can use the properties of mean and variance and regroup the terms so that

$$Var[\bar{h}_n(X)] = Var\left[\frac{1}{n}\sum_{i=1}^{n}Y_i\right] = \frac{1}{n^2}\left\{E\left[\sum_{i=1}^{n}Y_i\right]^2 - \left[E\sum_{i=1}^{n}Y_i\right]^2\right\}$$

$$= \frac{1}{n^2}\left(EY_1^2 + \cdots + EY_n^2 + 2E(Y_1 Y_2) + \cdots + 2E(Y_{n-1}Y_n)\right)$$

$$- \frac{1}{n^2}\left((EY_1)^2 + \cdots + (EY_n)^2 + 2E(Y_1)E(Y_2) + \cdots + 2E(Y_{n-1})E(Y_n)\right)$$

$$= \frac{1}{n^2}\left(\sum_{i=1}^{n}\sigma_i^2 + 2\sum_{i=1}^{n}\sum_{j=i+1}^{n}\sigma_{ij}\right).$$

$\blacksquare$

Notice that in general case the overall variance of (3.22) and hence the precision of Monte Carlo estimation is determined both by variances σ_i and covariances σ_{ij} so statistic (3.18) expressed only through the first and not the second is no longer a good measure of $Var[\bar{h}_n(X)]$, therefore Figures 3.7 and 3.8 are no longer helpful in assessing the precision of estimation.

The general idea of using correlated samples $\sigma_{ij} \neq 0$ is to achieve higher precision via variance reduction in (3.27) without increasing n. Methods of variance reduction suggest special sampling techniques providing mostly negative covariances $\sigma_{ij} \leq 0$ and lower value of (3.27). Three of these methods will be illustrated by a simple example in the following subsection.

3.4.4 Variance Reduction Methods

The principal example we will use to demonstrate different approaches to variance reduction is the case $X \sim Unif[0, 1]$. We will recall the following formulas:

$$EX^k = \int_0^1 x^k dx = \frac{1}{k+1}, \tag{3.28}$$

hence

$$Var(X^k) = EX^{2k} - (EX^k)^2 = \frac{1}{2k+1} - \frac{1}{(k+1)^2} = \frac{k^2}{(2k+1)(k+1)^2}, \tag{3.29}$$

in particular, $Var(X) = 1/12$ and $Var(X^2) = 4/45$. Also,

$$Cov(X, X^2) = EX^3 - EXEX^2 = \frac{1}{4} - \frac{1}{2} \times \frac{1}{3} = \frac{1}{12} \tag{3.30}$$

and

$$Cov(X, 1 - X) = E(X - X^2) - EXE(1 - X) = -EX^2 + (EX)^2 = -Var(X) = -\frac{1}{12}. \tag{3.31}$$

Now we will pretend to forget these facts for a while, and instead of using (3.28) directly will use Monte Carlo method to estimate $h^{(1)}(X) = EX$ and $h^{(2)}(X) = EX^2$.

The most straightforward way to do it is to create an i.i.d. sample

$$X = (X_1, \dots, X_n) \sim Unif[0, 1]$$

and use $\bar{X}$ to estimate EX and $\bar{X^2} = 1/n \sum_{i=1}^{n} X_i^2$ to estimate EX^2. This way we are assured to get $Var(\bar{X}) = \frac{1}{12n}$ and $Var(\bar{X^2}) = \frac{4}{45n}$. However, we can do much better than that in terms of precision. For that purpose we will suggest smarter sampling schemes involving correlated samples.

Antithetic Variables

The idea of this method is to append initial i.i.d. sample $(X_1, \dots, X_n)$ with an *antithetic* sample $(X_{n+1}, \dots, X_{2n})$ in such way that for some $i < n, j > n$ we obtain negative covariances $\sigma_{ij} = Cov[h(X_i), h(X_j)] < 0$, reducing variance (3.27) for the combined sample $(X_1, \dots, X_{2n})$.

For our example we append $X = (X_1, \dots, X_n)$, $X_i \sim Unif[0, 1]$ with antithetic values $X_{n+i} = 1 - X_i$ for all $i = 1, 2, \dots, n$. It is easy to check that the combined sample $(X_1, \dots, X_{2n})$ is still uniform on $[0, 1]$. Figure 3.10 shows initial sample size 10 in filled black circles, and antithetic values in unfilled circles. Notice that by nature of random number generation, the initial sample happens to shift to the right of the interval, containing several values on the extreme right, therefore by construction the antithetic sample shifts to the left containing some values on the extreme left.

If we consider $h^{(1)}(x) = x$, for most of the values $i < n, j > n, \sigma_{ij} = 0$. However for n pairs $(i, j = n + i), i = 1, 2, \dots, n$, as established in (3.31), $\sigma_{ij} = -\sigma_i$ and then according to (3.27), the variance of $\bar{h}^{(1)}_{2n}$ is zero and our Monte Carlo method is ideally precise even for sample size $n = 1$! If we consider $h^{(2)}(x) = x^2$, the advantage of antithetic sampling is not that striking but it is still there (see Exercises). A rather simplistic explanation for this method's efficiency is that the antithetic variables compensate for the initial sample being randomly shifted to the left or to the right, which affects the mean value and also the second moment.

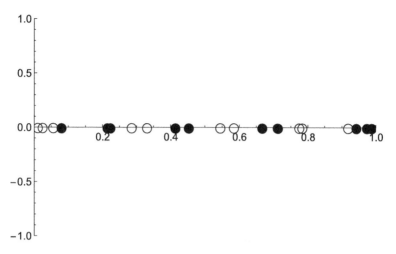

Figure 3.10 Antithetic sampling.

Control Variates

If our goal is to compute $E[h(X)]$, suppose that we can directly calculate $E[g(X)]$, where "control variate" $g(X)$ is in some sense close or related to $h(X)$. Then we can use the identity

$$E[h(X)] = E[h(X) - g(x)] + E[g(x)]$$

to define the ***control variate Monte Carlo estimator*** for $E[h(X)]$ as

$$\bar{h}_{CV} = \frac{1}{n} \sum_{i=1}^{n} [h(X_i) - g(X_i)] + E[g(X)].$$

Variance reduction may be achieved due to the fact that

$$Var(\bar{h}_{CV}) = \frac{1}{n^2} Var\left(\sum_{i=1}^{n} [h(X_i) - g(X_i)] \right)$$

$$= \frac{1}{n^2} \sum_{i=1}^{n} (Var[h(X_i)] + Var[g(X_i)] - 2Cov[h(X_i), g(X_i)]).$$

Control variate method has better precision than standard i.i.d. sampling

$$Var(\bar{h}_{CV}) \leq Var[\bar{h}_n],$$

if h and g are correlated strongly enough to compensate for additional variation in g:

$$Var[g(X_i)] \leq 2Cov[h(X_i), g(X_i)]. \tag{3.32}$$

Let us return to the uniform example with $g(x) = h^{(1)}(x)$, for which $E[g(X)]$ was directly calculated using antithetic variables, and $h(x) = h^{(2)}(x)$. Checking condition (3.32), we easily see that the left-hand side is $Var(X) = 1/12$ from (3.29), and the right-hand side is $2Cov(X, X^2) = 2/12$ from (3.30). Therefore instead of using $(\bar{X^2})$ to estimate $h^{(2)}(X)$, we suggest the control variate estimator

$$\bar{h}_{CV} = \frac{1}{n}\sum_{i=1}^{n}\left(X_i^2 - X_i\right) + \frac{1}{2} = (\bar{X^2}) - \bar{X} + \frac{1}{2}. \tag{3.33}$$

Its variance can be also calculated directly from (3.29) and (3.30) as $1/180n$, which is clearly a huge improvement with respect to i.i.d. sampling. This improvement can be explained by the fact that when the sample is randomly shifted to the left below the theoretical mean $1/2$, driving the estimator $(\bar{X^2})$ also down, the last two terms in (3.33) represent positive correction; when the sample is shifted to the right, the correction is negative.

Stratified Sampling
To increase the precision in estimating integral

$$I = \int_A h(x)f(x)dx,$$

let us partition the region of integration into k nonoverlapping subsets or strata A_j: $A = (A_1 \cup \cdots \cup A_k)$. We can create a *stratified sample*:

$$X_1, \ldots \ldots, X_{n_1} \in A_1$$
$$X_{n_1+1}, \cdots \ldots, X_{n_1+n_2} \in A_2$$
$$\cdots \cdots$$
$$X_{n_1+\cdots+n_{k-1}+1}, \cdots \ldots, X_n \in A_k,$$

where $n_1 + \cdots + n_k = n$. Suppose also $p_j = P(X \in A_j) = \int_{A_j} f(x)dx$.

We can define stratified estimator as

$$\bar{h}_{str} = \sum_{j=1}^{k} p_j \bar{h}_{n_j}, \quad \bar{h}_{n_j} = \frac{1}{n_j}\sum_{i=n_{j-1}+1}^{n_j} X_{i.} \tag{3.34}$$

With an appropriate choice of partition, $Var[\bar{h}_{str}]$ may be smaller than conventional $Var[\bar{h}_n]$ for simple i.i.d. sample.

Let us return to our example and evaluate $E(X)$ for $X \sim Unif[0,1]$. Let us choose strata as $A_1 = (0, 1/5), A_2 = (1/5, 2/5), \ldots, A_5 = (4/5, 1)$ and create a stratified sample according to (3.34) with $n_1 = \cdots = n_5 = 20$, $p_j = 1/5$. We will choose twenty points Y_j uniformly on every subinterval A_j and compare the variance of the stratified sample with a conventional i.i.d. sample of size

$n = 100$ drawn from $Unif[0, 1]$. It is easy to see that $Var(Y_j) = 1/(25 \times 12)$, so for $h^{(1)} = E(X)$

$$Var[\bar{h}_{str}] = \frac{1}{25}5Var[\bar{h}_{n_j}] = \frac{5}{25}\frac{1}{20 \times \times 25 \times 12} = \frac{1}{100 \times 25 \times 12}$$

and compare the variance of the stratified sample with a conventional i.i.d. sample of size $n = 100$ drawn from $Unif[0, 1]$:

$$Var[\bar{h}_n] = \frac{1}{100}Var(X_i) = \frac{1}{100 \times 12}.$$

Stratification achieves a substantial reduction in variance (see also Exercises).

Examples of variance reduction confirm that there may be a benefit in going beyond i.i.d. sampling. Some simple examples are contained in the **Appendices** in *varreduction.xlsx*. Monitoring variance and further variance reduction will be revisited in Chapter 4. Until then we will have to leave the discipline of simulation and delve into entirely different field of mathematics. Familiarity with this field will be necessary to introduce further development of sampling techniques and to further exploit the idea of dependent samples.

3.5   Markov Chains

3.5.1   Markov Processes

A stochastic process is a process developing in time according to certain probability laws. At each given point of time the state of a stochastic process can be represented by a random variable. The set of all these random variables determines a stochastic process. Time could be either continuous or discrete, as well as the state space: union of the ranges of values of these random variables. Trajectories or paths of a stochastic process are formed by numeric values of the corresponding random variables. Trajectories of a stochastic process are observable up to the present moment of time. Future allows for uncertainty, therefore the trajectory for the future is not unique: multiple paths are possible. For more general information on stochastic processes see, for instance, Ross [26].

One of the simplest examples of stochastic processes is *random process* also known as "white noise." In its discrete version it is defined by a sequence of independent normal random variables with zero mean and constant variance, see Figure 3.11. This process was briefly described in Section 1.7.

Another example also discussed in Section 1.7 is discrete *random walk.* This sequence of random variables as shown in Figure 3.12 is defined by the formula $X_{t+1} = X_t + \varepsilon_t$ where ε_t is the white noise.

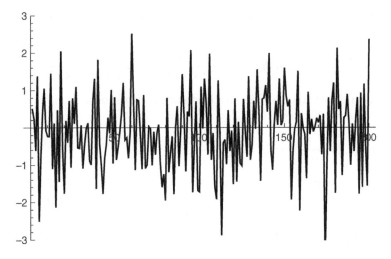

Figure 3.11 White noise.

A popular generalization of such a process is ***autoregressive process*** defined by

$$X_t = \sum_{k=1}^{p} \alpha_k X_{t-k} + \varepsilon_t, \tag{3.35}$$

where ε_t is the white noise (see also Section 1.7). The state of this process at any given point of time depends on several previous states and also contains a

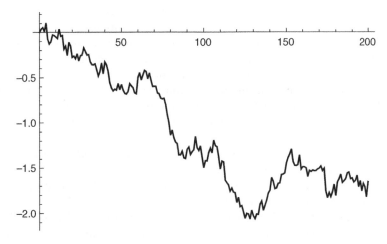

Figure 3.12 Random walk.

random component. We can say that autoregressive process has a "long memory."

However there exist stochastic processes whose next state is determined only by the present state and does not depend on the past:

$$P(X_t = x \mid X_1, \dots, X_{t-1}) = P(X_t = x \mid X_{t-1}). \tag{3.36}$$

This property of a stochastic process is known as **Markov property** named after Russian mathematician Andrey Markov. Stochastic processes satisfying the Markov property are known as **Markov processes** or, in discrete time, as **Markov chains.** For instance, random walk and autoregressive process of order 1 ($p = 1$) known as AR(1) are Markov processes.

Markov processes serve as good models for many stochastic processes one may encounter in applications. For example, the market efficiency hypothesis suggests that stock price changing with time is a Markov process. If you know the stock prices today, you have full information prescribing the future (random) states of the market, and the knowledge of the past stock prices adds nothing new. In other words, efficient markets assume that "everybody is doing their homework," and when you try to predict the future, the information regarding past prices is already incorporated in the present. In our further analysis we will try to avoid technical details. For a more rigorous treatment of Markov chains, please refer to Meyn and Tweedie [20], Norris [22], or to Haeggstroem [9] for more algorithmic applications.

3.5.2   Discrete Time, Discrete State Space

Unless otherwise specified, we will consider Markov chains, for which random states are assigned for some discrete times only. In this case we can assume that time takes on integer values only: $t = 0, 1, \dots, N, \dots$. Hence, Markov chains may be considered as sequences of random variables.

The union of the set of possible states of these random variables will be called the **state space** of the chain. We will begin with considering finite state spaces: at each point of time the process $\{X_t\}$ takes values from a finite set (state space) $S = \{S_0, \dots, S_k\}$. We will assume that $X_t = i$, if and only if at time t the process is at the state S_i. Extension to an infinite discrete state space does not require any serious additional efforts, though we will have to get more technical when we extend the state space to continuum.

3.5.3   Transition Probability

Here we will assume that the probability of transition from state i at time n into state j at time $n + 1$ (in one step) does not depend on time. Markov chains with such property are known as ***homogeneous***. One-step ***transition probabilities*** are denoted

$$P_{ij} = P(X_{n+1} = j \mid X_n = i). \tag{3.37}$$

Transition probabilities form one-step ***transition matrix***

$$P = \begin{pmatrix} P_{00} & \cdots & P_{0k} \\ \cdots & \cdots & \cdots \\ P_{k0} & \cdots & P_{kk} \end{pmatrix}, \sum_{j=0}^{k} P_{ij} = 1. \tag{3.38}$$

We will denote by $P_{ij}^{(n)}$ the probability of transition from state i into state j in n steps. We can build an n-step transition matrix $P^{(n)}$ from these probabilities. It is easy to verify by matrix multiplication that $P^{(n)} = P^n$.

3.5.4 "Sun City"

"Sun City," which we will consider for our basic example was described in a Russian children's book series, and later a cartoon, which was popular when both authors were still kids. The main hero of the books, Dunno (or Know-Nothing), a very charming though ignorant and lazy boy Shorty (a representative of a special race of little people) in his adventures happens to fall onto Sun City. The latter, named for its perpetually sunny weather, is a Shorty utopia noted for its incredible technological advances.

– Why do you call it Sun City? The houses are built of sun, are they? - asked Dunno.
– No, - the driver laughed. - It is called Sun City because the weather is always good and the sun shines at all times.
– Are there never any clouds? - Dunno was surprised.
– Why never? It happens, - the driver responded. - But our scientists developed a special powder: as soon as clouds gather, they toss some powder up, and the clouds disappear.
(Nikolay Nosov, Dunno in Sun City, [23]).

Things seem to be fairly deterministic in Sun City. We begin with this example because we need something dramatically simple to explain a few nontrivial ideas. Let us introduce a little randomness to allow for at least some rain to fall down. Otherwise, everything would stay as straightforward as it could be in only a synthetic city.

Let us consider a "mathematical model" of Sun City weather. Suppose that only two states are possible: either it is sunny or it rains. The weather does not change during the day. The weather tomorrow depends only on the weather today (the past is irrelevant): if it is sunny today, it will rain tomorrow with probability $1/6$; if it rains today, the probability of rain tomorrow is $1/2$. Therefore, the state space consists of only two states: S_0—sunny, S_1—rainy. The one-step (day-to-day) transition probability matrix is

$$P = \begin{pmatrix} \frac{5}{6} & \frac{1}{6} \\ \frac{1}{2} & \frac{1}{2} \end{pmatrix}. \tag{3.39}$$

The initial probabilities are chosen to allow for an easy hands-on simulation. The example was designed in such a way in order to avoid computer demonstrations. Namely, a simulation of day-to-day weather patterns in Sun City can be performed with one coin (two sided) and one die (six-sided cube).

3.5.5 Utility Bills

Let us consider another example bringing us back from the cartoon world to our daily rut. Consider the payment of utility bills (electricity, heat, water, etc.), which we are supposed to perform regularly, say, on a monthly basis. Suppose that as a utility customer we can find ourselves in one of three states: S_0—bills paid in full; S_1—bills paid partially; S_2—bills unpaid. Assume that the month-to-month transition matrix looks like this:

$$P = \begin{pmatrix} \frac{3}{4} & \frac{3}{16} & \frac{1}{16} \\ \frac{1}{2} & \frac{1}{4} & \frac{1}{4} \\ \frac{1}{8} & \frac{3}{4} & \frac{1}{8} \end{pmatrix}. \tag{3.40}$$

Analysis of this transition matrix, and especially of its second row corresponding to the partial bill payers, can confirm the old wisdom that too often temporary problems have an unfortunate tendency to become permanent. On the other hand, many customers in the database seem to make an earnest effort to come clean.

3.5.6 Classification of States

Two states S_i and S_j are ***communicating,*** if $P_{ij}^{(n)} > 0, P_{ji}^{(m)} > 0$ for some n, m. The subset of state space, where any two states are connected, forms a ***communicating class.*** A chain is ***irreducible*** if it consists of only one communicating class.

A state i is ***absorbing*** if the chain cannot leave this state once it gets there ($P_{ii} = 1$). Absorbing states do not communicate to any other states.

Consider a chain with transition matrix:

$$P = \begin{pmatrix} \frac{1}{2} & \frac{1}{2} & 0 & 0 \\ \frac{1}{2} & \frac{1}{2} & 0 & 0 \\ \frac{1}{4} & \frac{1}{4} & \frac{1}{4} & \frac{1}{4} \\ 0 & 0 & 0 & 1 \end{pmatrix}. \tag{3.41}$$

This chain can be split into two communicating classes $\{S_0, S_1\}$, $\{S_2\}$ and an absorbing state S_3.

Recurrence function for state S_i is defined as the probability to ever return to this state after leaving it. This function can be determined by the formula:

$$f_i = \sum_{m > n} P(X_m = i \mid X_n = i; X_t \neq i \text{ for } n < t < m). \tag{3.42}$$

The state is ***recurrent*** if this probability is equal to 1. It is ***transient*** if it is strictly less than 1. In a communicating class all the states belong to the same type: they are either recurrent or transient. In the following example we can split the chain into two communicating classes of recurrent states $\{S_0, S_1\}$, $\{S_2, S_3\}$ and a transient state S_4.

$$P = \begin{pmatrix} \frac{1}{2} & \frac{1}{2} & 0 & 0 & 0 \\ \frac{1}{2} & \frac{1}{2} & 0 & 0 & 0 \\ 0 & 0 & \frac{1}{2} & \frac{1}{2} & 0 \\ 0 & 0 & \frac{1}{2} & \frac{1}{2} & 0 \\ \frac{1}{4} & \frac{1}{4} & 0 & 0 & \frac{1}{2} \end{pmatrix}. \tag{3.43}$$

We say that state S_i is ***periodic with period*** $t > 1$, if $P_{ii}^{(n)} = 0$, when t does not divide n, and t is the largest number with this property. State S_i is ***aperiodic***, if such $t > 1$ does not exist.

Let a chain have transition matrix

$$P = \begin{pmatrix} 0 & 1 \\ 1 & 0 \end{pmatrix}. \tag{3.44}$$

Evidently, both states of the chain are periodic with period 2.

In a finite Markov chain any aperiodic recurrent state is called ***ergodic***. Markov chain is ***ergodic*** if it is irreducible and all its states are ergodic.

3.5.7 Stationary Distribution

Stationary distribution $\pi = (\pi_0, \ldots, \pi_k)$ can be defined for an ergodic chain. The components of this row-vector can be interpreted as probabilities of staying in corresponding states independent of the past or also as long-term proportions of time the chain spends in these states. It is easy to see that the stationary distribution should satisfy the following properties:

$$P^n \to \begin{pmatrix} \pi \\ \cdots \\ \pi \end{pmatrix} \ if \ n \to \infty, \ \pi P = \pi. \tag{3.45}$$

We can check that both states in the Sun City example are ergodic. Then we can look for the stationary distribution, as we know that it exists. First, we can use matrix multiplication to calculate several integer powers of the one-step transition matrix:

$$P^2 = \begin{pmatrix} \frac{5}{6} & \frac{1}{6} \\ \frac{1}{2} & \frac{1}{2} \end{pmatrix} \times \begin{pmatrix} \frac{5}{6} & \frac{1}{6} \\ \frac{1}{2} & \frac{1}{2} \end{pmatrix} = \begin{pmatrix} \frac{7}{9} & \frac{2}{9} \\ \frac{2}{3} & \frac{1}{3} \end{pmatrix},$$

$$P^3 = \begin{pmatrix} 0.76 & 0.24 \\ 0.72 & 0.28 \end{pmatrix}, \ P^4 = \begin{pmatrix} 0.75 & 0.25 \\ 0.74 & 0.26 \end{pmatrix}.$$

One might guess that the rows of this matrix seem to converge to the same vector

$(0.75, 0.25),$

which would indicate that the memory of the past is lost (after just 4 days, it does not matter if we started with a sunny or a rainy day). One can also use linear algebra to solve the system of two dependent linear equations (keeping in mind that the stationary probabilities of two states add up to 1)

$$\pi P = \pi \Rightarrow \begin{cases} \frac{5}{6}\pi_0 + \frac{1}{2}\pi_1 = \pi_0 \\ \frac{1}{6}\pi_0 + \frac{1}{2}\pi_1 = \pi_1 \end{cases} \Rightarrow \pi_0 = 3\pi_1 \Rightarrow \pi = (0.75, 0.25). \tag{3.46}$$

One can also determine the stationary distribution for the utility bills process. In this example we can get:

$$P^3 = \begin{pmatrix} 0.628 & 0.259 & 0.113 \\ 0.586 & 0.279 & 0.135 \\ 0.542 & 0.325 & 0.133 \end{pmatrix}, \ P^6 = \begin{pmatrix} 0.607 & 0.272 & 0.121 \\ 0.605 & 0.274 & 0.122 \\ 0.603 & 0.275 & 0.123 \end{pmatrix}. \tag{3.47}$$

A rough guess might suggest the stationary distribution of the form

$(0.61, 0.27, 0.12).$

Exact solution of the corresponding system of linear equations yields

$$\left(\frac{20}{33}, \frac{9}{33}, \frac{4}{33}\right).$$

We can see that in the long run on the average 61% pay in full, 27% pay partially, and 12% tend not to pay at all.

3.5.8 Reversibility Condition

If we look at a Markov chain reversing the course of time, we will see another Markov chain. Its transition probabilities for the reversed chain will be calculated according to

$$Q_{ij} = P(X_n = j \mid X_{n+1} = i) = \frac{P(X_n = j)P_{ji}}{P(X_{n+1} = i)}. \tag{3.48}$$

A Markov chain is **reversible** or *reversible in time*, if the transition matrices for direct and reversed chains coincide: $P_{ij} = Q_{ij}$.

Let the stationary distribution exist for a reversible Markov chain. Then the following relationship known as the **detailed balance equation**

$$\pi_i P_{ij} = \pi_j P_{ji} \tag{3.49}$$

follows from (3.2.7) when n tends to infinity. Let us check the detailed balance equation for the Sun City example.

$$\pi_0 P_{01} = \frac{3}{4} \cdot \frac{1}{6} = \frac{1}{4} \cdot \frac{1}{2} = \pi_1 P_{10}. \tag{3.50}$$

However, for the utility bills example (3.2.8) does not hold. For instance, $Q_{01} = \frac{9}{40} \neq P_{01}$.

3.5.9 Markov Chains with Continuous State Spaces

We will consider only the case when the state space is continuous and one dimensional. In this situation many of the definitions given above will require substantial modifications. Transition probability from a state to any single state in the continuous case is equal to 0, so we have to consider transition probabilities from a state x into a measurable state space A:

$$P(x, A) = P(X_{n+1} \in A \mid X_n = x). \tag{3.51}$$

Set of these probabilities for all x's and A's is called the **transition kernel** of a Markov chain. For a homogeneous chain the kernel does not depend on time. If $A = (-\infty, v]$ the transition kernel defines the cumulative distribution function of the transition

$$G(v \mid x) = P(X_{n+1} \leq v \mid X_n = x). \tag{3.52}$$

If this function has a continuous derivative

$$g(v \mid x) = \frac{\partial F(v \mid x)}{\partial v}, \tag{3.53}$$

then this function is called the one-step **transition density function**. n-step transition density function can be defined as a convolution:

$$g^{(n)}(v \mid x) = \int_{-\infty}^{+\infty} g(v \mid w) g^{(n-1)}(w \mid x) dw. \tag{3.54}$$

If the limit

$$\lim_{n \to \infty} g^{(n)}(v \mid x) = f(v) \tag{3.55}$$

exists and does not depend on x, then $f(v)$ is known as the **stationary state density**. If a Markov chain is reversible, it satisfies the **detailed balance equation**

$$f(x)g(v \mid x) = f(v)g(x \mid v). \tag{3.56}$$

We can see that many properties of Markov chains which we introduced for finite and discrete state spaces can be extended to the continuous state spaces. In our future development we will avoid exact proofs of the fine properties of the algorithms, therefore we will not need more precise definitions.

3.6   Simulation of a Markov Chain

Suppose we know the transition matrix for Sun City weather. Instead, we are interested in the stationary distribution, which would give us a long-term weather forecast for some distant future. Let us also pretend that we are very mathematically ignorant:

- We cannot multiply matrices.
- We cannot solve systems of liner equations.

This seems ridiculous for such an easy example. However, imagine a 10,000 × 10,000 transition matrix (not unusual in some applications). Without a computer we probably will not venture into using any of the two approaches suggested above. Is there another way we can follow? Actually, there is one suggestion which does not require computer help (and if you have a computer at hand, you might still prefer this approach to any of the two above). Just for the sake of clarity we will illustrate this approach using our simple example. It will still work for a 10,000 × 10,000 transition matrix, but will require a little more computing power. Have a fair coin and a fair die available.

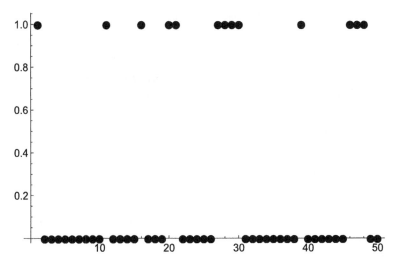

Figure 3.13 Sun City simulation.

Let us generate a long enough sequence of days using the following algorithm:

1. Choose one of the states (sun or rain) for Day 0. Do not think too hard, your choice would not matter.
2. For $n = 0, 1, 2, \ldots, N$ use the following rule:
   (a) If Day n is sunny, roll your die: if it turns on "six", at Day $n + 1$ move to "rain," otherwise stay at "sun."
   (b) If Day n is rainy, toss your coin: if it turns heads up, at Day $n + 1$ move to "sun," otherwise stay at "rain."

This simple algorithm will generate a Markov chain length N with correct transition probabilities. Then the long-term proportion of sunny days for large N will converge to the stationary probability of "sun," same for "rain."

One can also do the same in a more civilized way. In online Appendices we suggest computer implementation of this algorithm and graphical illustration using Microsoft Excel in **MCsimulation.xlsx**. We can also use R or other software instead of a coin and a die. Figure 3.13 demonstrates the behavior of a chain length 50 simulated in ***Mathematica*** jumping between two states: sun (0) and rain (1).

We see that in this particular case the probability of rain was estimated as 0.24 using the relative frequency of the rainy days in the simulated chain (12 out of 50 days). To increase the precision of estimation, take a longer run.

We can also generate a Markov chain for utility bills. See appendix for the algorithm in R, and in Figure 3.14 we suggest a graphical illustration for $N = 50$ with the chain moving from state to state: 0, 1, and 2. The probabilities of

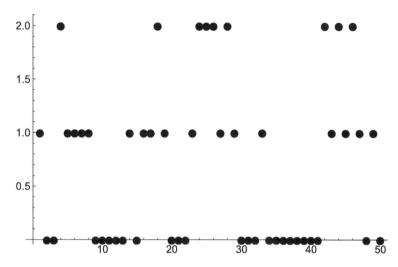

Figure 3.14 Utility bills simulation.

three states are estimated as 0.48, 0.34, 0.18. For better estimates we should substantially increase the number of steps.

Notice that this approach represents another modification of Monte Carlo integration. The goal is to estimate probabilities of the states using long-term relative frequencies of these states (similar to the example with estimation of π). What is different, instead of generating an independent sample directly from the target distribution, we generate a Markov chain converging to its stationary distribution.

3.7 Applications

3.7.1 Bank Sizes

Markov chains are used in many diverse applications: statistical process control, reliability, business, and insurance. There exist interesting applications to financial statistics. We will suggest one example from the authors' research experience.

The study involved Russian commercial banks during the period from 2007 to 2012. The objective was to build a reasonable model which would describe the distribution of bank size in Russia and describe its mobility (predict upward and downward movements in size).

Using available data, all Russian banks were split into four classes depending on their size (total capital). These classes could be roughly characterized as: small banks, medium banks, large banks, and giant banks. The boundaries for the classes were selected as in Table 3.2.

Table 3.2 Classification of banks

	Absolute values in 2011:Q4	Relative values with respect to total equity of the banking sector	Relative values, logarithmic scale	
Threshold level 1	0.3 billion rubles	0.006%	−5.070378	Minimum total equity for bank entering the industry (the restriction applies since 01/01/2012)
Threshold level 2	1 billion rubles	0.021%	−3.866405	Minimum total equity (the restriction is to be introduced in 2015)
Threshold level 3	19.6 billion rubles	0.411%	−0.890144	Top 30 banks

Belonging to a certain class at the end of the year is considered as being in *State 0* through *State 3*. In order to complete the state space two more states were introduced: *Entry* state (for the banks founded in a given year) and *Exit* state (for the banks ceasing their activity). In the assumption that the process of the bank size satisfies Markov property (probability of moving to a different class for the next year is determined only by the class where the bank currently is, and does not depend on the past), we can build a Markov chain estimating the transition probabilities, which are summarized in Table 3.3.

The constructed Markov chain is not irreducible and therefore is not ergodic. It contains an absorbing *Exit* state and a transient *Entry* state. Such a chain will not obtain a stationary distribution. We will not have much use for such chains

Table 3.3 Transition probabilities for bank sizes

	State in t					
State in $t-1$	Class 1	Class 2	Class 3	Class 4	Entry	Exit
Class 1	0.9675	0.0212	0.0002	0	0	0.0111
Class 2	0.0282	0.9455	0.0156	0	0	0.0107
Class 3	0	0.0183	0.9673	0.0025	0	0.0119
Class 4	0	0	0.0130	0.9805	0	0.0065
Entry	0.5806	0.1613	0.2581	0	0	0
Exit	0	0	0	0	0	1

Table 3.4 Distribution of bank sizes for 2012

	Class 1	Class 2	Class 3	Class 4
Number of active banks	379	299	269	30
Shares of size categories	38.80%	30.60%	27.49%	3.11%

in this book. However, this model itself might be useful for prediction of the structure of banking sector of economy.

Using statistical data we can estimate the distribution of bank sizes in 2012. It is summarized in Table 3.4.

3.7.2 Related Failures of Car Parts

Vehicle manufacturers do not have to make their failure reports public, thus historically very few data have been made available for statistical analysis. Recently, a lot of attention was attracted to consumer vehicle reliability. In 2005, Toyota faced lawsuits for withholding vehicle failure data, and many repair and recall data were collected in the aftermath. In response to these events, a rival company Hyundai made public a lot of their records, including a dataset of 58,029 manufacturer warranty claims on Hyundai Accent from 1996 to 2000. For each claim made, the following information was included: the vehicle identification number (VIN), the dates on which the vehicle was shipped, sold, and repaired, the amount of time (in days) from previous failure and from sale to repair, and the type of car usage (light, medium, or heavy). In [3] the author addressed the distribution models for this variable measured for various components, performing a preliminary analysis of each component repaired. In that study all failures were treated as independent events. However, the author concluded that failures of various components may be associated [11]. Our goal is to find ways to model this association. In this subsection we will touch on Markov chain approach, and in Section 7.6 we will return to the same problem with a very different toolkit [17].

Five main components that most frequently failed and caused warranty claims are chosen out of the full list of 60 detected in Hyundai Accent warranty claim records. They are: the spark plug assembly (A), ignition coil assembly (B), computer assembly (C), crankshaft position sensor (D), and oxygen heated sensor (E). The spark plug assembly brings power to the spark plug, which provides the spark for the motor to start. Spark plugs are typically replaced every 50,000 to 100,000 miles, but this rate can change depending on such factors as frequency of oil changes and quality of oil used. The ignition coil assembly regulates the current to the spark plugs, helping to ignite the spark. While these two parts are engineered to wear independently, they display such a high level of interaction that it can be difficult to tell which one needs repair. The computer assembly includes engine sensors, and it controls electronics for fuel-injection emission controls and the ignition. The crankshaft position sensor controls ignition system timings and reads rpms. Finally, the oxygen heated sensor determines the gas-fuel mix ratio by analyzing the air from the exhaust and adjusting the ratio as needed. The latter two components are controlled by the computer assembly. Once again, failure by one should not necessarily affect another, yet these three components are all closely related and could all need to be replaced if, for example, a single event caused the system to short out.

With regard to this dataset, data were recorded for only the cars that had at least one of the five main components fail within the warranty period. Thus, the main focus in this study is on the cars for which component failures are registered during this time. We also took into consideration the fact that with some cars, a main component needed to be replaced more than once. The assumption that the future need of replacements and repairs is most closely related to the most recent repairs already made reflects the Markov property. The ultimate goal in this section is to suggest a Markov chain model for the failures of five engine assembly components. Such a model would be helpful for better estimation of lifetimes and prediction of certain parts' failures given the car's history and recent replacements and repairs. The most important part of the modeling process is to describe the state space and transition matrix, and then to estimate its elements.

Seven states were identified: State S_0 corresponds to the state of a car before any repairs done. State S_i indicates: "Part (i) needs repair." Finally, state S_6 is the state: "No more repairs needed up to the end of the warranty period." We will assume that all repairs are done directly when diagnosed. Exact time of repair is not reflected in the transition matrix, only the right sequence of repairs. Simultaneous repairs of several parts are recorded as two or more consecutive repairs starting with the more common one (A to E order).

Transition probabilities P_{ij} will be estimated based on the relative frequency of such an event: a car detected at state S_i would at some point before the warranty expiration move directly to state S_j. These estimated probabilities are given in Table 3.5.

All the cars in the database will start at state S_0, but by definition no car will return to it after the first repair. Therefore S_0 is a transient state. States S_1-S_5 are communicating. After one repair is needed, another might be needed at a later date. State S_6 is the terminal or absorbing state. We get into this state only when no more status changes happen before the end of the warranty period. You cannot move directly from S_0 to S_6, because all cars in our database had at least one repair recorded.

What kind of information can be provided by the contents of Table 3.5? Let us consider, for instance, four highlighted numbers. A relatively high probability $P_{11} = 0.1814$ corresponds to more than one replacement of the spark plug

Table 3.5 Transition probabilities for car repairs

| | State in t | | | | | | |
State in t-1	S_0	S_1	S_2	S_3	S_4	S_5	S_6
S_0 (Initial)	0	0.3531	0.2486	0.1344	0.1414	0.1224	0
S_1 (A)	0	**0.1814**	0.0266	0.0100	0.0114	0.0209	0.7462
S_2 (B)	0	**0.0632**	0.0793	0.0303	0.0152	0.0109	0.8010
S_3 (C)	0	0.0292	**0.0425**	**0.0756**	0.0295	0.0153	0.8079
S_4 (D)	0	0.0216	0.0093	0.0195	0.0169	0.0087	0.9240
S_5 (E)	0	0.0286	0.0102	0.0076	0.0102	0.0203	0.9235
S_6 (Terminal)	0	0	0	0	0	0	1

assembly (A) during the warranty period, which can be considered unusual, but not totally impossible representing a relatively common maintenance operation. A potentially more troubling issue is a relatively high value of $P_{33} = 0.0756$, especially in conjunction with a relatively modest $P_{03} = 0.1344$, which means that more than half the replacements/repairs of expensive computer assembly (C) required an iterative repair within the warranty period. In general, all diagonal elements P_{ii} for $i = 1, \dots, 5$ correspond to multiple repairs of the same component. A couple of relatively high off-diagonal elements $P_{21} = 0.0632$ and $P_{32} = 0.0425$ indicate potentially interesting patterns: spark plug assembly (A) replacement following directly the replacement of ignition coil assembly (B), and ignition coil assembly (B) replaced directly after a problem with computer (C). Leaving further conclusions to technical experts, we may suggest that even some simple application of Markov chains might provide some insights to statistical dependence of components of complex engineering systems.

This short analysis above also makes it clear that a more thorough study should include such an important factor as time from repair to repair which is totally overlooked by our transition matrix. Also, there are such covariates as the type of the car usage, which should be taken into account.

References

1 Abramowitz, M., and Stegun, I. A. (1972). *Handbook of Mathematical Functions.* New York: Dover.
2 Asmussen, S., and Glynn, P. (2007). *Stochastic Simulation: Algorithms and Analysis. Stochastic Modelling and Applied Probability.* Berlin: Springer.
3 Baik, J. (2010). Warranty analysis on engine: case study. Technical report, Korea Open National University, Seoul, 1–25.

4 Cover, T. (1987). Pick the largest number. In: T. Cover and B. Gopinath (editors), *Open Problems in Communication and Computation.*, Springer Verlag.

5 Devroye, L. (1986). *Non-Uniform Random Variate Generation.* New York: Springer.

6 Fontenrose, J. E. (1978). *The Delphic Oracle, its Responses and Operations, With a Catalogue of Responses.* Berkeley: University of California Press.

7 Gamerman, D., and Lopes, H. F. (2006). *Markov Chain Monte Carlo.* Chapman & Hall/CRC.

8 Glasserman, P. (2004). *Monte Carlo Methods in Financial Engineering.* New York: Springer.

9 Haeggstroem, O. (2003). *Finite Markov Chains and Algorithmic Applications*, No. 52, Student Texts. London: London Mathematical Society.

10 Hammersley, J. M., and Handscomb, D. C. (1964). *Monte Carlo Methods.* Boca Raton FL: Chapman & Hall, CRC Press.

11 Heyes, A. M. (1998). Automotive component failures. *Engineering Failure Analysis*, 5(2), 698–707.

12 Karsten, R. (1926). *The Civilization of the South American Indians With Special Reference to Magic and Religion.* New York: Alfred A. Knopf.

13 Knuth, D. E. (1998). *The Art of Computer Programming. Volume 2 (Seminumerical Algorithms)*, 3rd ed. Reading, MA:Addison-Wesley.

14 Korn, R., Korn, E., and Kroisandt, G. (2010). *Monte Carlo Methods and Models in Finance and Insurance.* Chapman & Hall/CRC.

15 L'Ecuyer, P. (1994). Uniform random number generation, *Annals of Operations Research*, 53, 77–120.

16 L'Ecuyer, P., and Simard, R. (2002). TestU01: A software library in ANSI C for empirical testing of random number generators, Available at: http://www.iro.montreal.ca/~lecuyer.

17 Kumerow, J., Lenz, N., Sargent, K., Shemyakin, A., and Wifvat, K. (2014). Modelling related failures of vehicle components via Bayesian copulas, *ISBA-2014, Cancun, Mexico*, § 307, 195.

18 Marsaglia, G. (1996). The Marsaglia random number CDROM including the Diehard battery of tests of randomness. Available at: http://stat.fsu.edu/pub/diehard.

19 Metropolis, N., and Ulam, S. (1949). The Monte Carlo method. *Journal of the American Statistical Association*, 44, 335–341.

20 Meyn, S. P., and Tweedie, R.L. (2005). *Markov Chains and Stochastic Stability.* Cambridge University Press.

21 Mikhailov, G. A., and Voitishek, A. V. (2006). *Numerical Statistical Modeling. Monte Carlo Methods.* Moscow: Akademiia, (in Russian).

22 Norris, J. R. (1998). *Markov Chains.* Cambridge University Press.

23 Nosov, N. (1958) *Dunno in the Sun City.* Moscow, Mir (in Russian).

24 Robert, C. P., and Casella, G. (2004). *Monte Carlo Statistical Methods*, 2nd ed. Springer-Verlag.

25 Robert, C. P., and Casella, G. (2010). *Introducing Monte Carlo Methods with R*, Springer-Verlag.

26 Ross, S. M. (1996). *Introduction to Stochastic Processes*. New York: Wiley.

27 Rubinstein, R. Y. (1981). *Simulation and the Monte Carlo method*. New York: Wiley.

28 Rukhin, A., Soto, J., Nechvatal, J., Smid, M, Barker, E., Leigh, S., Levenson, M., Vangel, M, Banks, D., Heckert, A., Dray, J, and Vo, S. (2001). A statistical test suite for random and pseudorandom number generators for cryptographic applications. Special Publication 800–22, National Institute of Standards and Technology.

Exercises

3.1 Generate random sample size 100 from the distribution with density $f(x) = 2\exp(-2x), x \geq 0$. Check the feasibility of the obtained data using: histogram, mean, variance, EDF.

3.2 Generate a sample from double exponential distribution with variance 1: $f(x) = Ce^{-k|x|}, -\infty < x < \infty$. Implement accept–reject method to transform this sample into a sample from standard normal distribution. What is the resulting sample size? How does it confirm the theoretical results?

3.3 Use Monte Carlo method with $f(x) \sim Exp(\alpha)$ to evaluate integrals

$$\int_0^\infty \alpha x e^{-\alpha x} dx$$

and

$$\int_0^\infty \alpha \ln(x) e^{-\alpha x} dx$$

for $\alpha = 3$ and $\alpha = 4$.

3.4 Generate a sample $(x_1, \ldots x_n), n = 100$ from Poisson distribution $X \sim Poiss(\lambda = 1.5)$. Then use it to estimate $P(X > 2)$ using techniques from Section 3.3.

3.5 Using Monte Carlo method with a generated sample $x_i \sim Gamma(4, 2)$, estimate the integral

$$\int_0^\infty x^5 e^{-2x} dx.$$

In order to generate a sample $x_i \sim Gamma(4, 2)$, apply:

(a) Representation of gamma distribution $Gamma(4, 2)$ as a sum of exponentials.

(b) Accept–reject technique with instrumental density *Gamma*(2, 1).
(c) Importance sampling with instrumental density *Gamma*(2, 1).
Compare the results.

3.6 Use the method of antithetic variables to estimate $E(X^2), X \sim Unif[0, 1]$. Obtain exact numerical value for the variance $Var(\bar{h}_{2n}), h(x) = x^2$ and compare it with the simulation results (use 100 simulations with $n = 100$).

3.7 Use the method of control variates to estimate $E(X^3), X \sim Unif[0, 1]$. Obtain exact numerical value for the variance $Var(\bar{h}_{CV}), h(x) = x^3$ with control variate $g(x) = x^2$ and compare it with the simulation results (use 100 simulations with $n = 100$).

3.8 Use the method of stratified sampling to estimate $E(X^2), X \sim Unif[0, 1]$. Estimate the variance $Var(\bar{h}_{str}), h(x) = x^2$ from the simulation results (use 100 simulations with $n = 100$).

3.9 For the Markov chain with transition matrix

$$P = \begin{pmatrix} 0.3 & 0.3 & 0.4 \\ 0.2 & 0.5 & 0.3 \\ 0.1 & 0.3 & 0.6 \end{pmatrix}. \tag{3.57}$$

(a) Calculate the third, the fourth, and the fifth powers of P.
(b) Using linear algebra, evaluate the stationary distribution of the chain. Work by hand or use R, if you like. Submit all relevant code.
(c) Develop Monte Carlo algorithm for generating a chain with given transition probabilities. Estimate stationary probabilities. Use a chain of such length that will provide precision to within 0.01.

3.10 Are all the states in Table 3.3 communicating? Which states are transient, recurrent, or absorbing?

3.11 Characterize states S_1–S_5 in Table 3.5 as transient or recurrent. Prove your point using the definitions.

4

Markov Chain Monte Carlo Methods

In our introduction to Markov chain Monte Carlo (further often abbreviated as MCMC) methods we will try to avoid all details which are not immediately necessary to understand the main ideas of the algorithms. Thus we sacrifice mathematical rigor and computational convenience. We also avoid detailed descriptions of more modern and more complicated MCMC methods. For a more systematic exposure to MCMC we can recommend excellent texts of Gamerman and Lopes [9] and Robert and Casella [26]. Computational details are treated with attention by Bolstad [3] and Kruschke [17], and also with special reference to R in [25].

4.1 Markov Chain Simulations for Sun City and Ten Coins

In this section we will ask our readers to pretend being even more ignorant about the Sun City weather situation than in Sections 3.5 and 3.6. Here we assume that not only we are unable to analytically derive the stationary distribution for the Markov chain of daily weather changes from its transition matrix, but that this matrix itself is not known to us.

Therefore, in order to determine the long-term proportion of rainy days (which can be also treated as the mean value of the binomial variable taking two values: 1 if it rains, and 0 if it is sunny), we can use neither the analytic tools based on solving linear equations developed in Section 3.5 nor the direct Markov chain simulation described in Section 3.6.

For such a weird situation we may introduce the following two-stage algorithm, which combines Markov chain simulation with accept–reject techniques of Section 3.3. To begin with, we will propose a simulation from a Markov chain with a totally different transition matrix. The choice of this "proposal" matrix is very basic: first, it should have the same state space as the "target" chain and second, it should be easy to simulate. Tossing a fair coin each day independently

Introduction to Bayesian Estimation and Copula Models of Dependence, First Edition.
Arkady Shemyakin and Alexander Kniazev.
© 2017 John Wiley & Sons, Inc. Published 2017 by John Wiley & Sons, Inc.
Companion Website: http://www.wiley.com/go/shemyakin/bayesian_estimation

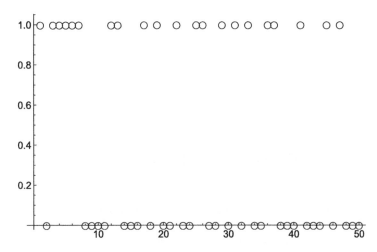

Figure 4.1 Simulation with the proposed transition: $N = 50$.

for rain or for sun will correspond to sampling from a Markov chain y_t with the "proposal" transition matrix

$$Q = \begin{pmatrix} 1/2 & 1/2 \\ 1/2 & 1/2 \end{pmatrix},$$

and will serve the declared purposes well enough. Two states, which in the "target" chain with correct probabilities (3.39) would be "light" or less probable (rain) and "heavy" or more probable (sun), are equally likely to occur in the proposal chain. Therefore if we count rainy and sunny days in the sample $(y_1, \ldots, y_N)$ from the proposal, their proportion will be approximately 50-50, which is substantially different from the proportion expected for the Sun City.

Figure 4.1 shows a simulated chain where hollow circles correspond to the states of the proposal chain. There are 21 points in the "rainy" state 1, and 29 points in the "sunny" state 0, bringing about the proposed proportion of rainy days of 0.42. This is far away from the Sun City theoretical proportion 0.25, so the sampling is not very effective in the determination of the long-term frequency of rainy days.

Thus follows the necessity of the second stage of the algorithm or an accept–reject procedure. Let us observe the proposal chain day-by-day and update it exactly like this: nothing changes in the proposal chain if it proposes a next day change to a heavier or equally heavy state (from sun to sun, rain to rain, or rain to sun). Only if a change from sun to rain is suggested by the proposal chain, we accept this "jump" to a lighter state with probability 1/3 and reject it with probability 2/3. Rejecting in this context means that instead of jumping into a rainy state, we stay with sun for one more day.

It is easy to verify that with the arbitrary choice of initial state X_0, the resulting chain X_t after applying the accept–reject procedure to the proposal chain for $t = 1, \ldots, N$ will have transition probabilities

$$p_{01} = P(X_t = 1 \mid X_{t-1} = 0) = q_{01} \times 1/3 = 1/2 \times 1/3 = 1/6,$$

$$p_{00} = P(X_t = 0 \mid X_{t-1} = 0) = q_{00} + q_{01} \times 2/3 = 5/6,$$

$$p_{10} = q_{10} = p_{11} = q_{11} = 1/2,$$

as is supposed to be for the target: the Sun City. Mission accomplished! It should be mentioned though that the secret of our success is in being able to use the magic number 1/3, which helped us to successfully restore the target transition matrix

$$P = \begin{pmatrix} 5/6 & 1/6 \\ 1/2 & 1/2 \end{pmatrix},$$

from the simulation which started with pretty arbitrary proposal

$$Q = \begin{pmatrix} 1/2 & 1/2 \\ 1/2 & 1/2 \end{pmatrix}.$$

As it happens, this magic number 1/3 expresses some partial knowledge of the stationary distribution of the target chain. So instead of using the target transition matrix to directly simulate a chain, we use information regarding its stationary distribution to alter a chain with the proposal transition so that it moves toward the target. What is the point of such simulation? Same as before, the most evident result is obtaining a numeric estimate for the mean of the target distribution (in other words, the long-time proportion of rainy days in the Sun City).

The simulation with $N = 50$ from Figure 4.1 with an additional accept–reject procedure applied at each time step is illustrated in Figure 4.2, where the filled circles correspond to the accepted states of the proposal chain, and hollow ones correspond to the rejected ones. The proposal chain spent 21 days at the "rainy" state 1 and 29 at the "sunny" state 0. However, 9 circles in state 1 stayed unfilled and the chain at these times dropped down to 0, which means that 9 proposed rainy dates were rejected, and on these 9 days the resulting chain stayed in the sunny state. The resulting chain includes only 12 rainy days, which brings the estimate of the long-term proportion of rainy days down to $\bar{X} = \frac{1}{N} \sum_{t=1}^{N} X_t = 0.24$, which is much better than the proposed proportion 0.42.

The procedure we implemented is a special case of so-called independent Metropolis algorithm (IMA). It is a representative of **MCMC: Markov chain Monte Carlo** methods. For the Sun City example it was used to introduce a Markov chain simulation, which required neither exact knowledge of the stationary distribution, nor even the knowledge of the transition matrix, however brought about a plausible numerical result.

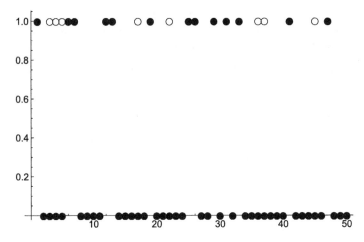

Figure 4.2 Two-step Markov chain Sun City simulation: $N = 50$.

In more general Bayesian framework, we may want to sample from the target posterior distribution, whose density, as we might remember from Chapter 2, is proportional to the product of prior and likelihood so typically is known to within a constant. For this purpose we build a Markov chain with the state space coinciding with the parametric set such that its stationary distribution is indeed the posterior. The transition matrix or, in case of infinite state space, the transition kernel of this chain might not be directly available. Therefore we might have to start with the proposal chain, which will be updated at every time step in order to converge to the target. This convergence may take a long time, thus we will need chains much longer than the one we used in the previous example.

Ten Coin Tosses

To further illustrate this version of Metropolis algorithm, we will revisit the ten coins example introduced in Chapter 2 and then reviewed in Chapter 3. Suppose that the probability of success θ is known to have $Beta(12, 12)$ prior distribution and the data x consists of 7 successes in 10 trials. We have derived the posterior distribution with density $\pi(\theta \mid x) \sim Beta(19, 15)$, and the Bayesian estimator we are looking for is the posterior mean $E(\theta \mid x)$. Section 2.6 contains the analytical solution. In Section 3.3 we have suggested several methods to generate independent random samples directly from the posterior and estimate the posterior mean via Monte Carlo approach. Now we want to estimate the posterior mean applying Markov chain simulation.

We will try to estimate the mean value of the posterior distribution $\pi(\theta \mid x) \sim Beta(19, 15)$ using Markov chain Monte Carlo simulation instead of analytical formula

$$E(\theta \mid x) = \frac{\alpha}{\alpha + \beta} = \frac{19}{19 + 15}.$$

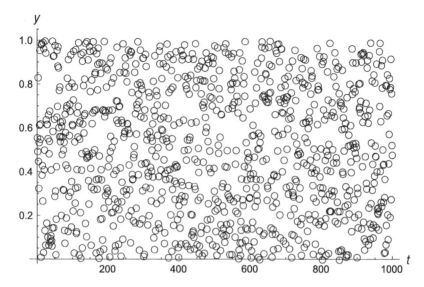

Figure 4.3 Ten coins. Proposal sample.

Here we will suggest a procedure similar to the one we just used for the Sun City, which will generate a Markov chain whose stationary distribution is the desired *Beta*(19, 15). This procedure is certainly of no practical use being much less efficient than analytical calculations in Section 2.6 or independent Monte Carlo simulation in Section 3.3. However, it demonstrates the potential of Markov chain simulation for Bayesian estimation purposes.

The first stage of the algorithm requires us to choose a transition kernel with support on the same state space as the target (posterior) and generate a Markov chain with this transition kernel. The parametric set $\Theta = [0, 1]$ forms an infinite state space for this proposal. The easiest choice of distribution to sample is the uniform *Unif*[0, 1]. We will build the proposal chain by generating $t = 1, 2, ..., N$ independent copies $y_t \sim Unif[0, 1]$ using independent transition kernel $g(y_t \mid y_{t-1}) = g(y_t) \equiv 1$ for any values $y_{t-1}, y_t \in [0, 1]$. These copies will uniformly fill up the horizontal band $\{1, ..., N\} \times [0, 1]$ in Figure 4.3, where time t on horizontal axis is discrete, and the parametric state space on y-axis is continuous.

The normalized histogram of the proposal sample is shown in Figure 4.4 along with the posterior density $\pi(\theta \mid x) \propto \theta^{18}(1 - \theta)^{14}$. Notice, that this figure represents the static summary of the chain, and thus the axes are different from the previous graph: horizontal axis corresponds to the parametric state space and the values on vertical axis serve to compare the empirical distribution of the proposal sample to the density of the posterior values in the state space. As we can clearly see, the histogram corresponds to the uniform distribution, and its shape does not even remotely resemble the target. Too many points in the proposal sample are located toward the ends of the interval [0, 1], and too few

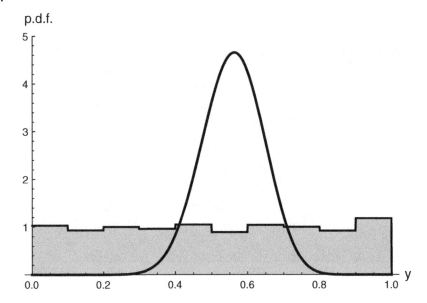

Figure 4.4 Ten coins. Histogram of the proposal and target density.

are grouped around the mode. The second stage of the algorithm will result in "thinning" of the proposal chain: values will drift toward the center of the state space corresponding to the higher posterior.

We will establish a natural ordering of "weight" of states $\theta_i \in \Theta$:

$$\theta_1 \preceq \theta_2 \text{ iff } \pi(\theta_1 \mid x) \leq \pi(\theta_2 \mid x)$$

which will simplify the following deliberations.

The second stage of the algorithm will require setting up an accept–reject procedure which will allow the proposal chain to gravitate from the outskirts of Θ toward the "heavier" states corresponding to higher target density which are located around its mode. For that purpose we will consider the ratio

$$R(\theta_1, \theta_2) = \frac{\pi(\theta_1 \mid x)}{\pi(\theta_2 \mid x)} = \left(\frac{\theta_1}{\theta_2}\right)^{18} \times \left(\frac{1 - \theta_1}{1 - \theta_2}\right)^{14}$$

for any two states in Θ. Then we can write down the natural ordering condition in terms of ratio R:

$$\theta_1 \preceq \theta_2 \text{ iff } R(\theta_1, \theta_2) \leq 1,$$

and define the *acceptance probability* as

$$\alpha(\theta_1, \theta_2) = \min\{1, R(\theta_1, \theta_2)\}. \tag{4.1}$$

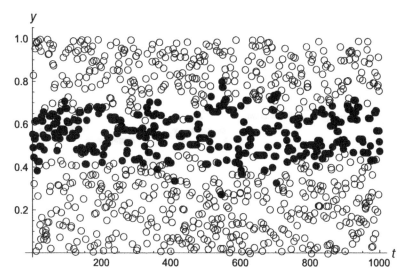

Figure 4.5 Ten coins. Markov chain simulation, $N = 1000$.

We will then generate a supplementary uniform sample $u_1, \ldots, u_N \sim Unif[0, 1]$ and construct the new Markov chain z_t starting with an arbitrary initial value $z_0 \in [0, 1]$ and at time t using proposal chain to update:

$$z_t = \begin{cases} y_t, & \text{if } R(y_t, z_{t-1}) \geq u_t \text{(acccept);} \\ z_{t-1}, & \text{otherwise(reject).} \end{cases} \tag{4.2}$$

The role of the supplementary sample (u_t) is to establish the correct acceptance probability

$$\alpha(y_t, z_{t-1}) = P(z_t = y_t | z_{t-1}) = \begin{cases} 1, z_{t-1} \leq y_t; \\ R(y_t, z_{t-1}) \text{ otherwise} \end{cases}$$

so that a proposed heavier state is always accepted, while a lighter state is accepted only with a certain probability based on the ratio R.

The resulting chains are shown in Figure 4.5 for $N = 1000$ and in Figure 4.6 for $N = 100$. In both pictures the filled circles correspond to accepted states of the proposal chain, while the hollow circles correspond to the rejected states. One may see that while the desired thinning of the edges of the state space is achieved in both and the corrected chain concentrates around the target mean, zooming-in with the smaller sample size makes it evident that the corrected chain often gets stuck at certain states which happens due to high rejection rate. This highly undesirable effect leading to high positive autocorrelations in the final chain will be discussed later.

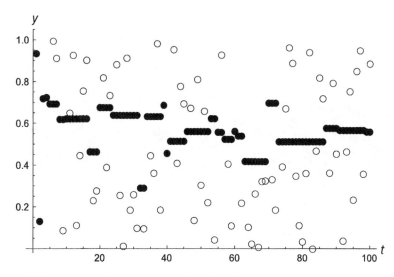

Figure 4.6 Ten coins. Markov chain simulation, $N = 100$.

Numerical estimates for the posterior mean, which we know to be approximately 0.559, are obtained as sample means for the resulting sample: $\bar{z} = 0.558$ for Figure 4.5 with $N = 1000$, and clearly less precise $\bar{z} = 0.568$ for Figure 4.6 with $N = 100$. It is also worthwhile to notice that for the resulting chain z_t not only the sample mean approximates the posterior mean, but its histogram also resembles the posterior density. This can be observed in Figure 4.7.

The procedure we described above is a special case of independent Metropolis algorithm introduced above for the Sun City simulation. More formal definition of the algorithm and its more general version are defined in the next section.

4.2 Metropolis–Hastings Algorithm

Let us describe the general scheme which will allow us to solve two related problems:

- Generate random sample from the distribution with posterior density $\pi(\theta \mid x)$.
- Estimate posterior mean of θ, $E(\theta|x)$.

Both problems are open when no analytical formula exists for posterior mean, and direct sampling from the posterior is not possible. Moreover, as usual in Bayesian practice, posterior density $\pi(\theta|x)$ may be known only to within a constant multiple. In this case, a Markov chain simulation may become not only feasible but also practical.

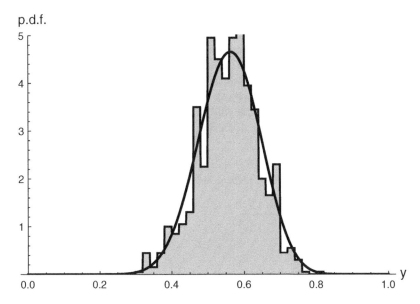

p.d.f.

Figure 4.7 Ten coins. Histogram of the resulting chain and target density.

Consider a function $f(\theta) \propto \pi(\theta \mid x)$ defined on Θ. Choose a valid **proposal** p.d.f. $g(\theta)$ on the same domain Θ such that it allows to define an easy sampling procedure. Then we will build a Markov chain z_t, $t = 1, \ldots, N$ whose stationary distribution coincides with the posterior playing the role of the target distribution.

Independent Metropolis Algorithm (IMA)

For $t = 0$ choose any z_0 from Θ. Define the ratio

$$R(\theta_1, \theta_2) = \frac{f(\theta_1)}{f(\theta_2)} \bigg/ \frac{g(\theta_1)}{g(\theta_2)} = \frac{f(\theta_1) \times g(\theta_2)}{f(\theta_2) \times g(\theta_1)}.$$

For $t = 1, 2, \ldots$

- Generate a random proposal $y_t \sim g(\theta)$;
- Generate a random $u_t \sim Unif[0, 1]$;
- Define

$$z_t = \begin{cases} y_t, & \text{if } R(y_t, z_{t-1}) \geq u_t \text{(acccept)}; \\ z_{t-1}, & \text{otherwise(reject)}. \end{cases}$$

The obtained Markov chain z_t will have posterior $\pi(\theta|x)$ as its stationary distribution therefore sample mean $\bar{z} = \frac{1}{N} \sum_{t=1}^{N} z_t$ will serve as a reasonable estimate for the posterior mean. Putting aside for now an important issue of the rate of convergence to the stationary distribution and even the very fact of this convergence which has to be proven yet, we may report the successful solution of both problems posted in the beginning of the section. Let us also introduce the following natural generalization of this algorithm, allowing one to introduce the step-by-step changes in the proposal.

Suppose that for any $\theta_1 \in \Theta$ we can sample from the conditional density $g(\theta \mid \theta_1)$ defined on Θ, playing the role of the proposal kernel. Then we will build a Markov chain z_t, $t = 1, \ldots, N$ using the modified algorithm below.

Metropolis–Hastings Algorithm (MHA)
For $t = 0$ choose any z_0 from Θ. Define the ratio

$$R_g(\theta_1, \theta_2) = \frac{f(\theta_1) \times g(\theta_2 \mid \theta_1)}{f(\theta_2) \times g(\theta_1 \mid \theta_2)}. \tag{4.3}$$

For $t = 1, 2, \ldots$

- Generate a random proposal $y_t \sim g(\theta \mid z_{t-1})$;
- Generate a random $u_t \sim Unif[0, 1]$;
- Define

$$z_t = \begin{cases} y_t, & \text{if} R_g(y_t, z_{t-1}) \geq u_t \text{(acccept)}; \\ z_{t-1}, & \text{otherwise (reject).} \end{cases}$$

Avoiding fine details of the actual proof, we can still suggest a sketch of the justification of general MHA. It is based on the verification of the detailed balance equation (3.56). Suppose that we redefine the ordering on Θ using (4.3):

$$\theta_1 \preceq \theta_2 \text{ iff } R_g(\theta_1, \theta_2) \leq 1.$$

Evidently, the detailed balance equation will not work with the proposal without a further accept/reject step:

$$f(\theta_1) \times g(\theta_2 \mid \theta_1) \neq f(\theta_2) \times g(\theta_1 \mid \theta_2),$$

but if we define acceptance probability $\alpha(\theta_1, \theta_2) = \min\{1, R_g(\theta_1, \theta_2)\}$ and transition kernel $g^*(\theta_1 \mid \theta_2) = \alpha(\theta_1, \theta_2) \times g(\theta_1 \mid \theta_2)$, reflecting possible rejection of the proposal g, it is easy to see that

$$f(\theta_1) \times g^*(\theta_2 \mid \theta_1) = f(\theta_2) \times g^*(\theta_1 \mid \theta_2). \tag{4.4}$$

Indeed, let us assume first that $\theta_1 \preceq \theta_2$. In this case by definition

$$g^*(\theta_2 \mid \theta_1) = g(\theta_2 \mid \theta_1),$$

$$g^*(\theta_1 \mid \theta_2) = R_g(\theta_1, \theta_2) \times g(\theta_1 \mid \theta_2) = \frac{f(\theta_1)}{f(\theta_2)} g(\theta_2 \mid \theta_1),$$

and both the left-hand side and the right-hand side in (4.4) are equal to the product $f(\theta_1) \times g(\theta_2 \mid \theta_1)$. On the contrary, if $\theta_2 \leq \theta_1$, then

$$g^*(\theta_2 \mid \theta_1) = R_g(\theta_2, \theta_1) \times g(\theta_2 \mid \theta_1) = \frac{f(\theta_2)}{f(\theta_1)} g(\theta_1 \mid \theta_2),$$

$$g^*(\theta_1 \mid \theta_2) = g(\theta_1 \mid \theta_2),$$

and both sides in (4.4) are equal to $f(\theta_2) \times g(\theta_1 \mid \theta_2)$. This means that a Markov chain with the transition kernel $g^* = \alpha g$, constructed by accepting the proposal with kernel g with probability α, whatever the proposal is, will converge to the stationary target f. For an elegant geometric interpretation of the MHA, see Billera and Diaconis [2].

There are publications (see, e.g., Dongarra and Sullivan [5]) expressing an opinion that the Metropolis–Hastings algorithm is one of the 10 most influential computer algorithms of the twentieth century. One can argue exact rankings like this, but the story of the MHA is indeed impressive. It can be dated back to the early 1950s. The early history of the algorithm is fairly well known and is associated with Los Alamos, NM. As many other inventions influencing the development of the discipline of statistics, it is closely tied to its applications. Out of the five authors of the paper [20], which introduced and justified the algorithm in its independent version (Metropolis, Rosenbluth, Rosenbluth, Teller, and Teller), the name of Edward Teller is by far the most notorious. Teller, along with Stanislaw Ulam, was one of the two most prominent physicists associated with the H-bomb project.

A certain cloud of controversy surrounds the invention and real authorship of MHA. Besides Teller, other physicists such as Ulam and even Enrico Fermi have been mentioned as possible sources of its idea (Teller [31]). Another honorary mention is due to the emergence of a brand new (then) computer MANIAC (1952), which made it possible to implement the algorithm numerically. Nevertheless, it is probably fair that the algorithm was named after Nicholas Metropolis, which seems to be justified by more than the pure alphabetic order of the authors of the above mentioned seminal paper [20]. His contribution to the paper and his role in the development of both MANIAC and MHA well deserves this level of appreciation (for more details, see Robert and Casella [27]). The second name associated with MHA is that of a Canadian statistician W. K. Hastings, who further generalized Metropolis algorithm and put it in its current mathematical framework in his later paper [15], which was further developed in Peskun [21].

MHA was historically the first of the sampling methods organizing simulation of a random sample in a Markov chain fashion, which were given the name of MCMC: Markov chain Monte Carlo. Most of the popularity of MCMC methods nowadays is related to their use in Bayesian statistics, and their modern development is to a large extent connected with the needs and demands of Bayesian statistical models.

The initial independent version of MHA was successfully complemented by further development including such modifications as random walk algorithms considered in Section 4.3. It was also closely followed by a very different idea of Gibbs sampling discussed in Section 4.4. The rest of the chapter deals with the diagnostic issues and provides a brief survey of the modern development of MCMC.

4.3 Random Walk MHA

When we consider any problem of parametric estimation, we can identify two main sources of error. One is the bias and another is the variance of the estimators as a random variable. The bias, which is detrimental for the accuracy of estimation, is defined as the systematic error or $b(\phi) = E(\phi^*) - \phi$, where ϕ is the parameter of interest and ϕ^* is its estimator. The variance is responsible for the estimation's precision. The mean squared error of estimation, which is routinely considered a natural measure of error or the most popular loss function can be defined as

$$E(\phi^* - \phi)^2 = b(\phi)^2 + Var(\phi^*).$$

If we consider estimation of posterior mean $\phi = E(\theta \mid x)$ to be the main purpose of the MHA, it is logical to try to accomplish two tasks: to keep the bias at bay, and also to minimize the variance of the estimator of ϕ obtained by Markov chain Monte Carlo sampling. Let us start with the variance considerations. Keeping in mind that according to the principle of Monte Carlo methods, the MHA estimator of the posterior mean is just the sample mean of the constructed chain, we can analyze the variance of the chain in the same fashion as we did for correlated samples in Section 3.4. Theorem 3.2 still works and identifies positive correlation between the sample elements as the enemy: the main source of variance.

Looking back at Figure 4.6, we can detect several flat segments, where the chain does not move and gets stuck at certain states for a few steps. The same picture could be observed for a longer chain in Figure 4.5, it is just harder to visualize. The most obvious reason for that is a too low **acceptance rate**, defined as the overall proportion of the new proposed states, which are accepted by the algorithm. Low acceptance rate by construction of independent Metropolis algorithm may be caused by a poor choice of the proposal bringing about too many rejections. If the proposed states are very "light" or improbable from the target point of view, they are likely to be rejected by the MHA due to their low acceptance probability (4.1). What is the problem with flat segments? They indicate high positive autocorrelation, which was defined in Section 1.7 as the correlation between the adjacent elements of the chain. Thus the overall variance in the chain stays high according to Theorem 3.2.

The simplest solution to the problem of flat segments and high positive auto-correlation would be a more adequate choice of the proposal. Better proposals could be sought for in the framework of IMA. However they are not that easy to find. We do not want a proposal to be **underdispersed** or tightly concentrated at certain regions of the state space. Underdispersed proposals tend to ignore other regions, which is especially dangerous if these other regions are target-heavy and need to be visited in order to build a representative sample from the target. The evident advantage of the uniform proposal used in ten coins example is its being **overdispersed** in the meaning of allowing greater variability than the target. It runs around the state space, and the low acceptance rate is just the price we have to pay for overdispersion. The curse of IMA is having to choose a proposal avoiding both extremes: underdispersion and low acceptance rate due to overdispersion (the latter being a lesser evil).

An opportunity to maneuver between these two extremes lies in a more flexible general MHA structure. Instead of wild up-and-down jumps between the states dictated by an overdispersed independent proposal often rejected by the MHA, we can try to propose each new state relatively close to the previous one guaranteeing a more pedestrian traveling style across the state space.

This approach is epitomized by the **random walk Metropolis–Hastings algorithm (RWMHA)**. In its most popular version the conditional proposal in (4.3) is chosen as

$$g(\theta_2 \mid \theta_1) = \frac{1}{\sigma\sqrt{2\pi}} e^{-\frac{(\theta_2 - \theta_1)^2}{2\sigma^2}} \sim N(\theta_1, \sigma^2). \tag{4.5}$$

An important property of this proposal is its symmetry: $g(\theta_2 \mid \theta_1) = g(\theta_1 \mid \theta_2)$, which reduces expression (4.3) to

$$R(\theta_1, \theta_2) = \frac{f(\theta_1)}{f(\theta_2)}.$$

Other choices of symmetric proposals are certainly possible, but normal distribution with varying σ usually provides sufficient flexibility. An important calibration or **scaling** issue is the choice of parameter σ [28]. Too large σ can bring about excessive overdispersion and in some cases even force the proposal to leave the target state space, which is undesirable. Too small σ leads to a chain with a very high acceptance rate, which is not necessarily good news. Such a chain may take a long time to travel across the state space, and guess what: with a small enough variance, it can exhibit a substantial bias!

The main problem can be stated as follows: convergence of an underdispersed random walk sample to the target is guaranteed, but it can be slow. The first values of the chain (and it is not clear, how many of these) may have a distribution

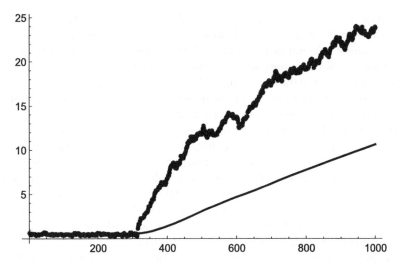

Figure 4.8 Ten coins RWMHA: $\sigma = 0.2$.

far away from the target, and it is not clear how long it takes to get the chain into the stationary mode.

Figures 4.8, 4.9, and 4.10 illustrate the performance of RWMHA for the ten coins example (compare with IMA chains in Figures 4.5 and 4.6). Filled circles represent the accepted values and the solid line corresponds to the cumulative

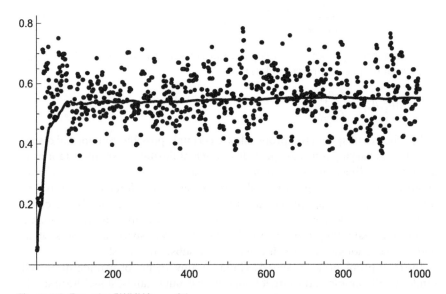

Figure 4.9 Ten coins RWMHA: $\sigma = 0.1$.

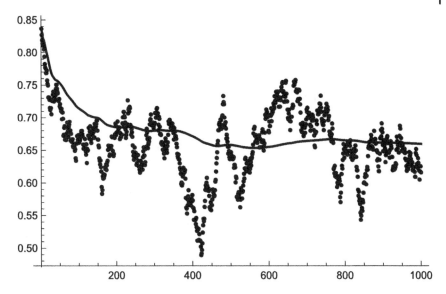

Figure 4.10 Ten coins RWMHA: $\sigma = 0.01$.

sample means providing the current estimates for the posterior mean as the chain runs on. Clearly, the choice of $\sigma = 0.2$ in Figure 4.8 with no qualifying conditions proves unfortunate, because the chain leaves the state space after 300 steps and drifts out. The chain is way overdispersed though its calculated acceptance rate 0.806 is rather high.

The moderate value of $\sigma = 0.1$ in Figure 4.9 brings about good ***mixing***, which is the standard term for a nice travel pattern around the state space and a reasonable numerical estimate of 0.566 for the posterior mean equal to 0.559 with the acceptance rate of 0.66. Convergence to the posterior mean is decent even after 200 or 300 steps of the chain.

On the other hand, too small value of $\sigma = 0.01$ in Figure 4.10 causes an overly high acceptance rate of 0.977, and while the final numerical estimate of 0.599 for the posterior mean is sufficiently bad, just look what a disaster it could have been if we stopped at $N = 300$ instead of moving on to 1000 iterations. A poor choice of initial value for an RWMHA chain causes a substantial bias of the sample mean as the estimate of the posterior mean. Also, high acceptance rate with low σ values does not prevent positive autocorrelations.

We will revisit the issues of acceptance rate, bias, and variance reduction in a more general context. At this point we have to admit that there is a number of issues with the choice of proposal for the MHA implementation. These issues are magnified in multidimensional parameter spaces causing slow or even breaking convergence of the chain to the target. In the next section we consider a different MCMC algorithm, which is specifically devised for multiparametric case and in many instances proves to be more effective than the MHA.

4.4 Gibbs Sampling

Triangular Uniform Distribution

Let us begin with a simple example. Consider state space $S = [0, 1] \times [0, 1]$ and the target distribution with density $f(x, y) = 2$, if $x + y \leq 1$ (uniform distribution on $A = \{(x, y) : x + y \leq 1\}$). How can we generate a sample from this distribution, which is just uniform on A? In a fashion similar to the Monte Carlo estimation of the number π in Section 3.3, we can generate a sample on the unit square S, and then use an accept/reject procedure to throw out extra points not satisfying condition $x + y \leq 1$. The only problem consists in losing roughly half of the initial sample during the accept/reject procedure. We can save some time by using the fact that the bivariate target distribution obtains two simple one-dimensional conditional distributions with densities $f(x \mid y) \sim Unif(0, 1 - y)$ and $f(y \mid x) \sim Unif(0, 1 - x)$. Drawing from each of these uniform distributions is easy.

Let us define an iterative process:

For $t = 0$ choose x_0 uniformly from $[0, 1]$. Then choose y_0 uniformly from $[0, 1 - x_0]$.

For $t = 1, 2, \ldots$

- Generate a random value $x_t \sim Unif[0, 1 - y_{t-1}]$;
- Generate a random value $y_t \sim Unif[0, 1 - x_t]$;
- Use pairs (x_t, y_t) as a sample from $Unif(A)$.

The resulting sample will neatly fill out the triangular region A. This procedure involving dimension reduction was possible because of our knowledge of conditional distributions and their relatively simple form. It allowed us to sample from univariate conditionals rather than directly from two-dimensional joint distribution. This idea lies in the foundation of Gibbs sampling. In general, Gibbs sampling scheme for two parameters could be described as follows.

Gibbs Sampling

Suppose $f(\theta, \phi) \propto \pi(\theta, \phi \mid x)$ is (to within a multiplicative constant) the target density with data x (assumed to be known) and two parameters θ and ϕ. Let $f_1(\theta)$ and $f_2(\phi)$ be two marginal densities, and

$$g_1(\theta \mid \phi) = \frac{f(\theta, \phi)}{f_2(\phi)}, g_2(\phi \mid \theta) = \frac{f(\theta, \phi)}{f_1(\theta)}$$

are two conditional densities. If direct sampling from conditional is feasible, the algorithm is as follows:

For $t = 0$ choose θ_0 arbitrarily. Then draw ϕ_0 from $g_2(\phi \mid \theta_0)$.
For $t = 1, 2, \ldots$

- Generate a random value $\theta_t \sim g_1(\theta \mid \phi_{t-1})$;
- Generate a random value $\phi_t \sim g_2(\phi \mid \theta_t)$;
- Use pairs (θ_t, ϕ_t) as a sample from $f(\theta, \phi)$.

The name associated with the algorithm is somewhat misleading, because Gibbs himself had nothing to do with its idea. Geman and Geman [12] introduced the term developing the sampling schemes for Gibbs random fields while working on image processing problems. Gelfand and Smith [10] were probably the first to point out the full potential of Gibbs sampling (and MCMC in general) to Bayesian statistics.

By construction, each step t of the algorithm is determined by the state at the previous step $t - 1$, therefore the resulting sample forms a Markov chain. Convergence to the target is achieved by satisfying the detailed balance equation. It is easy to see that updates of individual parameters with proposals g_1 and g_2 are reversible as $f_1(\theta)g_2(\phi \mid \theta) = f_2(\phi)g_1(\theta \mid \phi)$, but entire tth step might not be, and it is much harder to prove that the stationary distribution of the chain is indeed the target. It was first done by Besag [1] under certain additional conditions. For a more detailed discussion see Gamerman and Lopes [9].

Normal Distribution

The following example (see Robert and Casella [25]) addresses the problem of sampling from bivariate normal distribution. This distribution was defined in Section 1.6 and will play important role in further discussions in Chapters 5 and 6.

Suppose $(\theta, \phi) \sim BN(0, \Sigma)$, with zero means and unit covariance matrix

$$\Sigma = \begin{pmatrix} 1 & \rho \\ \rho & 1 \end{pmatrix},$$

where $\rho = \rho(\theta, \phi)$ measures the correlation between two components of the random vector.

The algorithm below is a Gibbs-style Markov chain sampling emphasizing the role of conditionals. An alternative direct sampling scheme not involving conditionals is introduced later in Section 5.5 and used in Section 6.6.

It is known from Chapter 1, that both marginals $f_1(\theta)$ and $f_2(\phi)$ are standard normal, and the conditionals are also normal. They can be represented as

$$g_1(\theta \mid \phi) \sim N(\rho\phi, 1 - \rho^2), \ g_2(\phi \mid \theta) \sim N(\rho\theta, 1 - \rho^2).$$

Using standard procedures for sampling from univariate normal distributions, we will apply Gibbs sampling.

For $t = 0$ make an arbitrary choice of $\theta_0 = c$. Then draw ϕ_0 from $N(\rho c, 1 - \rho^2)$. By construction, $\rho(\theta_0, \phi_0) = \rho$.

For $t = 1, 2, \dots$

- Generate a random value $\theta_t \sim N(\rho\phi_{t-1}, 1 - \rho^2)$;
- Generate a random value $\phi_t \sim N(\rho\theta_t, 1 - \rho^2)$;
- Use pairs (θ_t, ϕ_t) as a sample from $BN(0, \Sigma)$.

The correlation structure is maintained at every step, since by construction we get $\rho(\theta_t, \phi_t) = \rho$. Notice that due to arbitrary choice of the initial point $\theta_0 = c$ this algorithm does not immediately start at the stationary distribution and it takes some time to get there. However, it is easy to see that two substeps at step t may be combined to yield the following one-step $t - 1$ to t conditionals:

$$g(\theta_t \mid \theta_{t-1} = c) \sim N(\rho^2 c, 1 - \rho^4),$$

and by recursion

$$g(\theta_t \mid \theta_0 = c) \sim N(\rho^{2t} c, 1 - \rho^{4t}),$$

so that for $t \to \infty$ the distribution of θ_t (and similarly of ϕ_t) converges to standard normal.

Gibbs sampling algorithm can be clearly generalized to the case of three or more parameters. It could be technically considered to be a version of the Metropolis–Hastings algorithm for multiple parameters without the rejection part. As one can see, the main benefit of this method is avoiding any accept/reject steps and not losing any generated sample elements. The key is: we need to know how to sample directly from conditional distributions, which is fortunately possible in many interesting applications. However, when posteriors are too complicated and their conditionals are not analytically tractable, as happens to be the case with most applications in the second part of the book, Gibbs sampling cannot be productively used and the MHA or its modifications seem to be the last resort. "Metropolis within Gibbs" is a hybrid sampling scheme, which can be applied when some of the conditional distributions are tractable, and some are not, necessitating application of the MHA for some particular components of the vector parameter. In this book we will avoid a too detailed discussion of Gibbs sampling, concentrating instead on the diagnostics issues which are common for all MCMC methods.

4.5 Diagnostics of MCMC

Discussing the rate of convergence of general Monte Carlo methods in Chapter 3, we made the main emphasis on variance reduction. Monte Carlo estimates were obtained as the sample means of the samples drawn directly from the posterior distribution. Sample means converge to the theoretical posterior mean by virtue of the Law of Large Numbers. For Markov chain Monte Carlo methods we have to consider an additional source of error: the sample does not come directly from the posterior distribution, it just converges to the posterior, and the rate of this convergence has to be taken into account. Arbitrary choice of the initial point and the behavior of the chain in its randomly long initial segment before it gets into the stationary distribution cause a possible bias of the MCMC estimators. Rate of convergence for various MCMC procedures including the MHA and Gibbs sampler can be estimated by rigorous mathematical tools, but such results are often very technical and complicated [26]. Therefore monitoring and reducing both bias and variance based on the chain output plays an important role in practical applications of MCMC techniques.

4.5.1 Monitoring Bias and Variance of MCMC

The main visual tool which can be used for monitoring the behavior of an MCMC chain is its time series plot or *trace plot*. In case of multiple parameters trace plots can be drawn for each individual component of the vector parameter. Trace plots are equally applicable to the MHA chains and Gibbs sampler chains. Without using this exact name, we have seen many trace plots already for most of the examples in Chapter 4. For instructive purposes we tended to show both accepted and rejected states for the MHA in Figures 4.5 and 4.6. To avoid unnecessary confusion, one may want to drop the rejected points from consideration (recording the overall acceptance rate as an additional monitoring tool) and show the time series of all the accepted states against the time scale appended with the path of accumulated sample means. Examples of trace plots were presented in Figures 4.8, 4.9, and 4.10.

The visual clue suggesting good convergence is the stabilization of the time series of accepted states around the centerline of accumulated means, achieved in the case of Figure 4.9. It is evidently an imprecise and subjective criterion of convergence and it has to be confirmed by more definitive statistical procedures.

Another standard graphical summary of a Markov chain is its histogram. It should be mentioned that histograms do not preserve the order of observations in the chain and thus lose potentially important information which is captured in trace plots. However, histograms provide a valid insight on the shape of the sampling distribution of the chain which can be compared to the posterior density if its form is known. Such features as multimodality, skewness, and fat tails visible at an MCMC histogram can be considered a serious diagnostic warning if they are not expected from the posterior density. Histograms of the chains from Figures 4.9 and 4.10 are demonstrated in Figures 4.11 and 4.12, the latter looking decent but centered at a completely wrong point.

As we have noticed, high variance in an MCMC sample may be caused by positive autocorrelations in the chain, and in case of multiple parameters also by cross-correlations. Therefore the sample autocorrelation function (SACF) is

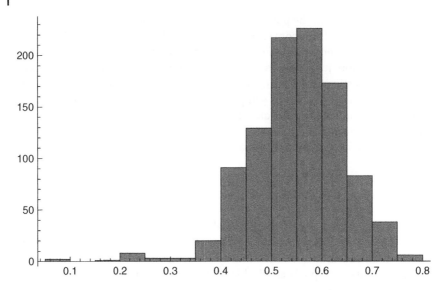

Figure 4.11 Ten coins RWMHA histogram: $\sigma = 0.1$.

a valuable additional tool for monitoring variance. Sample autocorrelation lag k for a finite time series y_t, where $t = 1, \ldots, N$ is defined as

$$r(k) = \frac{\sum_{t=k+1}^{N}(y_t - \bar{y})(y_{t-k} - \bar{y})}{\sum_{t=1}^{N}(y_t - \bar{y})^2}, \bar{y} = \frac{1}{N}\sum_{t=1}^{N} y_t. \tag{4.6}$$

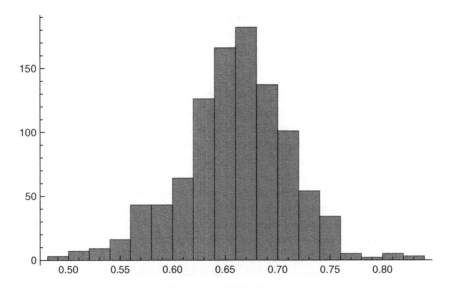

Figure 4.12 Ten coins RWMHA histogram: $\sigma = 0.01$.

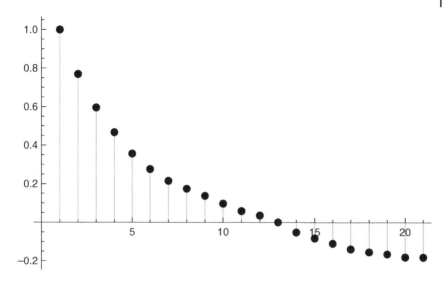

Figure 4.13 Ten coins RWMHA SACF: $\sigma = 0.1$.

Graphs of SACF for the chains from Figures 4.9 and 4.10 are shown in Figures 4.13 and 4.14.

The vertical lines on the graphs correspond to the values of sample autocorrelation $r(k)$ with increasing lag k. Ideally we want to see the autocorrelations

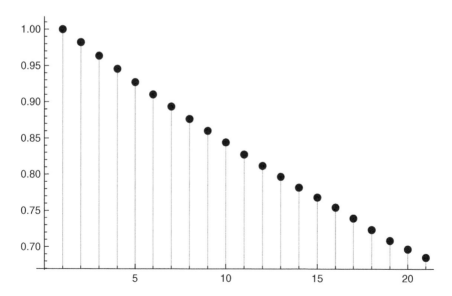

Figure 4.14 Ten coins RWMHA SACF: $\sigma = 0.01$.

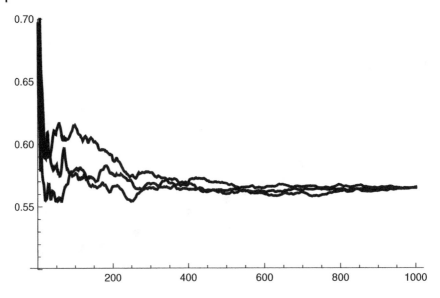

Figure 4.15 Multiple chains for $\sigma = 0.1$.

to be negligible or at least cut off or die down when the lag increases, as it happens in Figure 4.13. Sometimes negative autocorrelations can bring about faster convergence as stated in Theorem 3.2, but positive correlations as observed in Figure 4.14 inevitably slow it down.

A different approach to monitoring convergence requires running multiple chains simultaneously and projecting them on one multiple chain plot using the same time scale. Intuitively, when chains starting at several distant states stick together, this is a clear indication of the convergence to the stationary state. To prevent multiple chain diagrams from becoming too messy, it might be reasonable to drop the plots of accepted states and picture only multiple time series of accumulated means as done in Figures 4.15 and 4.16. Pay attention to the difference in vertical scales of these figures!

Though multiple chain plots are often very illustrative, clearly exposing the lack of convergence, there is always a question: is it worthwhile spending time on running m small chains of length N rather than running one long chain of length mN, which certainly has more time to converge than each of the smaller chains?

4.5.2 Burn-in and Skip Intervals

All of the trace plots or multiple chain graphs in the previous subsection exhibit very erratic behavior for the first 100 steps or so, and stabilize (or not) later on.

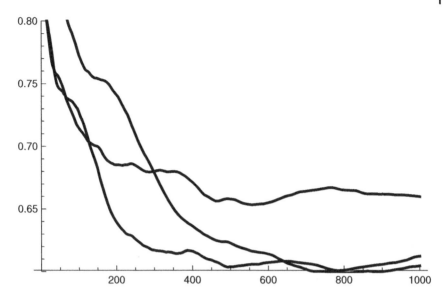

Figure 4.16 Multiple chains for $\sigma = 0.01$.

This is to be expected since the initial state was chosen in an arbitrary fashion causing potential bias of the estimation, and its effect lingers on in the chain until it is completely forgotten when and if the chain falls into the stationary state. One simple but hard decision is to cut the starting segment off in order to reduce the bias. This starting segment is known as the ***burn-in*** phase of the chain. When the decision is made, the first b sample elements are disregarded, and all further inference is based on the remaining sample size $N - b$.

This decision is hard for two reasons. First, it is always bad to lose data, even if they are somewhat compromised. The second reason is our very vague idea regarding the exact point where this cut-off has to be implemented. If we cut too early, we do not get rid of the bias, and if we cut too late, we throw out some completely valid data. The stochastic nature of the MCMC time series means that all inputs being equal, some chains will randomly behave better than the others, so no pre-fixed value b is going to ideally fit all possible trajectories of the process.

Some usual choices of b are between $0.1N$ and $0.5N$, which is not a definitive rule. In Figure 4.9, the choice of $b = 0.1N = 100$ seems to be plausible, cutting off most of the "garbage" observations in the beginning of the chain, and dramatically improving convergence. In Figure 4.6, the choice of $b = 0.1N = 10$ also improves convergence, while in Figure 4.5 it does not seem to be needed. Recommendation on appropriate burn-in phase is an important part of some diagnostic checks.

Using burn-in to reduce the bias, we also achieve a certain reduction of variance, which is likely to be higher in the initial segment of the chain. However, a more radical way to reduce the variance is to fight autocorrelation. If the adjacent values of the MCMC time series are too close to each other thus highly positively correlated, as it happens in Figure 4.10, which can be observed in Figure 4.14, or in case of flat segments due to low acceptance rate as observed in Figure 4.6, one can consider a possible rarefaction of the chain or skipping some of the adjacent values. Defining a ***skip interval*** k means taking in consideration only each kth element of the sample: $t = 1, k + 1, 2k + 1, \ldots$, and throwing out everything else. This approach brings about a radical reduction in the sample size: from N to N/k, which may be costly.

There exist some general results describing conditions under which no burn-in or skip interval are beneficial at all (see [18]). However both procedures are of a certain value and can be implemented when they are called for as a result of diagnostic checks.

4.5.3   Diagnostics of MCMC

In this section we will briefly describe several most popular techniques assessing the rate of convergence of MCMC sampling, addressing both bias and variance. All of these methods can be applied to many different MCMC settings including both the MHA and Gibbs sampler and can be used for vector parameters. This description is far from complete, and there exist many other methods of convergence diagnostics. Justification of most of the diagnostic tools discussed in this subsection requires a serious level of mathematical competence and is far beyond the level of the book. Nevertheless their basic ideas can still be presented in an intuitive way, which we are trying to do. Even if the readers are not given enough instruction to run specific diagnostic checks on their own, they will have some familiarity with the topic when invited to run these checks built into special software packages (e.g., BUGS, CODA, JAGS, Stan, and various R tools [17]).

Heidelberger–Welch: Stationarity Test

Using the general theory of time series, we can describe an MCMC chain as stationary (no drift, constant variance, and autocorrelations) when it is in or close to its stationary state. On the contrary, if a chain is not stationary, especially in its initial states, it might be beneficial to introduce a burn-in period, cutting off its nonstationary part. Heidelberger and Welch [16] propose to apply a stationarity test based on Cramer–von Mises statistic to the entire chain length N. If the chain passes the stationarity test, no burn-in is needed. If the test fails, introduce the burn-in of $b = 0.1N$, and repeat. If failed again, increase the burn-in to $0.2N$ and if needed, to $0.5N$. If the test still fails, it is necessary to increase the total length N of the initial chain, say, to $1.5N$ and repeat the process. The procedure itself is transparent other than the determination of the critical values of Cramer–von Mises statistic. It helps to detect a bias in the chain and determine optimal length of the burn-in period.

Geweke: Comparing Two Segments of the Chain

If the ultimate goal of the MCMC sampling is to estimate the posterior mean, it is natural to compare its estimates based on two different segments of the chain. Geweke in [13] suggested using the first $0.1N$ and the last $0.5N$ sample elements and compare the sample means for these two subsamples, which are supposed to be close enough. The test is based on the difference of the subsample means. As with the Heidelberger–Welch's method, the procedure is very simple, and the devil is in the detail: how to determine the critical value of the test statistic? Geweke's method helps to detect bias and address the variance issues, suggesting the reasonable length of the burn-in.

Raftery–Lewis: Quantile Estimation

Suppose the objective of running MCMC chain is to estimate a posterior quantile (most popular being the median, but other percentile points might be of interest too). In the case of quantile estimation, an MCMC time series can be treated as a binary Markov chain with two states: above and below a certain bound. This simplification makes it possible (see [24]) to use properties of binomial distribution in order to estimate such characteristics as the chain length N, burn-in b, and skip interval k after one single trial run of the chain.

Gelman–Rubin: Multiple Chains

The most popular MCMC diagnostic tool so far, is the **shrink factor** R introduced by Gelman and Rubin in [11], based on normal approximation for Bayesian posterior inference and involving a multiple chain construction. Gelman–Rubin's method requires first to draw initial values for multiple chains from an overdispersed estimate of the target distribution. The number of chains should be substantial: at least 10 for a single parameter and more for multiple parameter cases. All chains run for time N, and the second halves of the chains ($0.5N$ observations per chain) are used for the estimation of pooled within-chain variance W and between-chain variance B depending on the number of chains. Overall estimate of variance is $V = (1 - 2/N)W + (2/N)B$ and the shrink factor is determined as

$$\hat{R} = \sqrt{\frac{V}{W}}.$$

For a good convergence of MCMC process all chains are supposed to shrink together and the shrink factor should be close to 1.

Main criticism of Gelman–Rubin diagnostic relates to such issues as the sensitivity of results to the choice of starting points, validity of normal approximation, and necessity to discard a huge volume of observations (mainly, first halves of multiple overdispersed chains), which may not be further used in the inference.

The Metropolis–Hastings algorithm and Gibbs sampling are just two instances of general MCMC sampling methods. The main problem of these methods: the sample does not start at the target distribution, and it needs time to get there. Modern development of MCMC is driven to a large extent by two objectives: suppressing the bias and reducing the variance. In order to accomplish the first objective, one needs to be able to get to the stationary distribution as fast as possible. The second objective in case of the MHA is usually achieved by the choice of the proposal. In the next section we can briefly describe several directions of modification of basic MCMC algorithms in view of these two objectives.

4.6 Suppressing Bias and Variance

4.6.1 Perfect Sampling

The problem of determination of the length of the burn-in period would be resolved if we only knew the exact time when the chain gets into its stationary state, achieving the target distribution. The general properties of Markov chains state that after a chain gets into the stationary state, it stays there forever. Convergence has been achieved and since that moment on we can draw directly from the target distribution. The intuitive justification of this fact is the understanding that in its stationary state the chain has completely forgotten its past.

Returning to the Sun City example with just two states forming the state space, the bias is related to a difference between two possible starting points. Beginning simulation at a rainy day or a sunny day matters. However, if we start two chains: one from sun and one from rain, as soon as they hit the same state X at time T, which is known as ***coupling*** (and sooner or later this is going to happen), the past becomes irrelevant. The future behavior of both chains is determined only by the common state they share at time T, and not by the previous history. Therefore there is no necessity to run two chains any more, we can run one chain starting from the state X, and as soon as the observations prior to time T are discarded (burn-in), the bias is eliminated.

Imagine now that a chain has the state space with a finite number k of states. Then if we start k chains at time 0 from all possible states, and when any two chains couple, proceed with just one chain instead of these two, so that the total number of chains to run is reduced by 1. Sooner or later all the chains are going to couple together at some state X. The time T when it is achieved is our new starting point, and from there on we proceed with one chain which is known to stay at the stationary distribution creating a ***perfect sampling*** or exact sampling scheme. This procedure introduced by Propp and Wilson [23] is known as the ***coupling from the past*** (CFTP).

The price we pay for getting a perfect sample from the target distribution is high. A total of kT chain elements are being discarded. Finite state spaces are also not the most interesting ones. There exist modifications of CFTP to infinite state spaces.

An alternative technique such as backward sampling or backward coupling suggested by Foss and Tweedie in [7] using fine mathematical properties of Markov chains is one such modification. Instead of running multiple chains from the past into the future, it requires running one path backward using the reversibility condition, and after achieving a specific state in the past at time $-T$, running a forward chain from that state. It can be proven that under rather wide assumptions, this one backward/forward path will end up at the stationary state at time 0. However, this backward/forward run may take an unlimited amount of time, thus rendering the entire procedure hardly workable without additional adjustments. It seems like perfect sampling has a lot of promise but hardly has proven yet to be superior to other MCMC methods. For more recent development we can refer to two excellent surveys [4] and [14].

4.6.2 Adaptive MHA

An idea of variance reduction in MCMC can be addressed in many different ways. One of the most obvious ideas is related to the fact that "bad mixing" and slow convergence of chains due to high variance can be caused by a bad proposal distribution, which is too far from the target. Bad proposals are inevitable when we do not know much about the target. However, by observing a Monte Carlo chain, we may learn more while the chain still runs. Can we modify proposal "on the run," adjusting it as the chain goes?

Adaptive algorithms will do exactly that, using information obtained from the past values of the chain to modify the proposal for the future. The main problem with this approach is the loss of Markov property (according to this property, the future of a chain given present state cannot depend on the past). However, there exist many sophisticated ways to prove that under certain conditions either the Markov property still holds or its loss does not prevent the chain from converging to the target. A combination of proposal adaptations with perfect sampling may be promising [29] but has not proven to be very useful yet. For more references, see also [14].

4.6.3 ABC and Other Methods

The last decades gave rise to many different MCMC techniques including particle filtering, slice sampling, Hamiltonian Monte Carlo, quasi-Monte Carlo, and so on. One of the more promising directions is ***approximate Bayesian computation*** or ABC, which provides for some solutions when likelihood is not completely specified and thus traditional MCMC methods fail [14]. Modern MCMC methods can be much faster, more precise, and more efficient than the basic Metropolis–Hastings algorithm and even Gibbs sampling. Unfortunately, trying to keep things simple, we have to avoid further discussion of modern MCMC tools, suggesting to the readers to use the multiple reference sources provided above.

4.7 Time-to-Default Analysis of Mortgage Portfolios

In this section we will get back to basics and provide an illustration of the use of the RWMHA. We will apply it to a practical problem of mortgage defaults and discuss some interesting diagnostic advantages of Bayesian approach when implemented via MCMC.

4.7.1 Mortgage Defaults

Mortgage defaults played a critical role in the first financial crisis of the twenty-first century. Indeed, the majority of these were initiated in the subprime sector of the mortgage market. Accordingly, the practice and foundations of subprime lending have received great attention in the literature. Crucial to these conversations is an understanding of the inherent risk in subprime lending. In this section we discuss a Bayesian reliability model for the ***probability of default*** (PD) in mortgage portfolios. This approach was first suggested by Soyer and Xu [30], and then developed by Galloway et al. [8]. In the latter work, survival analysis methods were applied to a subprime mortgage portfolio of a major commercial bank. We will briefly discuss some results of this study.

Subprime mortgages are typically higher interest contracts considered riskier and not complying to the "prime" status, which would bring about a lower interest rate. The subprime portfolio in question is not homogeneous, including mortgages initiated during a 10-year period, representing a variety of terms, conditions, interest rates, and different sectors of the mortgage market. Available data are de-personalized, stripped of any identification and aggregated. The month of inception and the month of default (if it happened) are the only variables recorded. Early payoffs or refinancing as other causes for early termination could be considered as competing risks, but the data on these are unavailable. A large number of unobserved factors makes it necessary to find a way to reflect the latent (hidden) heterogeneity in default rates among the mortgage holders.

Out of the many possible approaches to modeling PD, one suggests defining a ***time-to-default*** (TTD) variable measuring the life of the mortgage from the inception to the default. Constructing a parametric model for the distribution of the TTD variable for the entire portfolio will provide estimates of PD for any given period of time. TTD models are popular in many applications including insurance, reliability, and biostatistics. We will first refer to one recent application from a seemingly unrelated field of business statistics which has suggested a new approach to modeling mortgage PD.

4.7.2 Customer Retention and Infinite Mixture Models

A long-standing problem of "customer retention" measuring the length of the period of customer's loyalty is especially important in the context of subscriptions which could be renewed or canceled, either at regular time intervals (e.g., monthly or annually) or in continuous time. This problem, historically related to journal subscription and customer loyalty programs, has attracted attention recently in connection with online subscription services. Customer retention time T is defined for a population of customers as a random variable measuring time until cancellation. It may serve as an indicator of success of marketing policy. A simple model for T can be derived from the discrete-time Beta-geometric (BG) model proposed by Fader and Hardie [6] and is very successful in describing subscription patterns. This model is based on the following assumptions.

(BG1) Cancellations are observed at discrete periods, that is, $T \in \{1, 2, 3, \dots\}$.

(BG2) The probability of cancellation for an individual customer, θ, remains constant throughout their holding. Thus T is described by the geometric distribution defined in (1.7) with distribution function

$$F_G(t|\theta) = P(T \le t|\theta) = 1 - (1 - \theta)^t, t \in \{1, 2, \dots\}.$$

(BG3) The probability of default, θ, varies among subscribers and can be characterized by the *Beta*(α, β) distribution (1.25) with parameters $\alpha, \beta > 0$ and probability density function

$$f_B(\theta|\alpha, \beta) \propto \theta^{\alpha-1}(1 - \theta)^{\beta-1}, \theta \in [0, 1].$$

This is an example of ***infinite mixture model***, where population heterogeneity is expressed by risk parameter θ pertaining to individual customers and randomly distributed across the population. Integrating θ out brings about the distribution of T depending on population parameters α and β. This use of conditional and marginal distributions resembles Bayesian approach, but its purpose may be considered as opposite: while in Bayesian approach we may use data $T = t$ with prior f_B to develop the posterior distribution of individual customer's risk θ given $T = t$, infinite mixture approach leads us to the marginal distribution of T unconditioned on θ characterizing entire population of customers. This requires exactly the calculation of the integral in the denominator of the Bayes formula (2.26), which we were so happy to avoid in most of the examples of Chapter 2.

The discrete-time Beta-geometric model assumptions oversimplify the reality of the mortgage default setting. To begin, the TTD for mortgages is typically measured on a continuous-time scale, that is, defaults can occur at any time. This is not a big problem, since one can use a continuous-time exponential-gamma model, where exponential distribution with parameter θ measures the individual TTD and gamma distribution characterizes the distribution of θ over the population of customers. This model was also considered in [6]. In case of exponential distribution, the hazard rate as well as the probability of cancellation θ in BG model does not change with time.

Further, one has to take into account that while penalties for subscription cancellation may be substantial, they rarely get as dramatic as the consequences of a mortgage default. Depending on the circumstances of the mortgage portfolio and its holder, the hazard rate may increase or decrease throughout the duration of the holding. To accommodate these features, Fader and Hardie also proposed a continuous-time Weibull-gamma (WG) model for customer retention. We can translate it into the language of mortgage defaults, treating T as the TTD.

(WG1) Defaults can occur at any time, that is, $T \geq 0$.

(WG2) The risk of default for an individual may increase or decrease throughout their holding. Thus T can be characterized by the Weibull distribution with probability density function $f_W(t|\theta, \tau)$ and corresponding distribution function

$$F_W(t|\theta, \tau) = P(T \leq t|\theta, \tau) = 1 - e^{-\theta t^{\tau}}, t \geq 0 \, .$$

Note that the parameter $\tau > 0$ represents the magnitude and direction of change of the *hazard rate* (risk of default) over time with $\tau = 1$ indicating a constant hazard rate and $\tau > 1$ ($\tau < 1$) indicating an increase (decrease) in the hazard rate. Further, $\theta > 0$ represents the transformed scale parameter of Weibull distribution with larger θ reflecting a larger risk of default.

(WG3) The risk of default varies among mortgage holders. To this end, heterogeneity in scale parameter θ is modeled by a Gamma(α, λ) distribution with parameters $\alpha, \lambda > 0$ and probability density function

$$f_G(\theta|\alpha, \lambda) \propto \theta^{\alpha-1} e^{-\lambda\theta}, \theta > 0 \, . \tag{4.7}$$

It follows that the joint Weibull-gamma distribution of (t, θ) is characterized by density function

$$f_{WG}(t, \theta|\alpha, \lambda, \tau) = f_W(t|\theta, \tau) f_G(\theta|\alpha, \lambda) \, .$$

Finally, integrating θ out produces a form of the Burr Type XII distribution:

$$F_{WG}(t|\alpha, \lambda, \tau) = \int_0^t \int_0^\infty f_{WG}(z, \theta|\alpha, \lambda, \tau)d\theta dz = 1 - \left(\frac{\lambda}{\lambda + t^\tau}\right)^\alpha. \quad (4.8)$$

4.7.3 Latent Classes and Finite Mixture Models

The Weibull-gamma framework under assumptions (WG1)–(WG3) provides a foundation for modeling TTD. However, it does not reflect the segmentation of portfolios into groups with different default risk patterns. For instance, in (4.8) parameter τ determines the hazard rate, which is increasing or decreasing with time for the entire population. There exist two controversial intuitive justifications for these two patterns. First, with time the equity of the mortgage holder grows and therefore the losses associated with default also grow. That would cause a sensible mortgage holder to become with time less and less default prone. Second, as it is known, for many households in shaky financial situations, mortgage payments themselves represent the main burden driving down the family finance. This means that the later in the mortgage history with more payments being made, the higher becomes the risk of default. That would suggest the existence of at least two different segments of the population with different values of parameter τ of the WG model. Let us start with dividing population into two different groups.

To this end, let $p \in (0, 1)$ be the segmentation weight reflecting the proportional size of the first group of mortgages and $1 - p$ be the size of the second group of mortgages. Further, assume default rates may fluctuate throughout the duration of the holding and may vary by group status. Then TTD T can be characterized by a Weibull segmentation model with probability density function $f_{WS}(t|\theta, \tau_1, \tau_2, p)$ and corresponding distribution function

$$F_{WS}(t|\theta, \tau_1, \tau_2, p) = 1 - pe^{-\theta t^{\tau_1}} - (1 - p)e^{-\theta t^{\tau_2}}, t \geq 0 \quad (4.9)$$

where τ_1 and τ_2 represent the magnitude and direction of change of hazard rates for two segments, respectively.

This is a ***finite mixture model***, corresponding to two homogeneous Weibull segments of the population. If risk rates expressed by parameter θ are assumed to be constant across mortgage holders in each segment, then this is a suitable TTD model. Otherwise, heterogeneity in θ can be modeled for each of the two segments by a separate *Gamma*(α, λ) distribution. In conjunction with the previous equation, it follows that a Weibull-gamma mixture model (WGM) for T has distribution function

$$F_{WGM}(t|\alpha, \lambda, \tau_1, \tau_2, p) = 1 - p\left(\frac{\lambda}{\lambda + t^{\tau_1}}\right)^\alpha - (1 - p)\left(\frac{\lambda}{\lambda + t^{\tau_2}}\right)^\alpha \quad (4.10)$$

with corresponding density function $f_{WGM}(t|\alpha, \lambda, \tau_1, \tau_2, p)$.

With these properties in mind, WGM is presented, which recognizes the segmentation of portfolios into two distinct groups while assuming heterogeneity in default risks among individual mortgage holders in each group. Similar strategies have been used to model heterogeneous survival data in various contexts. In fact, the WGM can be treated as an extension of the models proposed by Peter Fader and Bruce Hardie in a series of papers for predicting customer retention in subscription settings. Although mortgage defaults and subscription cancellations differ in nature and consequences, there are some similarities between these two. On the other hand, WGM can be considered to be an extension of TTD approach by Soyer and Xu [30].

Specification of the WGM requires the selection of the vector of model parameters $(\alpha, \lambda, \tau_1, \tau_2, p)$. The convention in financial and marketing applications is to estimate parameters using available data by maximum likelihood estimation (MLE) approach. However, this ignores potentially powerful prior knowledge and beliefs about model parameters. This knowledge may be based on the past experience and field expertise as expressed in the form of prior distribution. Also, MLE for mixture models often lacks robustness, being too sensitive to kinks in data. In contrast, we present a Bayesian alternative that evaluates parameters through a combination of observed data and prior knowledge. Though far from the standard, Bayesian applications in survival analysis settings are increasingly popular. We can refer to a study on Bayesian treatment of Weibull mixtures [19]. Another recent paper [22] applied Bayesian techniques to problems of loan prepayment, somewhat related to default analysis.

We illustrate the frequentist and Bayesian applications of the WGM using empirical data for a major American commercial bank [8]. These data include the monthly number of defaults for the subprime mortgage portfolios, pooled over 10 years from 2002 to 2011. We will see that in comparison to its MLE competitor, the Bayesian specification of WGM utilizing random walk Metropolis algorithm enjoys increased stability and helps to detect overparameterization. We consider two strategies below: the maximum likelihood method and Bayesian estimation.

4.7.4 Maximum Likelihood Estimation

In most financial and marketing applications, maximum likelihood methods of parametric estimation are the most widely used. Let $t_1, t_2, ..., t_n$ be the observed TTD for a random sample of n mortgage portfolios. Further, define likelihood function

$$L_{WGM}(\alpha, \lambda, \tau_1, \tau_2, p | t_1, t_2, ..., t_n) = \prod_{i=1}^{n} f_{WGM}(t_i | \alpha, \lambda, \tau_1, \tau_2, p).$$

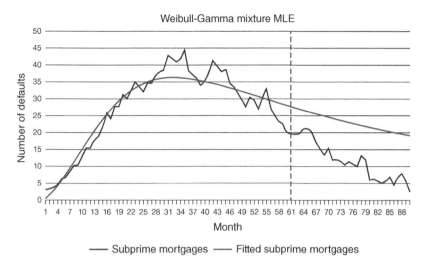

Figure 4.17 WGM model fit with MLE.

Taking into account aggregation (defaults recorded at discrete times—monthly) and right censoring of the data (finite period of observation), we will rewrite the likelihood functions. Let n be the total number of mortgages in the portfolio, m the observation period in months, and k_j the number of defaults recorded at time (month) j for $j = 1, \ldots, m$. Then

$$L_{WGM}(\alpha, \lambda, \tau_1, \tau_2, p | m, n, k_1, k_2, \ldots, k_m)$$

$$= \prod_{j=1}^{m} f_{WGM}^{k_j}(j | \alpha, \lambda, \tau_1, \tau_2, p)(1 - F_{WGM}(L | \alpha, \lambda, \tau_1, \tau_2, p))^{n - \sum_{j=1}^{m} k_j}.$$

Maximum likelihood estimates for the parameters of the WGM are obtained by numerical maximization of the likelihood function. Figure 4.17 shows the fitted WGM parametric curve next to the smoothed histogram of the data. First 60 months of observations were used for parametric estimation ($m = 60$), and the rest of the observations ($j > 60$) were used for model validation. It is clear that the model demonstrates a very poor fit at the right-hand tail.

Will a Bayesian model perform better?

4.7.5   A Bayesian Model

Bayesian formulation of the WGM (4.10) requires a specification of prior distributions for the relevant parameters. These are chosen to reflect prior knowledge of the parameters based on prior experience and subject-matter expertise.

The objective priors will be selected to reflect the absence of reliable prior information and will allow for the use of Bayesian methodology without introducing too much subjectivity. The proper choice of objective priors, according to Yang and Berger [32], is as follows:

$$\pi(\theta) \propto \frac{1}{\theta}, \ \theta > 0;$$

$$\pi(\lambda) \propto \frac{1}{\lambda}, \ \lambda > 0;$$

$$\pi(\alpha) \propto \sqrt{\sum_{i=0}^{\infty}(\alpha + i)^{-2}}, \ \alpha > 0;$$

$$\pi(\tau_i) \propto \frac{1}{\tau_i}, \ \tau_i > 0, i = 1, 2;$$

$$\pi(p) \propto p^{-1/2}(1 - p)^{-1/2}, \ p \in (0, 1).$$

Note that the choice of noninformative priors for λ and θ is standard for most scale parameters, $\pi(p)$ is a $Beta(1/2, 1/2)$ prior, and the least intuitive and most exotic of all is the prior for the shape parameter for gamma distribution, which can also be expressed in terms of *polygamma function* $\pi(\alpha) \propto \sqrt{PG(1, r)}$. Assume also that these prior distributions are independent. Further, let $(n, m, k_1, k_2, \dots, k_m)$ be the observed mortgage default data. Then the posterior distributions for the parameters of the WGM conditioned on the data are characterized by density functions

$$\pi_{WGM}(\alpha, \lambda, \tau_1, \tau_2, p | n, m, k_1, k_2, \dots, k_m) \propto$$

$$L_{WGM}(\alpha, \lambda, \tau_1, \tau_2, p | n, m, k_1, k_2, \dots, k_m) \cdot \pi(\alpha)\pi(\lambda)\pi(\tau_1)\pi(\tau_2)\pi(p),$$

where L_{WGM} is the likelihood function defined above.

In the Bayesian setting, let us run a random walk Metropolis–Hastings chain with normal conditional proposal, which we have illustrated in Section 4.3 for the case of one scalar parameter. With five-component vector parameter the simplest solution, though not the most economical one, will be to apply **block updates**, or update all five parameter components simultaneously on each step. It is obvious that the acceptance rate will suffer, because rejection will follow if even one of the five parameters is to be rejected. Therefore a longer chain is required to provide good mixing, and a more careful work has to be done on the fine tuning of the proposal, which in this case is reduced to the choice of standard deviations σ for all parameters. The results of RWMHA with a very long run of 3 million steps are illustrated in Figures 4.18 and 4.19 and Table 4.1.

Figure 4.18 representing the model fit looks like a vast improvement relatively to MLE. However, Table 4.1 tells a different story.

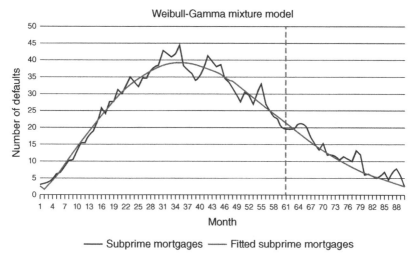

Figure 4.18 WGM model fit with Bayes estimates (RWMHA).

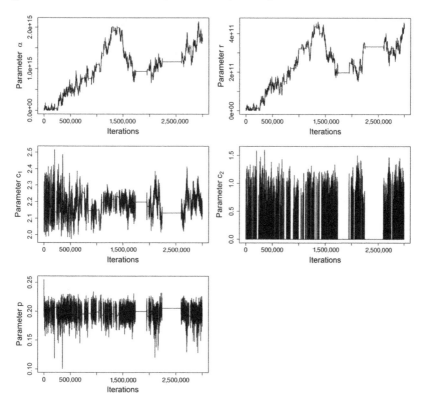

Figure 4.19 Trace plots for WGM model.

Table 4.1 Bayesian parameter estimates for WGM

Parameter	Estimate	Error	Scaling	Value
α	2.422×10^{11}	2.824×10^{9}	σ_{α}	10^{9}
λ	1.022×10^{15}	1.241×10^{13}	σ_{λ}	5×10^{12}
τ_1	2.179	1.227×10^{-3}	σ_{τ_1}	0.05
τ_2	9.386×10^{-2}	3.231×10^{-3}	σ_{τ_2}	0.2
p	0.199	1.293×10^{-4}	σ_p	0.007

Parametric estimates for α and λ behave rather strangely. Table 4.1 suggests very large values for both, and scaling parameters are also very high. A good overall fit is achieved without clear convergence of MCMC estimates, which smells trouble. A closer look at the trace plots might provide additional insight.

As we see in Figure 4.19, traces for α and λ (two charts in the upper row) seem to move up and down in coordination (flat intervals correspond to long series of rejections). The other parameters (τ_i in the middle row and p in the lower row) behave reasonably well other than sharing the same flat intervals. The information provided by trace plots is valuable: parameters α and λ are closely dependent!

A possible interpretation of this phenomenon is strikingly simple: the model is over-parameterized. We know that if parameter θ of the Weibull distribution in our mixture has a gamma distribution $Gamma(\alpha, \lambda)$, then according to (1.19)

$$E(\theta) = \frac{\alpha}{\lambda}, \ Var(\theta) = \frac{\alpha}{\lambda^2},$$

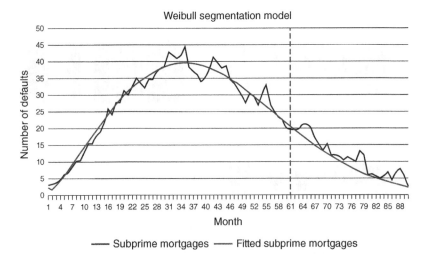

Figure 4.20 WS model fit with Bayes estimates (RWMHA).

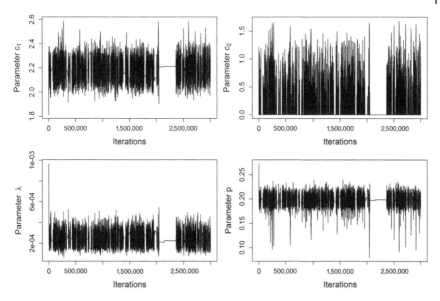

Figure 4.21 Trace plots for WS model.

and the case of both $\alpha \to \infty$ and $\lambda \to \infty$ so that $\alpha\lambda \sim c$ corresponds to $\theta = c$ being a positive constant, common for all individual observations. This consideration takes us back from WGM (4.10) to a model with fewer parameters (4.9) assuming homogeneity of risk rates θ in the population. We can characterize (4.9) as a **_Weibull segmentation model_** (WS) with no need to model heterogeneity within two segments. The results for this model (also using RWMHA) are demonstrated in Figure 4.20 (model fit), Table 4.2 (parameter estimates), and Figure 4.21 (trace plots). The overall results are much more attractive for a simpler model WS. The most interesting finding for the application might be the interpretation of parametric values from Table 4.2: two segments were detected. One including roughly 20% of the portfolio (value of p) may correspond to $\tau_1 > 1$ (increasing hazard rate). Another with roughly 80% includes

Table 4.2 Parameter estimates for the Bayesian WS model, subprime portfolios

Parameter	Estimate	Error	Scaling	Value
θ	2.381×10^{-4}	9.154×10^{-7}	σ_θ	3×10^{-5}
τ_1	2.187	1.286×10^{-3}	σ_{τ_1}	0.02
τ_2	0.115	5.286×10^{-3}	σ_{τ_2}	0.07
p	0.197	2.076×10^{-4}	σ_p	0.004

mortgage holders with $\tau_2 \approx 0.1 < 1$ (decreasing hazard rate), see [8]. It is interesting that this proportion has been found for some subprime loans by different authors [30].

Aside from some intresting practical results, from methodological point of view this example illustrates additional diagnostic opportunities (trace plots) provided by MCMC.

References

1 Besag, J. (1974). Spatial interaction and the statistical analysis of lattice systems. *Journal of the Royal Statistical Society, Series B*, 36(2), 192–236.

2 Billera, L., and Diaconis, P. (2001). A geometric interpretation of the Metropolis-Hastings algorithm. *Statistical Science*, 16(4), 335–339.

3 Bolstad, W. M. (2009). *Understanding Computational Bayesian Statistics*. New York: John Wiley & Sons.

4 Craiu, R., and Meng, X. L. (2011). Perfection within reach: exact MCMC sampling, In: *Handbook of Markov Chain Monte Carlo*, 199–226. London, New York: Chapman & Hall, CRC Press, Boca Raton.

5 Dongarra, J., and Sullivan, F. (2000). The top 10 algorithms. *Computer Science and Engineering*, 2, 22–33.

6 Fader, P. S., and Hardie, B. G. S. (2007). How to project customer retention. *Journal of Interactive Marketing*, 21, 76–90.

7 Foss, S. G., and Tweedie, R. L. (1998) Perfect simulation and backward coupling. *Stochastic Models*, 14(1–2), 187–208.

8 Galloway, M., Johnson, A., and Shemyakin, A. (2014). Time-to-Default Analysis of Mortgage Portfolios, MCMSki-V, Chamonix, France.

9 Gamerman, D., and Lopes,H. F. (2006). *Markov Chain Monte Carlo*. Chapman & Hall/CRC.

10 Gelfand, A. E., and Smith, A. F. M. (1990). Sampling-based approaches to calculating marginal denisities. *Journal of the American Statistical Association*, 85, 398–409.

11 Gelman, A., and Rubin, D. B. (1992). Inference from iterative simulation using multiple sequences (with discussion). *Statistical Science*, 7, 457–511.

12 Geman, S., and Geman, D. (1984). Stochastic relaxation, Gibbs distributions and the Bayesian restoration of images, *IEEE Transactions on Pattern Analysis and Machine Intelligence*, 6, 721–741.

13 Geweke, J. (1992). Evaluating the accuracy of sampling-based approaches to the calculations of posterior moments. In: J. M. Bernardo, J. Berger, A. P. Dawid, and A. F. M. Smith, (editors), *Bayesian Statistics*, 169–193, Oxford: Oxford University Press.

14 Green, P. J., Latusyznski, K., Pereyra, M., and Robert, C. P. (2015). Bayesian computation: a summary of the current state, and samples backwards and forwards. *Statistics and Computing*, 25(4), 835–862.

15 Hastings, W. K. (1970). Monte Carlo sampling methods using Markov chains and their applications, *Biometrika*, 57, 97–109.

16 Heidelberger, P., and Welch, P. D. (1983). Simulation run length control in the presence of the initial transient. *Operations Research*, 31, 1109–1144.

17 Kruschke, J. (2014). *Doing Bayesian Data Analysis. A Tutorial with R, JAGS and Stan*, 2nd ed. Elsevier, Academic Press.

18 MacEahern, S. N., and Berliner, L. M. (1994). Subsampling the Gibbs sampler. *The American Statistician*, 48, 188–190.

19 Marin, J. M., Rodriguez-Bernal, M. T., and Wiper, M. P. (2005). Using Weibull mixture distributions to model heterogeneous survival data. *Communications in Statistics Simulation and Computation*, 34, 673–684.

20 Metropoils, N., Rosenbluth, A. Rosenbluth, M., Teller, A., and Teller, E. (1953). Equations of state calculations by fast computing machines. *Journal of Chemical Physics*, 21, 1087–1091.

21 Peskun, P. (1973). Optimal Monte Carlo sampling using Markov chains. *Biometrika*, 60, 607–612.

22 Popova, I., Popova, E., and George, E. I. (2008). Bayesian forecasting of prepayment rates for individual pools of mortgages. *Bayesian Analysis*, 3, 393–426.

23 Propp, J., and Wilson, D. (1996). Exact sampling with coupled Markov chains and applications to statistical mechanics. *Random Structures and Algorithms*, 9, 223–252.

24 Raftery, A., and Lewis, S. (1992). How many iterations in the Gibbs sampler? In: J. M. Bernardo, J. Berger, A. P. Dawid, and A. F. M. Smith (editors), *Bayesian Statistics*, 763–773, Oxford: Oxford University Press.

25 Robert, C. P., and Casella, G. (2004). *Introducing Monte Carlo Methods with R*, 2nd ed. Springer-Verlag.

26 Robert, C. P., and Casella, G. (2010). *Monte Carlo Statistical Methods*. Springer-Verlag.

27 Robert, C. P., and Casella, G. (2011). A short history of MCMC: subjective recollections from incomplete data. *Statistical Science*, 26(1), 102–115.

28 Roberts, G.O., Gelman, A., and Gilks, W. R. (1997). Weak convergence and optimal scaling of random walk Metropolis algorithms. *The Annals of Applied Probability*, 7(1), 110–120.

29 Shemyakin, A. (2007). An adaptive backward coupling Metropolis algorithm for truncated distributions. *Model Assisted Statistics and Applications*, 2(3), 137–143.

30 Soyer, R., and Xu, F. (2010). Assessment of mortgage default risk via Bayesian reliability models. *Applied Stochastic Models in Business and Industry*, 26, 308–330.

31 Teller, E. (2001). *Memoirs: A Twentieth Century Journey in Science and Politics*. Perseus Publishing.

32 Yang, R., and Berger, J. O. (1998). A catalog of noninformative priors, Technical Report 42, Duke University Department of Statistical Science.

Exercises

4.1 Generate random samples size 100 and size 1000 for the Sun City example using Excel or R. First, use the "fair coin" proposal, and then apply accept/reject technique from Section 4.1. Compare the resulting estimates of the long-term proportion of rainy days with the theoretical probability 0.25.

4.2 Implement IMA (using any software) to estimate posterior mean with Poisson likelihood for the data $x = (3, 1, 4, 3, 2)$ and exponential prior with the mean 2, also known as $Gamma(1, 0.5)$. Choose an exponential proposal density with the parameter value you consider the best. Modify the proposal if you need it badly enough. Use sample size $N = 1000$. Draw relevant graphs: trace plot, histogram, and sample autocorrelation function.

4.3 Implement RWMHA (using any software) to estimate posterior mean with Poisson likelihood for the data $x = (3, 1, 4, 3, 2)$ and exponential prior with the mean 2, also known as $Gamma(1, 0.5)$. Choose a normal proposal density with the parameter value σ you consider the best. Modify the proposal if you need it badly enough. Use sample size $N = 1000$. Draw relevant graphs: trace plot, histogram, and sample autocorrelation function.

4.4 Calculate Geweke statistics for the chains obtained in Exercises 4.2 and 4.3. Compare these values.

4.5 Calculate Gelman–Rubin shrink factors for the chains obtained in Exercises 4.2 and 4.3. Compare these values.

4.6 Apply burn-in with $b = 0.1N$ and $b = 0.5N$ to the chain obtained in Exercise 4.2. Apply skip with $k = 10$. Does burn-in or skip or both improve your estimates?

4.7 Apply burn-in with $b = 0.1N$ and $b = 0.5N$ to the chain obtained in Exercise 4.3. Apply skip with $k = 10$. Does burn-in or skip or both improve your estimates?

4.8 Calculate Geweke statistics for the chains obtained in Exercises 4.6 and 4.7. Compare these values.

4.9 Calculate Gelman–Rubin shrink factors for the chains obtained in Exercises 4.6 and 4.7. Compare these values.

PART II

Modeling Dependence

5

Statistical Dependence Structures

5.1 Introduction

In this chapter we will briefly survey common statistical tools for modeling dependence between two or more random variables. In order to explain the problem in its most general form, let us consider a dataset consisting of n pairs (x_i, y_i), where x_i and y_i represent values of two random variables X and Y corresponding to the ith case (observation). A classical example introduced by Gary Alt, a famous black bear expert [1], contains measurements of length (X, measured in inches) and weight (Y, measured in pounds) of a sample of 143 black bears taken in the state of Pennsylvania. Figure 5.1 shows the scatterplot of this dataset.

Another example deals with a joint survival or joint mortality problem. We consider data on the length of human life, studying married couples, where X represents the wife's age at death, and Y represents the husband's. Figure 5.2 shows the scatterplot for about 11,000 pairs of spouses observed for a period of 5 years introduced in [7]. Evidently, not all of the observed died by the end of the period, so the actual number of points on the graph is much smaller than 11,000.

Why can X and Y in these two examples be considered dependent? In the case of bears, it is natural to believe that for a variety of reasons (such as age, sex, genetic predisposition, and life circumstance) some bears described as "big" tend to be both long and heavy, while the others described as "small" are short and lightweight. Both length and weight of a particular bear are the results of summary action of many factors and are likely to be positively correlated. In case of married couples, research seems to point out three major factors of positive dependence:

- common disaster,
- common lifestyle,
- broken heart syndrome.

Introduction to Bayesian Estimation and Copula Models of Dependence, First Edition.
Arkady Shemyakin and Alexander Kniazev.
© 2017 John Wiley & Sons, Inc. Published 2017 by John Wiley & Sons, Inc.
Companion Website: http://www.wiley.com/go/shemyakin/bayesian_estimation

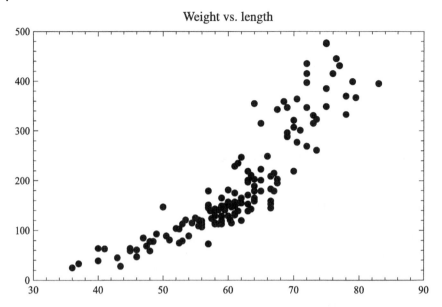

Figure 5.1 Black bears in Pennsylvania.

Common disaster factor corresponds to events such as car crashes or house fires, which are likely to shorten the lives of both spouses. *Common lifestyle* factor suggests that two spouses are likely to share a lifestyle which could be either healthy (physical exercise, healthy eating habits, marital harmony), extending their lives; or unhealthy (drug and alcohol abuse, stressful environment, frequent fights, etc.), making both lives shorter. Well-documented instances of *broken heart syndrome* suggest that after the death of a married person, especially in an elderly couple, the surviving spouse is more likely to die during the next 1 or 2 years. All these factors are important in joint life analysis [17, 18]. In a different context, for example, in engineering reliability or financial risk management problems, one may talk about common events or common shocks, event after action, short-term and long-term dependence patterns.

Building a mathematical model of dependence between X and Y for a situation like these two examples requires the study of both the strength and the character of dependence which would suggest an implicit functional representation $f(x, y) = c$ explaining the appearance of the scatterplots. The final objective is to fit the data to a functional model of relationship.

It is very tempting to measure the strength of dependence in terms of one numerical characteristic. One such characteristic is Pearson's linear correlation introduced in Section 1.6. In Section 5.2 we will discuss it in more detail, especially in the context of joint survival problems, for which there exist several ways to measure correlation. We will discuss some drawbacks of

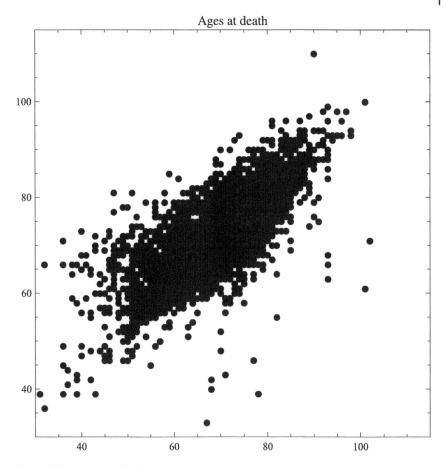

Figure 5.2 Joint mortality data.

correlation as an adequate measure of dependence, especially in the context of higher dimensions (more than two dependent variables), where correlation matrix is very convenient for modeling pairwise dependence but may miss more complex dependence patterns. We will also introduce alternative measures of strength of dependence such as Spearman's rank correlation and Kendall's concordance.

If we are not satisfied with measuring just the overall strength of dependence and are interested in the relationship between specific values of two variables X and Y, the next logical step is to consider entire conditional distribution. In this case it is commonplace to part with the symmetry and designate one of the variables, usually the one easily observed (say, X) as independent variable or predictor and the other (say, Y) as dependent variable or response. Then the conditional distribution of Y given X and its numerical characteristics such as conditional expected value $E(Y/X)$ can be estimated from the given data and used to predict the values of Y for any given $X = x$.

Regression models introduced in Section 5.3 suggest a functional relationship $y = f(x) + \varepsilon$, $f(x) = \mu_x = E(Y/X = x)$. Bayesian setting is considered in Section 5.4. These models will naturally fit the example with black bears, in which observations of length x_i may be quite logically treated as exactly measured values of independent variable X (nonrandom), which is used to explain and predict weight Y. However similar treatment of the joint mortality example would be more problematic. Considering that either observation x_i or y_i (ages at death) could precede the other one, taking one of the variables as a predictor and the other as a response is a logical stretch. Besides, when we look at the scatterplot in Figure 5.2, it does not exhibit behavior consistent with a meaningful explicit formula $y = f(x)$, rather suggesting an implicit representation by a two-dimensional joint density $f(x, y)$.

An alternative approach to modeling dependence between two random variables is based on the hazard rates. This will be appropriate for survival studies like husband/wife life length example. For instance, common lifestyle factor of joint mortality may be addressed by Cox proportional hazard models generalizing regression models. If a change in one variable brings about a change in the other one, we may consider regime switching models or multistage models where Markov chain theory becomes handy. Broken heart syndrome can be conveniently described in terms of these models and common disaster factor could be incorporated in this framework. These and other methods of dependence analysis in survival models, such as shared frailty, are briefly discussed in Section 5.5. For more in-depth coverage see [11] and [12].

All that being said, we see a point in building more general "holistic" models, which could address a variety of factors affecting the dependence between two variables. The most general approach would suggest constructing a model for the joint distribution function of two variables, which will allow for the calculation of all correlation and concordance measures, will provide all the conditional distributions considered in regression framework, and will include proportional/competing hazard and multistage models as particular cases.

The simplest method of modeling joint distributions will suggest using such universal framework as the multivariate normal distribution introduced in Section 1.6, or survival analysis options such as multivariate exponential or multivariate Weibull distribution families. Modeling whole distribution will be considered in Section 5.6. However we will see that the distribution families discussed there are often too restrictive, and will not allow for a good fit of the data. A good distribution model should provide a contour plot of joint density function consistent with the data scatterplots such as those presented in Figures 5.1 and 5.2. Section 5.7 contains an example demonstrating the dangers of misusing multivariate normal models relying on correlation. This gives us a sufficient motivation to move on to a more general class of models which will be introduced in Chapter 6 and will be analyzed in the rest of the book.

5.2 Correlation

Let us review the black bears' example from the previous section. Figure 5.1 gives us a clear presentation of dependence between the length and the weight of a bear, which also agrees with the common sense. Nevertheless, two questions arise: how strong is this dependence and how can it be expressed in a functional form? In this section we will address the strength of dependence.

5.2.1 Pearson's Linear Correlation

In Chapter 1 and further on we have discussed the idea of correlation as a numeric measure of relationship between two variables. We can recollect that if we suppose that the dependence between random variables X and Y is linear, then Pearson's correlation coefficient is an appropriate measure for the strength of dependence. The coefficient is calculated according to formula:

$$\rho(X, Y) = \frac{Cov(X, Y)}{\sqrt{Var(X)Var(Y)}}, \tag{5.1}$$

where

$$Cov(X, Y) = E((X - E(X))(Y - E(Y))) = E(XY) - E(X)E(Y), \tag{5.2}$$

and

$$Var(X) = E((X - E(X))^2), Var(Y) = E((Y - E(Y))^2). \tag{5.3}$$

A natural estimator for $\rho(X, Y)$ constructed by an independent paired sample $s = ((x_1, y_1), \ldots, (x_n, y_n))$ is *sample correlation*

$$r(s) = \frac{\sum (x_i - \bar{x})(y_i - \bar{y})}{\sqrt{\sum (x_i - \bar{x})^2 \sum (y_i - \bar{y})^2}} = \frac{\sum x_i y_i - n\bar{x}\bar{y}}{\sqrt{\sum (x_i - \bar{x})^2 \sum (y_i - \bar{y})^2}},$$

where $\bar{x}$ and $\bar{y}$ are sample means for the first and the second component correspondingly. In the black bears' example introduced in Section 5.1, $r = 0.87$.

This can be considered quite strong. The standard test for nonzero correlation performed in R brings about the following output

---

Pearson's product-moment correlation
data: length and weight
$t = 21.4$, df $= 141$, p-value $< 2.2e - 16$
alternative hypothesis: true correlation is not equal to 0
95 percent confidence interval: 0.83; 0.91
sample estimates: cor $= 0.87$.

---

rejecting the null hypothesis $\rho(length, weight) = 0$. However, this test has to be used with caution. The context of a problem will always matter, suggesting an important difference between statistical significance and practical significance. Recollect the lady tasting tea example from Section 2.5.

It is easy to show that $|\rho(X, Y)| \leq 1$, and $|\rho(X, Y)| = 1$ if and only if X and Y are linearly related: $Y = aX + b$. We can also prove that $\rho(X, Y) = 0$ for independent X and Y. The converse is not true. This can be demonstrated by a simple example $X \sim N(0, 1)$ and $Y = X^2$, where $\rho(X, X^2) = 0$, though these two variables are perfectly related by a quadratic formula.

The concept of correlation may be easily extended to the case of $d > 2$ random variables $X_1, \ldots, X_d$. Pairwise correlations between the variables are defined as

$$\rho_{ij} = \rho(X_i, X_j) = \frac{\sigma_{ij}}{\sigma_i \sigma_j},$$

where $\sigma_i^2 = Var(X_i)$ and $\sigma_{ij} = Cov(X_i, X_j)$.

Matrix $\Sigma = \{\sigma_{ij}\}$ of size $d \times d$ is known as *covariance matrix* and $R = \{\rho_{ij}\}$ as *correlation matrix* of the random vector $X = (X_1, \ldots, X_d)$, defining its dependence structure completely. Notice that in this context pairwise correlations fully determine all linear dependence patterns on the set of d variables.

Pearson's correlation ρ can be considered parametric in the sense that it can be fully determined by such basic distribution parameters as means and variances. In order to go beyond the linear dependence though, one might want to get rid of this parametric structure. The simplest nonparametric correlation coefficient is Spearman's ρ^*. We will use the asterisk to separate it from Pearson's ρ. For two variables X and Y with corresponding distribution functions F and G, we can define new random variables $U = F(X)$ and $V = G(Y)$. Then we can define

$$\rho^*(X, Y) = \rho(U, V).$$

Keeping in mind that U and V, as we have shown in Section 3.2, have uniform distribution on $[0, 1]$ with mean $1/2$ and variance $1/12$, we can rewrite

$$\rho^*(X, Y) = \frac{E(UV) - 1/4}{1/12} = 12E(UV) - 3.$$

5.2.2 Spearman's Rank Correlation

Spearman's correlation is also known as rank correlation, because its sample estimate can be calculated by the ranked observations. Consider a paired sample $s = ((x_1, y_1), \ldots, (x_n, y_n))$. If we define $R_i(x)$ as the rank (exact order by ascendance) of the ith element in the sample $x = (x_1, \ldots, x_n)$, and $R_i(y)$ as the rank (exact order by ascendance) of the ith element in the sample $y = (y_1, \ldots, y_n)$, sample Spearman's correlation serving as an estimator of ρ^* is

$$r^*(s) = \frac{\sum(R_i(x) - \bar{R}(x))(R_i(y) - \bar{R}(y))}{\sqrt{\sum(R_i(x) - \bar{R}(x))^2 \sum(R_i(y) - \bar{R}(y))^2}} = 1 - 6\frac{\sum(R_i(x) - R_i(y))^2}{n(n^2 - 1)}.$$

In the black bears' example, $r^* = 0.93$, as we can see from R output following the execution of R command *cor.test(length, weight, method = "spearman")*.

Spearman's rank correlation rho
data: length and weight
S = 33142, p-value $< 2.2e-16$
alternative hypothesis: true rho is not equal to 0
sample estimates: cor $= 0.93$.

5.2.3 Kendall's Concordance

Another measure of nonparametric rank correlation is Kendall's rank correlation (also known as **concordance**) τ [13]. When (X_1, Y_1) and (X_2, Y_2) are two independent pairs of variables with the same joint distribution as (X, Y),

$$\tau(X, Y) = P\Big((X_1 - X_2)(Y_1 - Y_2) > 0\Big) - P\Big((X_1 - X_2)(Y_1 - Y_2) < 0\Big). \qquad (5.4)$$

Sample concordance $\hat{\tau}(s)$ for a paired sample $s = ((x_1, y_1), \ldots, (x_n, y_n))$ can be defined using the notion of concordant and discordant pairs (i, j) of pairs $(x_., y_.)$. The total number of pairs of subscripts (i, j) such that $i, j = 1, \ldots, n; i < j$ is equal to $n(n - 1)/2$, so this is the number of ways to select i and j to choose a pair (x_i, y_i) and (x_j, y_j). Such a pair is **concordant** if either both $x_i < x_j$ and $y_i < y_j$ or both $x_i > x_j$ and $y_i > y_j$. A pair is **discordant** if either $x_i < x_j$ and $y_i > y_j$ or

$x_i > x_j$ and $y_i < y_j$. The possibility of "ties" (pairs which are neither concordant nor discordant) is not discussed so far, but must be important for discrete samples with repeating observations (see Exercises). The number of concordant pairs C and the number of discordant pairs D add up to $n(n-1)/2$, if there are no **ties**. To address ties if $x_i = x_j$ or $y_i = y_j$ we have to create special rules, which usually do not make a big difference for continuous variables X and Y.

Ignoring ties,

$$\hat{\tau}(s) = \frac{C-D}{C+D} = \frac{2(C-D)}{n(n-1)} = \frac{4C}{n(n-1)} - 1. \tag{5.5}$$

In the black bears' example, in a concordant pair of bears one of the bears is both longer and heavier than the other. Counting all concordant pairs, we obtain $\hat{\tau} = 0.79$, which is supported by the **R** output following the execution of **R** command *cor.test(length, weight, method = "kendall")*.

Kendall's rank correlation tau

data: length and weight

z = 13.769, p-value < 2.2e − 16

alternative hypothesis: true tau is not equal to 0

sample estimates: tau = 0.79.

For the same sample, values of Pearson's, Spearman's, and Kendall's correlation are not going to be radically different, though they also do not have to be too close. The following simple numerical example serves as an illustration of such discrepancy.

Example 5.1 Let us consider three small samples in Table 5.1 (size 5), where x_i and y_i were drawn independently, and $Z = e^{XY}$.

We will calculate Pearson's sample correlation coefficients $r(X, Z)$ and $r(Y, Z)$ and also Spearman's and Kendall's rank correlations r^* and $\hat{\tau}$ between these

Table 5.1 Three samples

X	Y	Z
−1.28	−2.14	15.45
−0.85	0.65	0.57
−0.40	0.62	0.78
−0.86	−1.37	3.23
−1.66	0.82	0.26

variables. We begin with rank correlations, which are easier to deal with numerically, and build the table of ranks:

$R(X)$	$R(Y)$	$R(Z)$
2	1	5
4	4	2
5	3	3
3	2	4
1	5	1

Calculations with ranks are simple enough.

$$r^*(X,Z) = 1 - 6\frac{((-3)^2 + 2^2 + 2^2 + (-1)^2 + 0^2)}{5 \times 24} = 1 - 6 \times 18/(5 \times 24) = 0.1.$$

Similarly,

$$r^*(Y,Z) = 1 - 6\frac{((-4)^2 + 2^2 + 0^2 + (-2)^2 + 4^2)}{5 \times 24} = 1 - 6 \times 40/(5 \times 24) = -1.$$

For Kendall's $\hat{\tau}$ we need to calculate the number of concordant pairs of rows in the table $(i,j), i \neq j$. For the pair of columns (X,Z) the concordant pairs of rows are: $(i,j) = \{(1,5),(2,3),(2,5),(3,5),(4,5)\}$. Their number is equal to 5, out of the total $n(n-1)/2 = 10$, so the number of discordant pairs is also 5. There are no concordant pairs of rows for the pair of columns (Y,Z). Therefore from the definition of $\hat{\tau}$ we conclude $\hat{\tau}(X,Z) = 0$ and $\hat{\tau}(Y,Z) = -1$.

More calculations are required to calculate Pearson's sample correlation. We need $\bar{x} = -1.01, \bar{y} = -0.284, \bar{z} = 4.058$, and also $\sum x_i z_i = -23.78$ and $\sum y_i z_i = -36.42$ to calculate the numerators in the definition. For the denominators, we need $\sqrt{\sum(x_i - \bar{x})^2} = 0.957$, $\sqrt{\sum(y_i - \bar{y})^2} = 2.745$, and $\sqrt{\sum(z_i - \bar{z})^2} = 12.953$. After that we obtain

$$r(X,Z) = \frac{-23.78 - 5 \times (-1.01) \times 4.058}{0.957 \times 12.953} = -0.265$$

and

$$r(Y,Z) = \frac{-36.42 - 5 \times (-0.284) \times 4.058}{2.745 \times 12.953} = -0.862.$$

As we see, the values of three coefficients are not too far from each other, but do not necessarily agree. Notice though, that Pearson's correlation is often more sensitive to changes in data: for instance, change of the last element in the first column of the data from -1.66 to -11.66 will not alter ranks and thus Spearman's and Kendall's sample correlations, while the value of Pearson's correlation will switch sign from negative to positive to $r(X, Z) = 0.276$. Also, if we change the first element in the second column from -2.14 to -12.14, it will not affect ranks and therefore rank correlations, but $r(Y, Z)$ will become -1.

5.3   Regression Models

What is the functional form of dependence between the length and weight of black bears? High value of Pearson's correlation coefficient suggests that there exists a strong linear dependence between these two variables. As suggested in Section 5.1, we assume that $X = length$ is a nonrandom explanatory variable and $Y = weight$ is the random response variable. We will look for a relationship

$$Y = \alpha + \beta X + \varepsilon, \varepsilon \sim N(0, \sigma^2) \tag{5.6}$$

where ε is the normal error reflecting all possible additional random factors contributing to a bear's weight and not related to bear's length. This model includes three unknown parameters: slope β (to mention first as the most important parameter, reflecting the dependence between X and Y), intercept α, and the nuisance parameter σ^2, measuring the variance of error. Using the sample $s = ((x_1, y_1), \dots, (x_n, y_n))$ containing lengths x_i and weights y_i of $n = 143$ bears whose scatterplot was shown in Figure 5.1, we want to build a reasonable model

$$\hat{y} = \hat{\alpha} + \hat{\beta}x, \tag{5.7}$$

which would approximate weight $\hat{y}$ of a bear with length x. The method of least squares, which is consistent if we assume that X is not a random variable, brings about the estimators of the slope and intercept parameters

$$\hat{\beta} = \hat{\beta}(s) = \frac{\sum(x_i - \bar{x})(y_i - \bar{y})}{\sum(x_i - \bar{x})^2}, \hat{\alpha} = \hat{\alpha}(s) = \bar{y} - \hat{\beta}\bar{x}.$$

Using R we obtain the output

Coefficients:	Estimate	Std. Error	t value	$Pr(> |t|)$
(Intercept)	−441.388	29.911	−14.76	$< 2e - 16$
length	10.338	0.483	21.43	$< 2e - 16$
Multiple R^2:	0.765,	Adjusted R^2:	0.763	
F-statistic: 459	1, 141 DF,	p-value:	$< 2.2e - 16$	

Using the results from the output we get $\hat{\beta} = 10.338$ and $\hat{\alpha} = -441.388$, so we can write down the dependence equation as $\hat{y} = 10.338x - 441.388$. The output also presents some quality indicators of the regression model. For example, we can see that standard errors (which are given in the third column) are much smaller than the absolute values of the estimates. For the ith observation from the sample (x_i, y_i) we can define the corresponding *residual* as $e_i = y_i - \hat{y}_i = y_i - 10.338x_i + 441.388$. Smaller residuals correspond to a better model, and the method of least squares estimators (5.7), indeed, minimizes the sum of squared residuals $\sum_{i=1}^{n} e_i^2$ providing the best overall fit of the model. The sum of squared residuals can be also used to estimate the parameter σ^2. *Coefficient of determination* R^2, which is an important measure of the model quality, is just the square of the Pearson's sample correlation, characterizing the strength of linear dependence.

5.3.1 Heteroskedasticity

Looking at Figure 5.3 we may notice that when the length is increasing from left to right, not only the weight itself is increasing, but the variation of sample values around the regression line is increasing, too. It corresponds to better fit expected for smaller bears, and the larger errors expected for larger bears. This phenomenon is known as ***heteroskedasticity*** and, strictly speaking, when observed it contradicts to the model assumption of homoskedasticity: variance of error σ^2 being constant. In order to check the feasibility of this assumption we need to perform a heteroskedasticity test. There exists a number of heteroskedasticity tests, and a particular method of heteroskedasticity correction is usually associated with the test. We will perform Glejser test [8]. To that end we will estimate the dependence of regression residuals' absolute values on the variable *length*.

Figure 5.3 Linear regression.

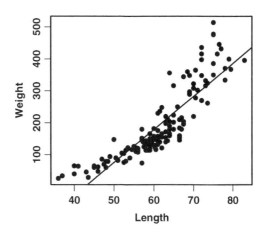

Coefficients:	Estimate	Std. Error	t value	$Pr(> |t|)$
(Intercept)	−1.975	18.373	−0.108	0.915
length	0.711	0.2964	2.400	0.018
Multiple R^2:	0.039,	Adjusted R^2:	0.032	
F-statistic: 5.759	1, 141 DF,	p-value:	< 0.01771	

Since the Glejser test is significant, observed heteroskedasticity does not allow us to accurately assess the quality of the model. It usually leads to the overstatement of standard errors. Let us introduce a method of correction for heteroskedasticity. To do this we will use the regression function $z_i = -1.975 + 0.711x_i$ from the test. Let us create new variables $y^* = y/z$, $invz = 1/z$, $x^* = x/z$. We will estimate the regression between new variables without a constant.

Coefficients:	Estimate	Std. Error	t value	$Pr(> |t|)$
$invz$	−351.781	24.326	−14.46	$< 2e - 16$
x^*	8.856	0.415	21.36	$< 2e - 16$
Multiple R^2:	0.932,	Adjusted R^2:	0.931	
F-statistic: 966	2, 141 DF,	p-value:	$< 2.2e - 16$	

The coefficient at the variable $invz$ corresponds to the constant in the original regression. We can see that the standard error of the slope coefficient is slightly smaller, and the quality of regression is somewhat better.

5.3.2 Nonlinear Regression

If we look at Figure 5.3 again, we may notice another serious problem. The assumption of linear nature of the dependence is doubtful. The scatterplot looks more like a parabola or even an exponential function. Let us suggest quadratic dependence between weight and length by introducing an additional variable, the square of length X^2.

Coefficients:	Estimate	Std. Error	t value	$Pr(> |t|)$
(Intercept)	376.748	114.907	3.279	0.0013
length	−17.674	3.858	−4.581	1.01e−05
length^2	0.234	0.032	7.303	1.94e − 11
Multiple R^2:	0.8298,	Adjusted R^2:	0.8274	
F-statistic: 341.3	2, 140 DF,	p-value:	$< 2.2e - 16$	

Figure 5.4 Quadratic regression.

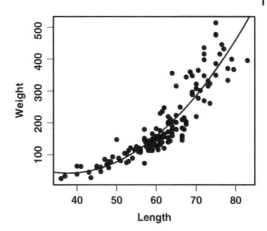

We can see that the coefficient at *length*^{2} is significantly different from zero, which supports our assumption of nonlinear dependence between the length and the weight of a bear. However, in general the improvement of the regression model is not really significant. The coefficient of determination increases only by 0.06. The graph of quadratic dependence is shown in Figure 5.4.

Rather than transforming the explanatory variable X, we can also transform the response Y. Let us estimate the dependence of the natural logarithm of weight $L = \ln(Y)$ on the length.

Coefficients:	Estimate	Std. Error	t value	$Pr(> |t|)$
(Intercept)	1.375	0.118	11.73	$< 2e - 16$
length	0.061	0.002	32.07	$< 2e - 16$
Multiple R^2:	0.8794,	Adjusted R^2:	0.8786	
F-statistic: 1028	1, 141 DF,	p-value:	$< 2.2e - 16$	

Thereby we obtain a linear model for the logarithm of weight $\hat{L} = 1.375 + 0.061x$. We can see that the quality of the model is quite satisfactory. Moreover, there is no heteroskedasticity detected, as can be observed in Figure 5.5. Transformations of response often help to reduce heteroskedasticity.

Another possible direction of the regression model development is to assume that a bear's weight depends not only on its length, but also on other parameters, for example, on the size of its head, the girth of its neck or its chest. Let us build a model for such dependence with multiple explanatory variables (multiple regression).

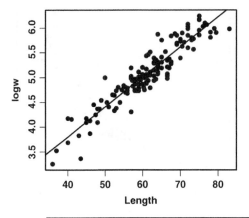

Figure 5.5 Logarithmic transform of weight.

Coefficients:	Estimate	Std. Error	t value	$Pr(> |t|)$
(Intercept)	−274.181	18.576	−14.76	$< 2e − 16$
length	0.802	0.628	1.276	0.204
head length	−4.292	2.78	−1.544	0.125
neck girth	6.963	1.356	5.134	$9.47e − 07$
chest girth	8.985	0.861	10.437	$< 2e − 16$
Multiple R^2:	0.9449,	Adjusted R^2:	0.9433	
F-statistic: 592	4, 138 DF,	p-value:	$< 2.2e − 16$	

We can see that the coefficient of determination in this multiple regression is much higher than in our initial simple linear regression model. However the coefficients corresponding to variables *length* and *head length* are not significantly different from zero. It may indicate a strong dependence between the explanatory variables. Their correlation coefficients are given in the following table.

	Head	Neck	Chest
Length	0.895	0.873	0.889

This phenomenon is known as ***multicollinearity***. We will remove the regression coefficients for *length* and *head length* from the equation in order to eliminate the multicollinearity effect.

Coefficients:	Estimate	Std. Error	t value	$Pr(> |t|)$
(Intercept)	−280.838	10.015	−28.041	$< 2e − 16$
neck girth	6.612	1.278	5.172	$7.84e − 07$
chest girth	9.142	0.787	11.616	$< 2e − 16$
Multiple R^2:	0.9439,	Adjusted R^2:	0.9431	
F-statistic: 1177	2, 140 DF,	p-value:	$< 2.2e − 16$	

We can see that the coefficient of determination virtually does not change, but now both explanatory variable coefficients are significantly different from zero. This functional form closer corresponds to our intuitive understanding of what are the biometric measurements related to the bear's weight.

5.3.3 Prediction

Let us assume that we observe a new bear, whose length is 72 inches. How heavy can it be? To answer this question we should make a prediction of the value of dependent variable $Y = weight$, provided that the corresponding value of the explanatory variable $X = length$ is known. Prediction is one of the basic applications of regression models. We will consider only the simplest linear model (5.6). Naturally, the process of prediction is much more complicated for nonlinear or multiple regression models. It is usually easy to make a point prediction. To do this we need to put a new value of the explanatory variable into the equation of the regression function. In our example, the predicted bear's weight turns out to be approximately 303 pounds. However it is not the complete answer to our question, as we understand that the bear's actual weight might be different from the predicted number. We should be able to figure out what is the range of prediction error, that is, we should make an interval prediction of the dependent variable. Therefore we need to solve two problems: to calculate the prediction error and to suggest a model for the distribution of the residuals. The prediction error is calculated under the following formula

$$\delta = s \sqrt{\left(1 + \frac{1}{n} + \frac{(x_{n+1} - \bar{x})^2}{\sum (x_i - \bar{x})^2}\right)}, \tag{5.8}$$

where x is the explanatory variable, s is the typical error defined as the square root of the mean square error $s^2 = \frac{1}{n-2} \sum_{i=1}^{n} e_i^2$, and n is the number of observations in the sample. In the bears' example this prediction error is 54.02. Further construction of the interval prediction is based mainly on the assumption that the random residuals are normally distributed. In this case the variable $(\hat{y}_{n+1} - y_{n+1})/\delta$ has Student's distribution with $n - 2$ degrees of freedom. In our example we obtain the following 95% prediction interval: (196.2, 409.8). As we can see, in our case it is quite large.

So far we hastily judged the performance of our models comparing the values of the coefficient of determination which estimates what fraction of the total variation in data is captured by the model. One may also perform some formal significant tests for the entire model (F-test) or particular explanatory variables (t-tests). Scatterplots also provide a good insight, often helping to detect heteroskedasticity and nonlinear patterns. However, residual analysis provides an important diagnostic toolkit determining the quality of regression models. One

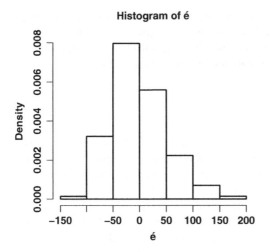

Histogram of é

Figure 5.6 Histogram of the residuals.

of the graphical parts of this toolkit is the histogram of the residuals, which for linear model (5.6) with estimates (5.7) is shown in Figure 5.6.

We can also formally test the normality of residuals, which is one of the model assumptions. We will perform Kolmogorov–Smirnov test in *R*. The result follows:

One-sample Kolmogorov–Smirnov test: D = 0.1021, p-value = 0.1012.

We can see that the hypothesis of residuals being normally distributed is not rejected. Therefore, one of the important assumptions of (5.6) is not clearly violated. However, looking at the evidently skewed histogram, we may assume that a different distribution of residuals might be more appropriate. If normality is violated, then the method of least squares is not necessarily optimal. Deviation from normality in residuals compromises the quality of the model. A strong pattern in residuals different from normality might indicate the presence of some additional effects, not addressed in the model. Therefore it suggests introducing additional explanatory variables.

5.4 Bayesian Regression

Our treatment of regression models in Section 5.3 was purely classical. Let us introduce a Bayesian version of linear regression. To avoid gruesome technical details, we will consider a simplified version of (5.6) with no intercept term $\alpha = 0$, which is known as RTO (regression through the origin):

$$Y = \beta X + \varepsilon, \varepsilon \sim N(0, \sigma^2). \tag{5.9}$$

To justify this simplification we can use two reasons: first, if we introduce matrix notation convenient for multiple regression models, where X is the matrix of explanatory variables, the intercept term can be included into the vector β of slopes and does not receive a special treatment. Second, there exist many applications where the regression line by construction passes through the origin. We will use this model here in order to be able to illustrate the principal ideas of Bayesian regression without getting into higher dimensions.

It is easy to see that the method of least squares for RTO applied to a sample $s = ((x_1, y_1), \ldots, (x_n, y_n))$ brings about a version of (5.7):

$$\hat{\beta} = \frac{\sum x_i y_i}{\sum x_i^2}. \tag{5.10}$$

We observe two unknown parameters: the slope β and the nuisance parameter σ^2. Full Bayesian approach requires us to set priors on both. However we will begin with treating σ^2 as a known constant, and will determine the prior on β conditional to σ^2. What will be a convenient form for a prior? From (5.9) we know that $y_i \sim N(\beta x_i, \sigma^2)$, so that the likelihood is

$$l(\beta; y, \sigma^2) = \frac{1}{(2\pi\sigma^2)^{n/2}} e^{-\frac{1}{2\sigma^2} \sum (y_i - \beta x_i)^2}. \tag{5.11}$$

Using the idea of conjugate priors leading to posteriors from the same family, which was investigated in Section 2.7, we arrive at the normal prior

$$\pi(\beta \mid \sigma^2) \sim N\left(\mu, \frac{\sigma^2}{g \sum x_i^2}\right), \tag{5.12}$$

introduced by Zellner [19] and known since then as g-prior. If $g = 0$, the prior is noninformative and improper (infinite variance), but values $g > 0$ do not cause any problems. The higher g corresponds to the more precise prior information. Formal expression for g-prior is

$$\pi(\beta \mid \sigma^2) = \left(\frac{g \sum x_i^2}{2\pi\sigma^2}\right)^{1/2} e^{-\frac{g \sum x_i^2}{2\sigma^2}(\beta - \mu)^2}. \tag{5.13}$$

Multiplying together expressions (5.11) and (5.13) and performing standard manipulations in the exponential terms (completing the square and ignoring terms not including β), we obtain conditional posterior

$$\pi(\beta \mid \sigma^2, y) \propto e^{-\frac{(g+1) \sum x_i^2}{2\sigma^2}(\beta - \beta^*)^2}, \tag{5.14}$$

where the posterior mean a.k.a. the Bayes estimator of β is

$$\beta^* = \frac{\sum x_i y_i + \mu g \sum x_i^2}{(g+1) \sum x_i^2} = \frac{1}{(g+1)}\hat{\beta} + \frac{g}{(g+1)}\mu, \tag{5.15}$$

and the posterior variance is

$$\frac{\sigma^2}{(g+1)\sum x_i^2}. \tag{5.16}$$

Therefore, if σ^2 is known, g-prior brings about a proper posterior, even in case $g = 0$. In this latter case the Bayes estimator coincides with the estimator (5.10).

However, if σ^2 is unknown, we have to address additional terms of the likelihood, which if treated as a function of σ^2 conditionally on β, assumes the form

$$l(\sigma^2 \mid y, \beta) \propto \left(\frac{1}{\sigma^2}\right)^{n/2} e^{-\frac{1}{2\sigma^2} A(y,\beta)},$$

and the conditional prior brings in

$$\pi(\beta \mid \sigma^2) \propto \left(\frac{1}{\sigma^2}\right)^{1/2} e^{-\frac{1}{2\sigma^2} B(\mu,\beta,g)},$$

where $A(y,\beta)$ and $B(\mu,\beta,g)$ are some functions containing no terms with σ^2. To obtain a conjugate prior on σ^2, one can suggest using a product of a power function and an exponential:

$$\pi(\sigma^2) \propto \left(\frac{1}{\sigma^2}\right)^{\alpha} e^{-\frac{1}{\sigma^2}\lambda},$$

which corresponds to $1/\sigma^2 \sim Gamma(\alpha, \lambda)$, or σ^2 having **inverse Gamma distribution**. Multiplying together three expressions for the likelihood and both priors, we obtain the joint posterior

$$\pi(\beta \mid \sigma^2, y)\pi(\sigma^2 \mid y),$$

and after factoring out the normal conditional posterior (5.14), obtain the inverse gamma expression for the second factor of this product, the posterior of σ^2:

$$\pi(\sigma^2 \mid y) \propto \left(\frac{1}{\sigma^2}\right)^{\alpha+1/2} e^{-\frac{1}{\sigma^2}(\lambda + C(y,\mu,g))},$$

where $C(y, \mu, g)$ is an additional term corresponding to the "leftovers" after completion of the square in (5.14):

$$C(y, \mu, g) = \frac{1}{2}\left(\sum y_i^2 + \frac{\sum x_i^2}{g+1}(g\mu^2 - 2\hat{\beta}g\mu - \hat{\beta})\right).$$

Bayesian estimation procedure suggests calculating the posterior mean of σ^2 first, and then the posterior mean β^*. If β is the parameter of interest, its estimator itself does not depend on σ^2, but its precision does. If we prefer using MCMC procedure to evaluate the joint posterior, the best solution is a Gibbs sampler requiring consecutive draws from Gamma distribution for $1/\sigma^2$, and then from normal for β using conditional posterior (5.14).

Coffee Shops

To illustrate the use of Bayesian regression, we may offer a toy example. Let us consider a small coffee-shop chain which was not in existence in 2010, opened 5 first shops in 2011, grew to 11 shops in 2012, 15 in 2013, and 18 in 2014. In order to model the growth of the chain (and possibly predict its future) we may want to use time series regression. Other methods of dealing with time series described in Chapter 1 could be more adequate, but we are not solving a real problem here. Setting the first year of the observation, 2010 as time 0, we obtain a paired sample

$$s = (x, y) = \{(0, 0), (1, 5), (2, 11), (3, 15), (4, 18)\}.$$

With origin included in the sample, we may consider RTO as a reasonable model. In this case $\sum x_i^2 = 30$, $\sum x_i y_i = 144$, and therefore according to (5.10), nonBayesian slope estimate (growth per year) is $\hat{\beta} = 144/30 = 4.8$. Let us also consider $\sigma^2 = 1$ known. A hint: if we average the squares of our residuals for $x = 1, 2, 3, 4$, we may arrive at a similar number for the variance of errors.

Now let us introduce a prior. Assume that a conservative assessment (preset goals) suggests the average growth per year at $\mu = 3$. If we apply g-prior with $g = 1$, according to (5.15) the Bayes estimate is

$$\beta^* = \frac{1}{2} \times 4.8 + \frac{1}{2} \times 3 = 3.9,$$

and its standard error from (5.16) is $\sqrt{1/2 \times 30} = 1/\sqrt{60} \approx .017$.

Changing the prior to $g = 2$ (more confidence in the prior mean) brings about $\beta^* = 3.6$ with the standard error of $1/\sqrt{90} \approx .011$.

5.5 Survival Analysis

In this section we will concentrate on dependence structures for random variables measuring the length of associated lives. For the distribution of such variables, hazard rate $h(t) = -\frac{d}{dt} \ln S(t)$ is a very convenient characteristic describing the survival pattern. It is easy to see that integrating the hazard rate over time, we obtain an expression for the survival function:

$$H(t) = \int_0^t h(s)ds = -\ln S(t) \Rightarrow S(t) = e^{-H(t)}.$$

5.5.1 Proportional Hazards

Let us review the joint mortality example from the introduction and think of the "common lifestyle" factor which influences both dependent lives. In this case one can separately address the hazard factor which is common for two lives, and the individual mortality patterns of husband and wife. For such problems, Cox [2] suggested a universal model platform based on the definition of the hazard function. Suppose we observe a group of n individuals. Let us define the hazard rate for the ith individual from the group as

$$h_i(t) = h_0(t)e^{\beta_1 x_{i1} + \cdots \beta_k x_{ik}}, \tag{5.17}$$

where $h_0(t)$ is the **baseline hazard function**, which is the same for the entire group, and $x_{.j}, j = 1, \ldots, k$ are k different additional factors defining the individuals. Both baseline hazard and coefficients β_j are to be estimated. Model (5.17) is known as *Cox proportional hazard model*.

In the context of joint mortality $h_0(t)$ may represent common lifestyle hazard of a heterosexual married couple, $n = 2$, and the only factor X defining the individual differences is the binary gender variable so that $k = 1$, $x_{11} = 0$ for the female partner and $x_{21} = 1$ for the male, so the female hazard rate is $h_1(t) = h_0(t)$, and the male is $h_2(t) = h_0(t)e^{\beta}$.

A very convenient simplification for the Cox proportional hazard model is provided by the choice of Weibull distribution to represent the common hazard. As we remember from Chapter 1, parameter λ of Weibull distribution is a linear factor in the expression for the hazard rate, so for different values of individual factors x_{ij}, all lifetimes in the group are modeled by Weibull distributions with the common shape parameter and individual values of λ_i. In this case $h_0(t) = \lambda_0 \alpha t^{\alpha-1}$, and both lives have Weibull distributions with $h_i(t) = \lambda_i \alpha t^{\alpha-1}$, where $\lambda_1 = \lambda_0$, $\lambda_2 = \lambda_0 e^{\beta}$, and α is the shape parameter, common for both distributions.

5.5.2 Shared Frailty

One may want to look at the proportional hazard model from the other side, considering two related lives. Suppose that two individuals have the hazard rates $Yh_1(t)$ and $Yh_2(t)$, where $h_i(t), i = 1, 2$ represent their independent baseline hazard rates, and Y is some random variable affecting both lives in the same way. Y in this context is called *shared frailty* or *common frailty*. It is easy to see that if we integrate in the definition of the hazard rate,

$$Yh_i(t) = -\frac{d}{dt} \ln S_i(t) \Rightarrow Y \int_0^t h_i(s)ds = YH_0(t) = -\ln S_i(t)$$
$$\Rightarrow S_i(t) = e^{-YH_i(t)}.$$

If we consider two lives sharing the same common frailty factor, we can take into account conditional independence of their hazard rates given $Y = y$ and write down the joint survival function conditional on Y as

$$S(t_1, t_2 | Y = y) = P(X_1 > t_1, X_2 > t_2 | Y = y) = e^{-yH_1(t_1)} \times e^{-yH_2(t_2)}$$
$$= e^{-y\left(H_1(t_1) + H_2(t_2)\right)}$$

and its expected value

$$S(t_1, t_2) = Ee^{-Y\left(H_1(t_1) + H_2(t_2)\right)} = L_Y(H_1(t_1) + H_2(t_2)) \tag{5.18}$$

is the Laplace transform of the variable Y at point $\left(H_1(t_1) + H_2(t_2)\right)$. If we want to choose a particular distribution model for Y, in order to cancel Y out and obtain a formula directly connecting the bivariate survival function with its margins, one of the simpler choices is gamma distribution $Y \sim Gamma(\alpha, \lambda)$. In this case the bivariate Laplace transform can be expressed as

$$L(H_1(t_1) + H_2(t_2)) = \left(\frac{\lambda}{\lambda + H_1(t_1) + H_2(t_2)}\right)^{-\alpha}.$$

Setting $t_2 = 0$, we can use the expression of the marginal survival function as

$$S_1(t_1) = S(t_1, 0) = \left(\frac{\lambda}{\lambda + H_1}\right)^{\alpha} \Rightarrow H_1 = \lambda(S_1^{-1/\alpha} - 1),$$

and using symmetric expression for H_2 and returning to (5.18) obtain

$$S(t_1, t_2) = \left(S_1^{-1/\alpha}(t_1) + S_2^{-1/\alpha}(t_2) - 1\right)^{\alpha}. \tag{5.19}$$

Notice that the inverse scale λ disappeared from the formula, and the final expression depends on shape of α only. An alternative to gamma is the so-called *positive stable* distribution with parameter $\beta, 0 < \beta \leq 1$. The only thing we need to know about this distribution is that its bivariate Laplace transform is

$$L(H_1(t_1) + H_2(t_2)) = e^{-(H_1(t_1) + H_2(t_2))^{\beta}},$$

and then setting $t_2 = 0$ we obtain

$$S_1(t_1) = S(t_1, 0) = e^{-H_1^{\beta}} \Rightarrow H_1 = \left(-\ln S_1\right)^{1/\beta},$$

and returning to (5.18) we get a different model

$$S(t_1, t_2) = e^{-\left(\left(-\ln S_1(t_1)\right)^{1/\beta} + \left(-\ln S_2(t_2)\right)^{1/\beta}\right)^{\beta}}. \tag{5.20}$$

For more details on derivation of two shared frailty models (5.19) and (5.20) see Hougaard [11]. What is the most important from our prospective, these two models will be recognized in the next chapter as special types of copulas.

5.5.3 Multistage Models of Dependence

Common lifestyle is not the only factor affecting joint mortality. Common disaster may lead to simultaneous failures of two lives, and the broken heart syndrome may strongly affect the life of one spouse after the death of the other. These factors of mortality can be modeled in the following fashion. Let us define a state space describing the status of joint lives of the spouses: X_0 corresponds to both spouses being alive, X_1 corresponds to wife being alive and husband dead, X_2 to husband being alive and the wife dead, and X_3 to both being dead. Clearly, the last one is an absorbing state. Transition matrix for a fixed period of time (say, 1 year), could be represented as

$$P = \begin{pmatrix} 1 - \sum_{i=1}^{3} p_{0i} & p_{01} & p_{02} & p_{03} \\ 0 & 1 - p_{13} & 0 & p_{13} \\ 0 & 0 & 1 - p_{23} & p_{23} \\ 0 & 0 & 0 & 1 \end{pmatrix}. \tag{5.21}$$

In this case $p_{03} > p_{01}p_{02}$ reflects both common lifestyle and common disaster factors, and $p_{01} \le p_{23}$ and $p_{02} \le p_{13}$ correspond to the broken heart syndrome. Probabilities in (5.21) can be calculated by integrating hazard rates corresponding to various settings. For instance, $p_{13} = \int_0^1 h_1(t)dt$, where $h_1(t)$ corresponds to the hazard rate for widowed women, while the calculation of p_{03} requiring two events happening within a year may be much more complicated. Moreover, homogeneous Markov model would not be adequate in this situation, because the physical age of the spouses is an important factor, which makes the 1-year transition matrix P clearly time dependent. For multistage models of joint lifetimes, see Hardy and Lee [10].

5.6 Modeling Joint Distributions

The idea of modeling dependence between two variables X and Y via building a model for entire joint distribution $F(x, y) = P(X \le x, Y \le y)$ is very tempting. Indeed, given a joint distribution or rather its density $f(x, y)$ such that $F(x, y) = \int \int f(x, y)dxdy$, it is relatively easy to determine the marginal distributions $F_1(x)$ and $F_2(y)$ by their densities

$$f_1(x) = \int_{-\infty}^{\infty} f(x, y)dy = \frac{\partial}{\partial x}F(x, \infty), \quad f_2(y) = \int_{-\infty}^{\infty} f(x, y)dx = \frac{\partial}{\partial y}F(\infty, y)$$

and conditional densities for $X = x$ and $Y = y$ as

$$g_1(x \mid y) = \frac{f(x, y)}{f_2(y)}, \quad g_2(y \mid x) = \frac{f(x, y)}{f_1(x)}.$$

The problem that often is in the center of attention, is determining conditionals, which help us to directly use one variable to draw conclusions regarding the other. For instance, regression models requiring the determination of $E(Y \mid X = x)$ or $E(X \mid Y = y)$ may be easily derived from the joint distribution of X and Y. All information on dependence structure, no matter what factors determine this dependence, should be contained in the formula for the joint distribution. All factors such as common disaster or shared frailty can be addressed. Both our introductory examples, black bears and joint lifetimes, can be modeled this way.

It would be especially nice to be able to construct the joint distribution directly based on given marginals. In most statistical applications it is much easier to model the marginals than the association structure. Independent case $F(x, y) = F_1(x)F_2(x)$ serves as a convenient starting point, and association structure can be introduced as a certain "deviation" from independence.

5.6.1 Bivariate Survival Functions

We will begin with presenting joint distribution models characteristic for survival analysis. Confining ourselves to the case of two nonnegative "lifetime" variables, it is common to consider joint survival function $S(x, y) = P(X > x, Y > y)$ rather than $F(x, y)$. Marginal survival functions can be defined as $S_1(x) = P(X > x) = S(x, 0)$ and $S_2(y) = P(Y > y) = S(0, y)$ and independence assumption yields $S(x, y) = S_1(x)S_2(y)$.

The simplest case to consider is the case of exponential marginals $S_1(x) = e^{-\lambda_1 x}$ and $S_2(y) = e^{-\lambda_2 y}$ for $x, y > 0$. How can we build a bivariate distribution function with these marginals allowing for dependence between X and Y? In 1960 Gumbel [9] suggested several options including

$$S(x, y) = e^{-x-y-\delta xy}, 0 \leq \delta \leq 1 \tag{5.22}$$

and

$$S(x, y) = e^{-\lambda_1 x - \lambda_2 y} \times \left(1 - \alpha(1 - e^{-\lambda_1 x})(1 - e^{-\lambda_2 y})\right), -1 \leq \alpha \leq 1. \tag{5.23}$$

Both formulas are simple enough, special cases $\delta = 0$ in (5.22) and $\alpha = 0$ in (5.23) correspond to independence, and both marginals are exponential. For model (5.22) conditional expected value is

$$E(Y \mid X = x) = \frac{1 + \delta + \delta x}{(1 + \delta x)^2},$$

so clearly the association between two variables X and Y is nonlinear and cannot be described by linear regression. The same can be said about (5.23), for which conditional expectation $E(Y \mid X = x)$ is an exponential function of x.

The good news is that these models capture some nonlinear dependence patterns. Unfortunately, the ability of these models to address strong linear association is limited. This fact is demonstrated by the inequalities for Pearson's correlation obtained for these two joint distributions in Gumbel [9]: in (5.22) $-0.40365 \leq \rho(X, Y) \leq 0$, and in (5.23) $-0.25 \leq \rho(X, Y) \leq 0.25$.

Bivariate exponential distributions (5.22) and (5.23) share another weakness which is typical for mathematically natural generalizations of popular models. There is no clear justification for these exact ways to generalize exponential distribution to multidimensional case, and there is no clear reason to choose one against the other. Such factors as common shock or shared frailty are addressed by the model only implicitly. To correct this situation, Marshall and Olkin suggested a different bivariate exponential model [14]:

$$S(x, y) = e^{-\lambda_1 x - \lambda_2 y - \lambda_{12} \max(x, y)}. \tag{5.24}$$

This model can be intuitively explained by introducing the idea of "fatal shock" events leading to separate failures of X or Y and also to a simultaneous failure of X and Y. If these three "fatal shock" events result from three independent Poisson flows with intensities λ_1, λ_2, and λ_{12}, the joint lifetime distribution of X and Y is given by (5.24). Mathematically, this joint distribution is not absolutely continuous, because $P(X = Y) \neq 0$, and requires some caution. However it naturally addresses common shock events. It also can be demonstrated to have the residual lives' distribution independent of age:

$$S(x, y) = P(X > x + t, Y > y + t \mid X > t, Y > t)$$

for any $t, x, y > 0$, which is a nice generalization of the memoryless property of the univariate exponential distribution.

It can be shown for (5.24) that

$$E(X) = \frac{1}{\lambda_1 + \lambda_{12}}, E(Y) = \frac{1}{\lambda_2 + \lambda_{12}}, Var(X) = E(X)^2, Var(Y) = E(Y)^2,$$

and

$$\rho(X, Y) = \frac{\lambda_{12}}{\lambda_1 + \lambda_2 + \lambda_{12}},$$

so that depending on parameters λ_1, λ_2, and λ_{12}, correlation can attain any values from 0 to 1. Therefore model (5.24) unlike (5.22) and (5.23) can describe the cases of strong linear dependence.

If (X, Y) is distributed according to (5.24), random vector $(X^{1/\alpha_1}, Y^{1/\alpha_2})$

$$S(x, y) = e^{-\lambda_1 x^{\alpha_1} - \lambda_2 y^{\alpha_2} - \lambda_{12} \max(x^{\alpha_1}, y^{\alpha_2})}$$

will have *bivariate Weibull distribution* with Weibull marginals. Further generalizations are possible. Some special cases of bivariate survival functions built on arbitrary marginals (5.19) and (5.20) were discussed earlier.

5.6.2   Bivariate Normal

Multivariate normal (Gaussian) distribution defined in Chapter 1 is by far the most popular choice of a model for multivariate joint distribution. It can be explained by the special role of normal distribution in general. Besides, it has many nice mathematical properties simplifying its technical treatment. In case of two variables, distribution density function of bivariate normal distribution with means (μ_x, μ_y), variances (σ_x^2, σ_y^2), and correlation ρ can be written down as

$$\varphi(x, y) = \frac{1}{2\pi\sigma_x\sigma_y\sqrt{1 - \rho^2}} \tag{5.25}$$

$$\times \exp\left\{-\frac{(x - \mu_x)^2/\sigma_x^2 - 2\rho(x - \mu_x)(y - \mu_y)/(\sigma_x\sigma_y) + (y - \mu_y)^2/\sigma_y^2}{2(1 - \rho^2)}\right\}.$$

As was stated in Chapter 1, if a vector (X, Y) is normally distributed, both marginals are normal, $X \sim N(\mu_x, \sigma_x^2)$ and $Y \sim N(\mu_y, \sigma_y^2)$. What is especially important, conditional distributions of Y given $X = x$ (and vice versa) are also normal for any x,

$$g_2(y \mid x) \sim N\left(\mu_y + \rho\frac{\sigma_y}{\sigma_x}(x - \mu_x), (1 - \rho^2)\sigma_y^2\right).$$

Seeing that conditional expected value $E(Y \mid X = x)$ is a linear function of x, we may conclude that linear regression model with response Y and predictor X is quite viable, and the dependence between X and Y is indeed linear. Therefore, Pearson's correlation is the number determining the strength of this dependence.

Reducing all dependence patterns to correlation is a nice simplification. However, it is not only a blessing but also a curse: when the character of association is essentially nonlinear, bivariate normal model may fail to capture this nonlinearity and will not be adequate.

5.6.3   Simulation of Bivariate Normal

First of all, in simulation problems it is sufficient to sample from standardized bivariate normal (W_1, W_2) with correlation ρ, zero means, and unit variances. Nonstandard normal for (X, Y) can be obtained by linear transformations

$X = \sigma_x W_1 + \mu_x$ and $Y = \sigma_y W_2 + \mu_y$. If we want to simulate a sample from bivariate normal distribution, we can either use the procedure described in Section 4.4 as an example of Gibbs sampling, or a more straightforward alternative, which we will also use in Section 6.6. This alternative works as follows:

1. Generate independently two standard normal variables $z_1, z_2 \sim N(0, 1)$.
2. Define correlated standard normal variables as $w_1 = z_1$ and $w_2 = \rho z_1 + \sqrt{1 - \rho^2} z_2$.

Why does this scheme work? It is easy to see that in this construction W_2 is also a standard normal variable:

$$E(W_2) = \rho \times 0 + \sqrt{1 - \rho^2} \times 0 = 0, Var(W_2) = \rho^2 \times 1 + (1 - \rho^2) \times 1 = 1,$$

and the correlation $\rho(W_1, W_2) = \rho$. Therefore the construction is complete.

5.7 Statistical Dependence and Financial Risks

The following example is a gross oversimplification of a serious real-life problem. This problem might lie in the foundation of the most recent financial crisis related to the US mortgage bubble which burst in the first decade of the twenty-first century. Misunderstanding the association between possible defaults on related securities brought about mispricing of credit derivatives. These financial products were novel and extremely popular at the turn of the century and became infamous later on. Abbreviations like credit default swaps (CDS) and collateralized debt obligations (CDO) became widely known after the credit derivative bubble burst. We will return to the story of credit derivatives in a different context in Chapter 7. For more serious discussion of the issues related to financial risks we can refer to Embrechts et al. [4], Engle's approach to dynamic correlation [5, 6], and also going beyond correlation in [15]. The latter along with [16] and [3] provide a good insight into the ways statistical dependence modeling meets financial risk management, and provide a smooth transition to Chapter 6.

5.7.1 A Story of Three Loans

Here we will start with a very simple illustration of problems with pricing credit derivatives. All the numbers we consider below are totally unrealistic and are chosen with the sole purpose to simplify the arithmetic of the underlying calculations.

Suppose you are a loan officer at a very small bank and you are going to issue three identical loans to three of your local clients. Each of them needs the principal of \$100,000, each of them is promising you to repay the loan in 1 year, and is comfortable with paying certain interest. After 1 year all the interest is collected, the loans are repaid, financial results are analyzed, and as the loan officer you do not care what will happen after that. Let us say you will retire?

We will assume that there is no risk involved with your customers making interest payments. You will be paid the interest for granted. However, some of the customers may default at some point during the year. In case of default your bank simply does not get back the principal amount of the loan (zero recovery rate). Your bank's credit risk department is doing their jobs very well, and you have a pretty good estimate of individual default probabilities for each loan. Assume that each of your clients may default during this year with probability $q = 0.1$.

How can you calculate the fair amount of interest you need to collect? To avoid lengthy calculations related to the time value of money, let us take into account only the future value of the total interest payments calculated at the end of the year. The bank needs to get something out of the deal, or at least not to lose any money. Therefore let us suggest $r = 10\%$ as the break-even condition. Any interest rate higher than that will guarantee a positive expected gain for the bank.

You calculate: in the end of the year each of three loans will pay you back the value of \$100,000 plus \$10,000 of interest with probability $1 - q = 0.9$ (this is the gain of \$10,000 with probability 0.9) or (in case of default) you get back the interest only and do not recover the principal with probability $q = 0.1$. This is a loss of \$90,000 with probability 0.1. Merging the two scenarios, you will get the expected gain of

$$\$10,000 \times 0.9 - \$90,000 \times 0.1 = \$9000 - \$9000 = 0.$$

Any additional interest you charge will constitute return on your \$100,000 investment. Say, you can charge 11% interest. This will bring you an expected return of \$1000 on each loan.

But what about the risks? They are rather substantial. The sheer amount of money you might be losing on the way to a modest financial gain is downright scary: without taking probabilities into account, your maximum loss on each loan is \$90,000. Multiplied by the factor of three it constitutes a possible (and probable) loss of \$270,000. Do you not want some kind of insurance on this deal which would cover you if the loans go sour?

Insurance policies which allow you to hedge your risks as a lender are known as credit default swaps or CDS. Your counterparty, CDS seller takes full responsibility for your losses in case of a default, expecting a reasonable fee in exchange. You can buy a CDS for each of the three loans and if all your risks are taken away, it is fair to expect to pay to the CDS seller the premium of \$10, 000 per loan equal to the collected interest. Then you end up with a zero gain but also have zero risk. If you manage to buy CDS at a cheaper price, you are in a no-risk positive gain situation known in finance as an arbitrage.

An alternative to buying individual CDS for all your loans would be to securitize entire package of liabilities. An asset-backed security (ABS) would require putting all three loans in one portfolio and than repackaging your risks and selling them in tranches. The buyer of the riskiest *equity tranche* for a designated premium takes the full responsibility to repay the principal in case of the first of the three loans (A, B, or C) going into default. The buyer of the *mezzanine tranche* will pay off the principal on the second default, if it happens during the year. And finally, the buyer of the *senior tranche* will pay off the principal on the third default, if all three loans (A, B, and C) default in one year, which is not very likely to happen. This payoff structure distributing the risks between the tranches is known as *waterfall*.

The payoff for each tranche is set at the full principal amount of one loan of $100,000, so the losses are quite clearly defined. However, the premiums paid to buyers of all three tranches have to be determined. The calculation of a fair premium (expected gain 0) must take into account both the fixed size of the loss and the probability of default.

5.7.2 Independent Defaults

If we assume that defaults happen independently and can be modeled by tossing three separate coins with "success" probability of 0.1, the expected losses for all three tranches can be determined from Venn's diagram in Figure 5.7. Denoting by A, B, and C the default events for loans A, B, and C, we can conclude that the probability of at least one default happening in 1 year can be calculated from the diagram as

$$P(A \ or \ B \ or \ C) = 0.081 \times 3 + 0.009 \times 3 + 0.001$$
$$= 0.243 + 0.027 + 0.001 = 0.271.$$

Evidently, there exist other ways to do this calculation, but let us stick to the diagram for its visual simplicity. We have calculated the probability of the loss

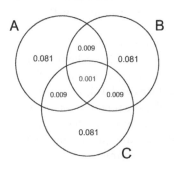

Figure 5.7 Independent defaults.

for the equity tranche. Thus the expected loss (EL) of the equity tranche

$$EL(Eq) = \$100,000 \times 0.271 = \$27,100.$$

Similarly, we can use Figure 5.7 to calculate the probability of double default (at least two defaults happening in 1 year):

$$P(AB \text{ not } C) + P(AC \text{ not } B) + P(BC \text{ not } A) + P(ABC)$$
$$= 0.009 \times 3 + 0.001 = 0.028,$$

and the expected loss for the mezzanine tranche is

$$EL(Mezz) = \$100,000 \times 0.028 = \$2800.$$

For the least risky senior tranche, the loss occurs in case of triple default (all three loans going down), so $P(ABC) = 0.001$, making

$$EL(Sen) = \$100,000 \times 0.001 = \$100.$$

Adding up all three expected losses, we arrive at the number

$$\$27,100 + \$2800 + \$100 = \$30,000,$$

which would coincide with your initial expected losses before interest charged, or also with the summary premium of three CDS if issued for each loan separately. Now they are very unevenly spread between the tranches. To determine fair premium to be paid to the tranche holders, we take expected losses as the basis. Anything we put on top of that will make the deal attractive and potentially profitable for the tranche buyer. For instance, if the senior tranche holders are offered a premium of \$1000, which exceeds their expected loss by an order of magnitude, they should be extremely happy with such a nice deal.

This approach, however, is very naive. The assumption of independence is not realistic: in actual markets, many factors influencing default probabilities would work for all three loans, what makes positive dependence of the default events very possible. Let us suppose that we want to take into account default correlation. This correlation is usually hard to assess, but we will imagine an ideal situation.

5.7.3 Correlated Defaults

Denote by X, Y, and Z three binomial variables associated with default events A, B, and C.

$X = 1$ in case of A, and $X = 0$ otherwise,
$Y = 1$ in case of B, and $Y = 0$ otherwise,
$Z = 1$ in case of C, and $Z = 0$ otherwise.

Suppose that $\rho(X, Y) = \rho(Y, Z) = \rho(X, Z) = \rho$. How will this assumption change the Venn diagram from what we saw in Figure 5.7? It is easy to see

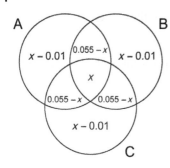

Figure 5.8 Correlated defaults with $\rho = 0.5$.

that since for binomial variables $E(X) = E(Y) = 0.1$ and $Var(X) = Var(Y) = 0.1 \times 0.9 = 0.09$,

$$\rho(X, Y) = \frac{E(XY) - E(X)E(Y)}{\sqrt{Var(X)Var(Y)}} = \frac{E(XY) - 0.01}{0.09}$$

and $E(XY) = 0.09\rho + 0.01$. Choosing for example $\rho = 0.5$, we obtain $P(AB) = E(XY) = 0.055$, and symmetrically $P(BC) = P(AC) = 0.055$, which completes the calculation of double default probabilities. Unfortunately, not much can be said regarding the probability of triple default $x = P(ABC)$ to be placed in the center of the Venn diagram corresponding to correlated defaults demonstrated in Figure 5.8. We can complete Venn's diagram "from inside out" as:

$P(ABC) = x.$
$P(AB) = 0.055 \Rightarrow P(AB \text{ not } C) = 0.055 - x.$
$P(A) = 0.1 \Rightarrow P(A \text{ not } B \text{ not } C) = 0.1 - (0.055 - x) - (0.055 - x) + x = x - 0.01.$

Two inequalities $x \leq 0.055$ and $x \geq 0.01$ are necessary conditions for probabilities of all disjoint events in the diagram to stay nonnegative.

The fact that correlations do not determine the probability of triple default does not allow us to evaluate the payoff probabilities and expected losses for the tranches. According to Figure 5.8, the default probabilities are:

$P(A \text{ or } B \text{ or } C) = 3(x - 0.01) + 3(0.055 - x) + x = 0.135 + x,$
$P(AB \text{ not } C) + P(AC \text{ not } B) + P(BC \text{ not } A) = 3(0.055 - x) + x = 0.165 - 2x,$
$P(ABC) = x.$

Expected losses can be calculated as

$EL(Eq) = \$100,000 \times (0.135 + x) = \$13,500 + 100,000x,$
$EL(Mezz) = \$100,000 \times (0.165 - 2x) = \$16,500 - 200,000x,$
$EL(Sen) = \$100,000x = 100,000x,$

Figure 5.9 Extreme scenario 1.

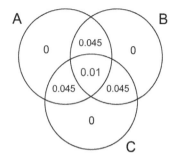

that once again add up to $30,000, but all three numbers depend on the value of x. We can observe two extreme scenarios.

1. $x = 0.01$ (min): $EL(Eq) = EL(Mezz) = \$14,000, EL(Sen) = \$1000$.
2. $x = 0.055$ (max): $EL(Eq) = \$19,500, EL(Mezz) = EL(Sen) = \$5500$.

Figure 5.9 demonstrates the Venn diagram for extreme scenario 1 and in Figure 5.10 we see the Venn diagram for extreme case scenario 2, two particular cases of correlated defaults defined by pairwise correlations $\rho = 0.5$.

Now let us return to financial markets. A popular credit derivative known as collateralized default obligation or CDO can be formed by a seller of three CDS for three separate loans, holding an equivalent of ABS package with a positive cash flow of $ 30,000 a year and the risks equivalent to the risks of a bank who issued the loans and bought our CDS. CDO seller can establish a tranche structure and sell the risks: the first $100K of losses (caused by the default of any of the three loans) is paid off by the buyer of the equity tranche; the second $100K, caused by the double default is paid off by the buyer of the mezzanine tranche, and the final $100K (caused by the triple default only) is paid off by the buyer of the senior tranche of a CDO.

We will concentrate on the risk of the senior tranche holder, which should help us to establish a fair price of this tranche. Independent defaults suggest the expected loss is $100. However, if defaults are correlated with $\rho = 0.5$, this number underestimates the expected loss from 10 times (scenario 1) to 55 times

Figure 5.10 Extreme scenario 2.

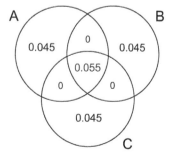

(scenario 2). So being paid $1000, as suggested above, is not such a good deal as it seemed before. Without further details, one of the main causes of the recent financial crisis was exactly the miscalculation of the risks for the CDO tranche holders leading to gross mispricing of CDO tranches.

This somewhat lengthy example was meant to serve as an illustration of one simple fact: even full knowledge of the correlation structure does not equate to the full assessment of dependence, especially in case of such extreme events as multiple defaults. That is why the idea of models based on multivariate normal distribution, which can be fully described by its means, variances, and correlations, becomes less and less attractive in modern financial mathematics and other applications dealing with extreme events and tails of joint distributions. The instruments of dependence analysis we will introduce in Chapter 6 will make it possible to go far beyond correlation.

References

1 Alt, G. L., Alt, F. W., and Lindzey, J. S. (1976). Home range and activity patterns of black bears in northwestern Pennsylvania. *Northeast Fish and Wildlife Conference*, 33, 45–56.

2 Cox, D.R. (1972). Regression models and life-tables. *Journal of the Royal Statistical Society, Series B*, 34(2), 187–220.

3 Dias, A., Salmon, S., and Adcock, C., (editors). (2013). *Copulae and Multivariate Probability Distributions in Finance*. London-New York: Routledge, Taylor and Francis.

4 Embrechts, P., McNeil, A., and Straumann, D. (2003). Correlation and dependency in risk management: properties and pitfalls. In: *Risk Management: Value at Risk and Beyond*, Cambridge University Press, 176–223.

5 Engle, R. (2002). Dynamic conditional correlation: a simple class of multivariate generalized autoregressive conditional heteroskedasticity models. *Journal of Business and Economic Statistics*, 20(3) 339–350.

6 Engle, R. (2009). *Anticipating Correlations: A New Paradigm for Risk Management*. Princeton Press.

7 Frees, E. W., Carreiere, J. F., and Valdez, E. (1996), Annuity valuation with dependence mortality. *Journal of Risk and Insurance*, 63, 2.

8 Glejser, H. (1969). A new test for heteroskedasticity. *Journal of the American Statistical Association*, 64(235), 315–323.

9 Gumbel, E. J. (1960). Bivariate exponential distributions, *Journal of the American Statistical Society*, 55, 292, 30–44.

10 Hardy. M., and Siu-Hang Li, J. (2011) Markovian approaches to joint life mortality. *North American Actuarial Journal*, 15(3), 357–376.

11 Hougaard, P. (2000). *Analysis of Multivariate Survival Data*, New York-Berlin-Heidelberg: Springer Verlag.

12 Joe, H. (1997). *Multivariate Models and Dependence Concepts*. London: Chapman & Hall.

13 Kendall, M. (1938). A new measure of rank correlation. *Biometrika*, 30(1–2), 81–89.

14 Marshall, A. W., and Olkin, I. (1967). A multivariate exponential distribution, *Journal of the American Statistical Society*, 62(317), 30–44.

15 Maschal, R., and Zeevi, A. (2002). Beyond correlation: Extreme co-movements between financial assets, Available at: http://dx.doi.org/10.2139/ssrn.317122

16 McNeil, A.J., Frey, R., and Embrechts, P. (2015) *Quantitative Risk Management: Concepts, Techniques and Tools*, 2nd ed. Princeton University Press.

17 Shemyakin, A., and Youn, H. (2006). Copula models of joint last survivor insurance. *Applied Stochastic Models of Business and Industry*, 22(2), 211–224.

18 Youn, H., and Shemyakin, A. (1999). Statistical aspects of joint life insurance pricing, 1999 Proceedings of the Business and Economic Statistics Section of the American Statistical Association, 34–38.

19 Zellner, A. (1986). On assessing prior distributions and Bayesian regression analysis with g prior distributions. In: P. Goel, and A. Zellner (editors), pp. 233–243. *Bayesian Inference and Decision Techniques: Essays in Honor of Bruno de Finetti. Studies in Bayesian Econometrics 6*. New York: Elsevier.

Exercises

5.1 Find such values c that for independent standard normal variables X and Y Pearson's correlation
(a) $\rho(X + cY, X - cY) = 0.5$;
(b) $\rho(X + cY, X - cY) = -0.5$.

5.2 Suggest an example of a sample (x, y) with four elements such that the sample Kendall's concordance $\hat{\tau}$ equals to 0 (number of concordant and discordant pairs are equal), while both Pearson's and Spearman's sample correlations r and r^* are positive.

5.3 Can you suggest an example of a sample (x, y) with three elements such that the sample Kendall's concordance $\hat{\tau}$ equals to 0 (number of concordant and discordant pairs are equal)? What needs to be done to make it possible?

5.4 Calculate sample Pearson's correlation, Spearman's rank correlation, and sample Kendall's concordance for
(a) Heating oil NYMEX futures and jet fuel price fluctuations *OilFuel-CDO.xlsx* file, columns A and B;

     (b) Lifetime of two tranches of synthetic CDOs ***OilFuelCDO.xlsx***, columns C and D.

     Comment on similarities and differences between these two situations.

5.5 Using the daily percentage returns on IGBM Madrid Stock Exchange and JSE Africa stock indexes in ***IGBMJSE.xlsx***, estimate the probability of simultaneous daily drop of these two indexes by more than by one standard deviation.

5.6 Suggest a linear regression model for JSE as a function of IGBM using the file ***IGBMJSE.xlsx***. Will multivariate normal distribution provide a good model for the joint distribution of these two indexes?

5.7 Using the model developed in the previous problem, estimate the probability of simultaneous daily drop of these two indexes by more than by one standard deviation. Use simulated bivariate normal variables (sample size 3000).

5.8 Modify the example of three loans in Section 5.7 for pairwise default correlations of 0.5 without changing mortgage terms, amounts, interest rates, or individual default probabilities of 0.1, so that the expected loss of the equity tranche is $16,000. What would be the expected losses of the

     (a) Senior tranche;

     (b) Mezzanine tranche?

6

Copula Models of Dependence

6.1 Introduction

The previous chapter contained a brief description of models of dependence between two random variables X and Y. We have agreed that modeling entire joint distribution of X and Y described by a function $H(x, y) = P(X \leq x, Y \leq y)$ for all possible values x and y would make it possible to address all sorts and sources of dependence.

However, simple bivariate distributions such as bivariate normal or bivariate exponential would rarely satisfy the needs of good data fit. Geometrically speaking, scatterplots of real bivariate data such as black bears' biometrics and joint mortality discussed in Section 5.1 exhibit types of behavior (skewness, asymmetry, multimodality, fat tails) which are impossible to describe within the confinements of most common bivariate distribution families.

Additional consideration should be given to our observation that in many applications researchers tend to obtain much more data and prior information regarding individual variables X and Y taken separately and possess relatively little information related to their interdependence. Therefore, as in the joint mortality example, there exist many good models for separate male and female life lengths based on extensive mortality databases; at the same time very few reliable models have been developed for joint mortality since little is known about dependent lives, and only a few paired datasets are commonly available.

Ideally, a statistical model for joint distribution of X and Y would include two aspects: estimating marginal distributions using all available data and prior information; and estimating the strength and character of dependence using paired data which are often limited. For both cases (estimation of marginals and estimation of dependence) Bayesian approach comes naturally. In the first case it is due to typically ample prior information. In the second case it is due to typically small sample sizes.

Introduction to Bayesian Estimation and Copula Models of Dependence, First Edition.
Arkady Shemyakin and Alexander Kniazev.
© 2017 John Wiley & Sons, Inc. Published 2017 by John Wiley & Sons, Inc.
Companion Website: http://www.wiley.com/go/shemyakin/bayesian_estimation

Leaving statistical aspects aside until Chapter 7, we will describe the class of mathematical models allowing for exactly this sequence of actions:

- modeling marginal distributions $F(x)$ and $G(y)$,
- modeling the joint distribution as

$$H(x, y) = C(F(x), G(y)), \tag{6.1}$$

"mixing" the marginals with the help of a special bivariate function C : $(u, v) \to C(u, v)$, reflecting the dependence pattern.

This class of functions C is known as *copulas* and the representation (6.1) is the result of the famous Sklar's theorem discussed in more detail in Section 6.2.

Section 6.3 contains the simplest examples of copula functions. The following two sections describe the methods of construction of joint distributions using two most popular subclasses of copula functions: elliptical copulas and Archimedean copulas.

One of attractive features of copula models is a natural way to organize simulation from joint distributions, whose dependence structure is defined by copulas. Simulation from the elliptical family and Archimedean copulas is considered in Section 6.6.

As in Chapter 5, we will mostly restrict ourselves to the case of two variables or *pair copulas*, though it is often possible to extend the suggested methodology to higher dimensions. Such extension comes naturally for elliptical copulas and requires some additional formalism in C-vine and D-vine construction for Archimedean copulas and more general classes. Elliptical copulas for dimensions higher than 2 and also vine and hierarchical copulas are discussed in Section 6.7.

Copula models are becoming a popular instrument of applied statistical analysis in such fields as finance [4], risk management [23], or engineering [33]. All statistical aspects of copula modeling such as estimation of parameters and model selection are relegated to Chapter 7. As we see, Bayesian approach comes very naturally. Certain problems recently encountered with practical application of copula models which brought them a rather bad publicity (see [10] and [24]) are also discussed there.

6.2 Definitions

In this section, we introduce the definition of a copula, which is one of the basic definitions in this book. We will also discuss the basic properties of copulas. Avoiding technical details, we will try to maintain a certain level of mathematical formalism.

Figure 6.1 Rectangular area.

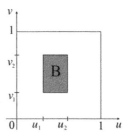

6.2.1 Quasi-Monotonicity

We introduce the following notation: $I = [0; 1]$ is the unit segment; $I^2 = [0, 1] \times [0, 1]$ is the unit square. For any $u_1 \leq u_2$, $v_1 \leq v_2$ $B = [u_1, u_2] \times [v_1, v_2]$ is a rectangular region in the plane as shown in Figure 6.1.

Let $A(u, v)$ be a function from I^2 to I and B be a rectangular region in the unit square. We will define A-volume of the region B as the value

$$V_A(B) = A(u_2, v_2) - A(u_1, v_2) - A(u_2, v_1) + A(u_1, v_1). \tag{6.2}$$

Let $\Delta_{u_1}^{u_2}$ and $\Delta_{v_1}^{v_2}$ represent the first consecutive difference:

$$\Delta_{u_1}^{u_2} A(u, v) = A(u_2, v) - A(u_1, v), \Delta_{v_1}^{v_2} A(u, v) = A(u, v_2) - A(u, v_1).$$

Then we may rewrite the definition of A-volume as follows:

$$V_A(B) = \Delta_{v_1}^{v_2} \Delta_{u_1}^{u_2} A(u, v).$$

Definition 6.2.1 Function $A(u, v)$ is called ***quasi-monotone*** if for any rectangular area B in the unit square its A-volume is nonnegative.

Example 6.2.1 Let $A(u, v) = uv$, then for rectangular area B we obtain $V_A(B) = (u_2 - u_1)(v_2 - v_1) \geq 0$ so this function is quasi-monotone.

Quasi-monotonicity often corresponds to the monotonicity in both arguments. However, the following two examples demonstrate that these two definitions of monotonicity are not equivalent, and none of them is more general than the other.

Example 6.2.2 Let $A(u, v) = max(u, v)$. It is obvious that this function increases in each argument, but it is not quasi-monotone. For instance, for rectangle with vertices $A(0.2, 0.4)$, $B(0.2, 0.9)$, $C(0.6, 0.9)$, $D(0.6, 0.4)$ A-volume is equal to -0.2.

Example 6.2.3 Let $A(u, v) = (2u - 1)(2v - 1)$, then for rectangular area B we obtain $V_A(B) = 4(u_2 - u_1)(v_2 - v_1) \geq 0$. This function is quasi-monotone, but for some values of u, for instance for $u = 0.1$, this function decreases in v.

Definition 6.2.2   Function $A(u, v)$ is **grounded** on I^2 if $A(0, v) = A(u, 0) = 0$ for any $u, v \in I$.

Lemma 6.1   *Any grounded nonnegative quasi-monotone function on I^2 is increasing in each argument.*

We will not prove this lemma. For a more rigorous mathematical exposure we can recommend Nelsen [25], [27], [28], or a recently published book by Durante and Sempi [8].

6.2.2   Definition of Copula

Definition 6.2.3   Function $C : I^2 \to I$ is called a **copula** if it is grounded, quasi-monotone, and for any $u, v \in [0, 1]$ $C(u, 1) = u$, $C(1, v) = v$.

In other words, a function $C(u, v)$ is a copula if it satisfies four conditions:

1. $C : I^2 \to I$.
2. For any $u, v \in [0, 1]$ $C(0, v) = C(u, 0) = 0$.
3. For any $u, v \in [0, 1]$ $C(1, v) = v$, $C(u, 1) = u$.
4. For any $0 \le u_1 \le u_2 \le 1$, $0 \le v_1 \le v_2 \le 1$, $V_C(B) = C(u_2, v_2) - C(u_2, v_1) - C(u_1, v_2) + C(u_1, v_1) \ge 0$.

   For any copula $C(u, v)$ partial derivatives $\frac{\partial C}{\partial u}$ and $\frac{\partial C}{\partial v}$ exist for almost all $u, v \in [0; 1]$. Let $\frac{\partial^2 C}{\partial u \partial v}$ and $\frac{\partial^2 C}{\partial v \partial u}$ exist and be continuous on I^2. Then function

$$c(u, v) = \frac{\partial^2 C}{\partial u \partial v} = \frac{\partial^2 C}{\partial v \partial u}$$

is a **copula density**.

From the definition, it is evident that if $u = F(x)$ and $v = G(y)$ are two distribution functions, then any copula $C(u, v) = C(F(x), G(y))$ is a valid bivariate distribution function. Joint probability density function of X and Y can be represented as

$$f(x, y) = \frac{\partial^2 C}{\partial u \partial v} \cdot \frac{dF}{dx} \cdot \frac{dG}{dy},$$

where $\frac{dF}{dx}$ and $\frac{dG}{dy}$ are marginal densities of X and Y.

What is more important though, is that the converse is also true. Every joint c.d.f. is a copula!

6.2.3   Sklar's Theorem

The following statement is known as Sklar's theorem [30], see also Sklar [31]. In the previous section we mentioned the converse. We will use the form of this theorem provided by Nelsen [25].

Theorem 6.1 *Let H be a joint distribution function with margins F and G. Then there exists a copula C such that for all x, y*

$$H(x, y) = C(F(x), G(y)). \tag{6.3}$$

If F and G are continuous, then C is unique.

The role of Sklar's theorem is that not just every copula function with marginal distributions as arguments is a valid bivariate distribution. It states that every valid bivariate distribution can be represented as a copula of its marginals. First of all, it allows for effective separation of marginal modeling from the modeling of dependence, which is a big procedural simplification. Then, it means that if we have to build a model for bivariate distribution with given marginals, the only problem is to find the proper copula which exists according to Sklar's theorem and is often unique. However, as we will see soon, this is not a trivial task.

6.2.4 Survival Copulas

Let $C(u, v)$ be a copula defining a joint distribution with marginals u and v. The following function

$$\bar{C}(u, v) = u + v - 1 + C(1 - u, 1 - v) \tag{6.4}$$

is called a ***survival*** copula. Survival copulas satisfy all four copula properties in Definition 6.2.3.

Let X be a random variable with distribution function $F(x)$ and Y be a random variable with distribution function $G(y)$. So $S_1(x) = 1 - F(x) = P(X > x)$ and $S_2(y) = 1 - G(y) = P(Y > y)$ are survival functions. Let $S(x, y)$ be a joint survival function: $S(x, y) = P(X > x, Y > y)$. Then one may check whether

$$S(x, y) = \bar{C}(S_1(x), S_2(y)).$$

This representation of the joint survival function as a survival copula based on marginal survival functions provides a useful alternative to (6.2.2) in cases when survival functions are easier or more natural to use than c.d.f.'s (e.g., exponential or Weibull distributions).

Let us also consider two transformations on the set of bivariate functions, where $C(u, v)$ is a copula:

$$\widetilde{C}(u, v) = u + v - C(u, v),$$
$$\widehat{C}(u, v) = 1 - C(1 - u, 1 - v).$$

The first transformation is called *dual* copula and the second is known as *co-copula*, but neither of these is strictly speaking a copula satisfying all four conditions of Definition 6.2.3.

6.3 Simplest Pair Copulas

In this section, we will consider some simple examples of copulas. First of all, we will establish upper and lower bounds on the class of copulas known as *Frechet–Hoeffding bounds* or the *maximum copula* and the *minimum copula*. If we imagine all copulas as surfaces spanned over the unit square I^2, these two bounds would represent the uppermost and the lowermost such surfaces, located so that all the other copula surfaces are squeezed in between these two. Then we will consider some special cases corresponding to certain types of dependence between the margins (such as *product copulas*) or convenient functional form of the copula representation $C(u, v)$ such as FGM or RLUF copula families.

6.3.1 Maximum Copula

Let us consider the function $M(u, v) = \min(u, v)$ on the unit square I^2. It is obvious that this function satisfies the first three conditions from the definition of a copula: $M(u, v) \in [0; 1]$, $M(u, 0) = M(0, v) = 0$, $M(u, 1) = u$, and $M(1, v) = v$ for any $u, v \in [0; 1]$. To check the fourth condition it is necessary to consider different cases of the location of a rectangular region B with respect to the diagonal $u = v$ of the unit square. There are three possible cases one should consider.

The first case assumes that all four vertices lie on one side of the diagonal as in Figure 6.2. The legend next to the vertices corresponds to the values of the function. In this case $V_M(B) = u_2 - u_2 - u_1 + u_1 = 0$.

The second case allows three vertices to lie on one side of the diagonal, and the fourth vertex to lie on the other side as in Figure 6.3. In this case $V_M(B) = u_2 - u_1 - v_1 + u_1 = u_2 - v_1 \geq 0$.

The third case is when two vertices lie on one side of the diagonal and the other two lie on the other side as in Figure 6.4. In this case $V_M(B) = u_2 - u_1 - v_1 + v_1 = u_2 - u_1 \geq 0$.

So function $M(u, v)$ represents a copula which is called *maximum copula*. Figure 6.5 depicts a three-dimensional image and contour plot corresponding to this copula surface.

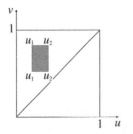

Figure 6.2 Maximum copula, case 1.

Figure 6.3 Maximum copula, case 2.

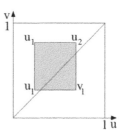

Figure 6.4 Maximum copula, case 3.

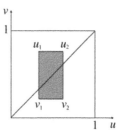

Let X be a continuous random variable. Let $\phi(x)$ be an increasing differentiable function and $Y = \phi(X)$. According to Sklar's theorem there exists a copula which models the dependence between X and Y. One may check that this copula is $M(u, v)$. Notice that, therefore, the maximum copula models a direct monotonic functional relationship between X and Y, which could be both linear (as in the case of perfect positive linear correlation $\rho = 1$) and nonlinear. This constitutes a significant difference from Pearson's correlation, which as discussed in Chapter 5 captures the linear dependence but fails to detect nonlinear relationships between the dependent variables.

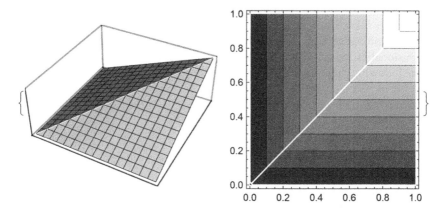

Figure 6.5 Maximum copula, three-dimensional surface, and contour plot.

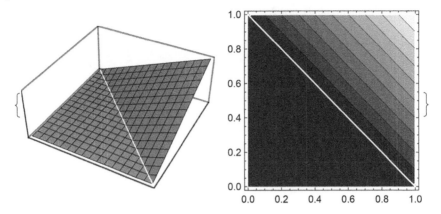

Figure 6.6 Minimum copula, three-dimensional surface, and contour plot.

6.3.2 Minimum Copula

Let us consider the function $W(u, v) = \max(u + v - 1, 0)$ on the unit square I^2 and call it the ***minimum copula***. It is easy to check that this function satisfies all copula properties. Figure 6.6 shows the three-dimensional surface and contour plot for this copula surface.

Let X be a continuous random variable. Also let $\phi(x)$ be a decreasing differentiable function and $Y = \phi(X)$. According to Sklar's theorem there exists a copula which models the dependence between X and Y. One may check that this copula is $W(u, v)$. Thus the minimum copula provides another model for direct negative relationship between X and Y, which could be both linear (as in the case of perfect negative linear correlation $\rho = -1$) and nonlinear.

The case of two independent random variables X and Y corresponds to the joint distribution function being the product of the margins. This copula $P(u, v) = uv$ is known as the ***product copula***. It is straightforward to check that this function satisfies all the copula properties. Figure 6.7 depicts the three-dimensional image and contour plot for the product copula.

The following theorem follows from the condition of quasi-monotonicity.

Theorem 6.2 *For any copula C and any $u, v \in [0, 1]$ the following inequality holds*

$$W(u, v) \leq C(u, v) \leq M(u, v). \tag{6.5}$$

Proof: Let us illustrate the idea of the proof by considering a rectangle B with vertices $A(u, v)$, $B(1, v)$, $C(1, 1)$, $D(u, 1)$. For this rectangle and for any copula $C(u, v)$ it is true that $V_C(B) = 1 - u - v + C(u, v) \geq 0$, so $C(u, v) \geq u + v - 1$, and of course $C(u, v) \geq 0$. So we obtain $C(u, v) \geq W(u, v)$. ∎

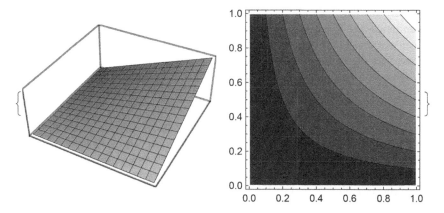

Figure 6.7 Product copula, three-dimensional surface, and contour plot.

6.3.3 FGM Copulas

What would be the simplest examples of copula construction? One of the most important classes of functions is the class of polynomials. An attractive feature of polynomials is the ease of basic calculus operations on polynomials: differentiation and integration. Moreover, it is important that many other classes of functions allow for a convenient polynomial approximation. Let us try to construct simplest polynomial copulas. From condition $C(u, 0) = C(0, v) = 0$ we obtain $C(u, v) = uvQ(u, v)$, where $Q(u, v)$ is a bivariate polynomial. If $Q(u, v)$ is a constant it must be equal to one, so the only result is the product copula. It is easy to check that $Q(u, v)$ cannot be a linear function. That would violate the definition. So let $Q(u, v)$ be the next in line: a quadratic polynomial. Using conditions $C(1, v) = v$ and $C(u, 1) = u$, we obtain

$$Q(u, v) = 1 + \alpha(1 - u)(1 - v),$$
$$C(u, v) = uv(1 + \alpha(1 - u)(1 - v)).$$

It is easy to check that for a rectangle B with vertices u_i, v_i

$$V_C(B) = (u_2 - u_1)(v_2 - v_1)(1 + \alpha(1 - u_2 - u_1)(1 - v_2 - v_1)).$$

Let us assume that $u_2 = v_2 = t$, $u_1 = v_1 = t^2$ and t approaches 0. Quasi-monotonicity yields $\alpha \geq -1$. Let us assume now that $u_2 = t, u_1 = t^2, v_2 = 1 - t$, $v_1 = (1 - t)^2$ and t approaches 0. Quasi-monotonisity results in $\alpha \leq 1$.

Thus we obtain the following class of copulas depending on parameter α

$$C(u, v) = uv(1 + \alpha(1 - u)(1 - v)), \quad -1 \leq \alpha \leq 1. \tag{6.6}$$

This parametric family of copulas is known as ***Farlie–Gumbel–Morgenstern*** class abbreviated as FGM copulas. This is an easy and practical choice for modeling weaker dependence, but it is not particularly suitable for stronger dependence (see Exercises at the end of the chapter). Comparing (6.6) to (5.23) we can see that FGM copulas may naturally appear in different approaches to multivariate distribution modeling.

We may consider some generalization of this class such as

$$C(u, v) = uv(1 + \alpha f(u)g(v)), \tag{6.7}$$

where parameter α and functions $f(u)$ and $g(v)$ must satisfy some additional conditions. This class is known as **Rodriguez Lallena–Ubeda Flores copulas** or RLUF copulas. It allows for asymmetry in u and v when different functions $f(u) \neq g(v)$ are used in representation (6.7). This is convenient for the description of the so-called directional dependence [32].

6.4 Elliptical Copulas

In this section, we consider one popular class of copulas which is often used in applications. These copulas are called *elliptical* copulas. The class of elliptical copulas includes such two important cases as Gaussian (normal) and Student t-copulas.

6.4.1 Elliptical Distributions

In the previous section we considered some examples of copula families. With the exclusion of the RLUF family, all of the copulas we considered were symmetric with respect to the main diagonal of the unit square $u = v$. In this section we will require the copulas to demonstrate elliptical symmetry, which will cause the copula functions to be also symmetric with respect to the diagonal $u = 1 - v$. This will result in the identity

$$C(u, v) = \bar{C}(u, v),$$

so that an elliptically symmetric copula coincides with its survival version.

We will consider the class of bivariate elliptical distributions $Q_\rho(s, t)$, defined by their density functions

$$q_\rho(s, t) = \frac{k^2}{\sqrt{1 - \rho^2}} g\left(\frac{s^2 - 2\rho st + t^2}{1 - \rho^2}\right).$$

Here $\rho \in (-1; 1)$, function $g : \mathbf{R} \to \mathbf{R}^+$ is such that $\int_{-\infty}^{\infty} g(t)dt < \infty$, and k is the normalizing constant. Consider also

$$Q(t) = \int_{-\infty}^{\infty} q_0(s, t)ds = \int_{-\infty}^{\infty} q_0(t, s)ds$$

as the marginal distribution of the first and second components of the vector (s, t) corresponding to $\rho = 0$ with symmetric density $q(t) = kg(t^2)$.

The elliptical class includes normal distribution family, Student t-distribution family including Cauchy distribution, logistic distribution, and Laplace distribution [25].

6.4.2 Method of Inverses

One of the possible methods of building an elliptically symmetric copula using an elliptic distribution $Q_\rho(s, t)$ is the method of inverses. Using the fact that for U and V independently uniformly distributed on $[0, 1]$, the inverse transforms $Q^{-1}(U)$ and $Q^{-1}(V)$ are two independent random variables with the same c.d.f. $Q(t)$, we will define an ***elliptical copula*** for any $u, v \in [0, 1]$ as

$$C_\rho(u, v) = Q_\rho(Q^{-1}(u), Q^{-1}(v)).$$

The copula density will assume the form

$$c_\rho(u, v) = \frac{q_\rho(s, t)}{q(s)q(t)},$$

where $s = Q^{-1}(u)$, $t = Q^{-1}(v)$.

This construction is attractive because it allows for effective separation of the marginal distributions $u = F(x)$ and $v = G(y)$, which could be chosen at will, from the structure of dependence which corresponds to a specific elliptical distribution. This way a bivariate joint distribution $H(x, y)$ with margins $F(x)$ and $G(y)$ can be modeled as

$$H(x, y) = Q_\rho(Q^{-1}(F(x)), Q^{-1}(G(y))), \tag{6.8}$$

where the parameter ρ is responsible for the strength of dependence.

6.4.3 Gaussian Copula

The Gaussian copula and the Student copula are the ones most frequently used in applications. We obtain the Gaussian copula if $H(x, y)$ in the formula (6.8) is constructed with the help of the distribution function of the bivariate normal distribution $\Phi_\rho(s, t)$ with zero means, unit variances, and correlation between the components ρ, and $\Phi(t)$ is the standard normal c.d.f. So we obtain

$$C_\rho(u, v) = \Phi_\rho(\Phi^{-1}(u), \Phi^{-1}(v)). \tag{6.9}$$

Denoting $s = \Phi^{-1}(u)$ and $t = \Phi^{-1}(v)$, we may write down the density of the Gaussian copula as

$$c_\rho(u, v) = \frac{\partial^2 C_\rho(u, v)}{\partial u \partial v} = \frac{\partial^2 \Phi_\rho(s, t)}{\partial s \partial t} \cdot \frac{\partial s}{\partial u} \cdot \frac{\partial t}{\partial v} = \frac{\phi_\rho(s, t)}{\phi(s)\phi(t)}, \tag{6.10}$$

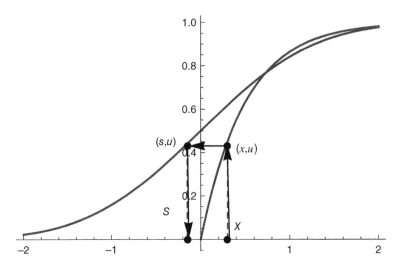

Figure 6.8 Transformation of variables for Gaussian copula.

where $\phi_\rho(s, t)$ is the density function of the bivariate normal distribution with zero means, unit variances, and correlation ρ, $\phi(t)$ is the density of the standard normal distribution. Then we can explicitly obtain

$$c_\rho(u, v) = \frac{1}{\sqrt{1 - \rho^2}} \exp \left(-\frac{\rho^2 s^2 + \rho^2 t^2 - 2\rho st}{2(1 - \rho^2)} \right). \tag{6.11}$$

For modeling joint distributions we can combine Gaussian copula with any marginal distributions $u = F(x)$ and $v = G(y)$. Figure 6.8 represents the transformation of variables $x \to u \to s$. Similarly one can transform $y \to v \to t$. As we can see from the picture, the idea of Gaussian copula is to transform two random variables X and Y with respective c.d.f.'s F and G into standard normal variables $S = \Phi^{-1}(F(X))$ and $T = \Phi^{-1}(G(V))$. Then the dependence between X and Y is expressed in terms of the dependence structure of their normal transformations S and T, therefore it can be reduced to linear correlation (see Section 5.4). With this approach, nonlinear dependence between X and Y is expressed through the linear dependence of their standard normal transforms.

Joint Life
A joint life study in [29] considers copula approach to modeling joint husband/wife mortality in the problem introduced in Section 5.1. The statistical analysis of more than 11,000 married couples, which will be discussed as a case

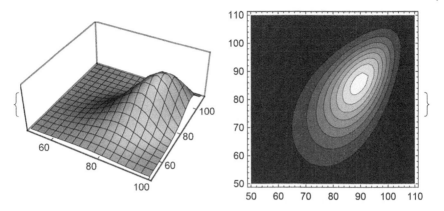

Figure 6.9 Density: Gaussian copula, Weibull margins (6.12) and (6.13).

study in Section 7.5, suggests Weibull marginal distributions for female mortality

$$F(x) = 1 - \exp\left(-(x/\theta_x)^{\tau_x}\right), \ \tau_x = 9.96, \theta_x = 89.51, \tag{6.12}$$

and male mortality

$$G(y) = 1 - \exp\left(-(y/\theta_y)^{\tau_y}\right), \ \tau_y = 7.65, \theta_x = 85.98. \tag{6.13}$$

In Figure 6.9 one may see a three-dimensional surface graph and contour plot of the p.d.f.

$$h(x, y) = \frac{\partial^2}{\partial x \partial y} H(x, y) = \frac{dF}{dx} \cdot \frac{dG}{dy} c_\rho(F(x), G(y))$$

corresponding to the joint distribution $H(x, y)$ defined as in (6.8) by Gaussian copula with $\rho = 0.575$ and Weibull margins $u = F(x)$ and $v = G(y)$.

To illustrate the results numerically, we can use (6.8) to obtain $H(65, 65) = 0.0139$ and $H(55, 55) = 0.0016$. In insurance-related joint life studies special attention is often paid to the left tails corresponding to relatively short lives of both spouses creating high insurance risks. Copulas help us to express the positive dependence of related lives. For the sake of comparison, the assumption of independence of two lives corresponding to product copula brings about much lower left tail probabilities $H(65, 65) = 0.0045$ and $H(55, 55) = 0.0003$.

6.4.4 The *t*-copula

In applications it is often necessary to model heavy-tailed multivariate distributions and tails of the joint distributions. In this situation, as in the one-dimensional case, the Student t-copula may be used instead of the Gaussian copula. If we use bivariate Student t-distribution with η degrees of freedom and correlation coefficient ρ in formula (6.8) we obtain the Student copula or *t-copula*. As we saw for the Gaussian copula, the choice of an elliptical copula model does not prescribe the choice of marginals. They might be chosen separately.

$$C_{\eta\rho}(u, v) = T_{\eta\rho}(T_\eta^{-1}(u), T_\eta^{-1}(v)). \tag{6.14}$$

For inverse transform we use two univariate t-distributions with η degrees of freedom. Let us denote $s = T_\eta^{-1}(u)$ and $t = T_\eta^{-1}(v)$. Similarly to the case of Gaussian copulas we obtain

$$c_{\eta\rho}(u, v) = \frac{\partial^2 C_{\eta\rho}(u, v)}{\partial u \partial v} = \frac{\psi_{\eta\rho}(s, t)}{\psi_\eta(s)\psi_\eta(t)}. \tag{6.15}$$

Here $\psi_{\eta\rho}(s, t)$ is bivariate t-distribution density and $\psi_\eta(x)$ is univariate t-distribution density with the same number of degrees of freedom. Then we obtain an explicit formula for the copula density

$$c_{\eta\rho}(u, v) = \frac{\Gamma\left(\frac{\eta+2}{2}\right)\Gamma\left(\frac{\eta}{2}\right)}{\sqrt{1-\rho^2}\Gamma^2\left(\frac{\eta+1}{2}\right)} \cdot \frac{\left(\left(1+\frac{s^2}{\eta}\right)\left(1+\frac{t^2}{\eta}\right)\right)^{\frac{\eta+1}{2}}}{\left(1+\frac{s^2+t^2-2\rho st}{\eta(1-\rho^2)}\right)^{\frac{\eta+2}{2}}}. \tag{6.16}$$

For modeling joint distributions we can combine Student copula with any marginal distributions $u = F(x)$ and $v = G(y)$. Figure 6.10 demonstrates the three-dimensional graph and contour plot of the model (6.8) corresponding to Student copula with $\eta = 1$, $\rho = 0.575$, and Weibull marginals (6.12) and (6.13) applied to the same joint life problem as before.

Numerical results yield $H(65, 65) = 0.0290$ and $H(55, 55) = 0.0058$. Both values corresponding to relatively short lives (left tails) are higher than in the case of Gaussian copula. This can be also observed in Figure 6.10, especially in the contour plot.

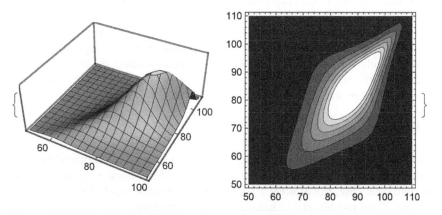

Figure 6.10 Student t-copula, Weibull margins (6.12) and (6.13).

It is possible to consider a generalization of the Student t-copula

$$C(u, v, \eta_1, \eta_2, \eta_3, \rho) = T_{\eta_1 \rho} \left(T_{\eta_2}^{-1}(u), T_{\eta_3}^{-1}(v) \right). \tag{6.17}$$

allowing for different numbers of degrees of freedom for two components. Another possibility to build a more flexible model based on t-copula structure is to consider "skewed t-distributions" which would allow for two asymmetric tails.

6.5 Archimedean Copulas

6.5.1 Definitions

We can think of a bivariate distribution function with the dependence structure represented by a copula $C(u, v)$ with marginals u and v as a certain deviation from the "ideal" independence case corresponding to a product copula $C(u, v) = uv$. Via logarithmic transform, we can suggest additive structure of transformed independence copula: $\ln C = \ln u + \ln v$. Since u, v, and C are all between 0 and 1, negative logarithms would bring about similar expression, where all terms are positive. It is tempting to determine a subclass of copulas using certain transformations ϕ which support additivity: $\phi(C) = \phi(u) + \phi(v)$. Mathematically, such copulas will be easier to manipulate, mostly because of attractively simple differentiation and integration of additive expressions. The following definitions will suggest such a construction, discussed more rigorously by Genest and MacKay in [11] and [12].

Let $\phi(t)$ be a continuous, strictly decreasing function from $[0, 1]$ to $[0, \infty)$ such that $\phi(1) = 0$. We determine the *pseudo-inverse* of ϕ as follows:

$$\phi^{[-1]}(t) = \max\{\phi^{-1}(t), 0\}. \tag{6.18}$$

So $\phi^{[-1]}(t) = \phi^{-1}(t)$ if $0 \leq t \leq \phi(0)$ and $\phi^{[-1]}(t) = 0$ if $t > \phi(0)$. Pseudo-inverse is additionally defined for the values out of the range of the original function, which makes it different from the regular inverse. If $\phi(t) \to \infty$ when $t \to 0$, then the pseudo-inverse function coincides with the inverse function. Pseudo-inverses serve to extend the inverse transformation to the functions of limited range. Figure 6.11 shows the graphs of function ϕ with range $[0, \infty)$ and its inverse function. Figure 6.12 depicts the graphs of function $\phi = 1 - t$ with restricted range $[0, 1]$ and its pseudo-inverse function.

Now we consider the main theorem in this section.

Theorem 6.3 [25] *Let $\phi : I \to [0, \infty)$ be a continuous, strictly decreasing function such that $\phi(1) = 0$. Then the function*

$$C_\phi(u, v) = \phi^{[-1]}(\phi(u) + \phi(v)) \tag{6.19}$$

is a copula if and only if ϕ is convex.

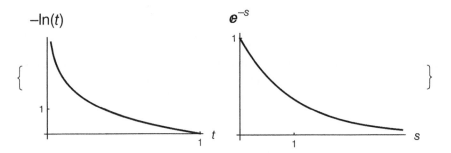

Figure 6.11 Graphs of $\phi(t) = -\ln(t)$ and its inverse.

The proof of this theorem can be found in [25]. If a copula $C_\phi(u, v)$ satisfies conditions of Theorem 6.7.1, it is called an ***Archimedean copula*** and the function $\phi(t)$ is its ***generator***.

How can one check whether an arbitrary copula is Archimedean? Let us denote by $\delta_C(u) = C(u, u)$ the diagonal projection of a copula. If for all $u \in (0, 1)$ $\delta_C(u) < u$ then $C(u, v)$ is an Archimedean copula. We will also notice two important algebraic properties of Archimedean class. It is obvious that Archimedean copulas are commutative: $C(u, v) = C(v, u)$. But it is much harder to verify that they are associative: $C(C(u, v), w) = C(u, C(v, w))$ [25], though this property will play an important role in some further constructions.

Let the second derivative $\phi''(t)$ exist. Then the density of an Archimedean copula can be expressed through its generator and its derivatives as

$$c(u, v) = \frac{\partial^2 C(u, v)}{\partial u \partial v} = -\frac{\phi''(C(u, v))\phi'(u)\phi'(v)}{(\phi'(C(u, v)))^3}. \tag{6.20}$$

6.5.2 One-Parameter Copulas

In this subsection we give simple examples of Archimedean copulas. We choose a simple generator and then apply Theorem 6.3 from the previous subsection.

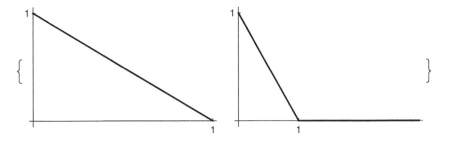

Figure 6.12 Graphs of $\phi(t) = 1 - t$ and its pseudo-inverse.

Example 6.5.1   Let us consider generator $\phi(t) = -\ln t$, then $\phi^{-1} = \exp(-t)$. Graphs of these two functions were shown in Figure 6.11. Using this generator we obtain the copula

$$C_\phi(u, v) = \exp\{-(-\ln u - \ln v)\} = uv = P(u, v).$$

This is the product copula from Section 6.3, corresponding to the case of two independent variables.

Example 6.5.2   Let us consider generator $\phi(t) = 1 - t$. Then the pseudo-inverse is $\phi^{[-1]}(s) = \max\{1 - s; 0\}$. Graphs of these two functions were shown in Figure 6.12. Using this generator we obtain the copula

$$C_\phi(u, v) = \max\{1 - (1 - u) - (1 - v); 0\} = \max\{u + v - 1; 0\} = W(u, v),$$

which we know from Section 6.3 as the minimum copula.

Example 6.5.3   We can also consider generator $\phi(t) = t^{-1} - 1$. Then the pseudo-inverse is $\phi^{[-1]}(s) = (1 + s)^{-1}$, which is positive for all $s > 0$, and no additional restrictions are required. Using this generator we obtain the copula $C_\phi(u, v) = (u^{-1} + v^{-1} - 1)^{-1}$.

Example 6.5.4   Let us consider the function $\phi(t) = \tan(\pi(1 - t)/2)$ and check whether this function is a generator. We calculate

$$\phi'(t) = -\frac{\pi}{\cos(\pi(1 - t)) + 1} < 0,$$

and

$$\phi''(t) = \frac{\pi^2 \sin(\pi(1 - t))}{(\cos(\pi(1 - t)) + 1)^2} > 0.$$

So function ϕ is a generator and its inverse function is $\phi^{-1}(t) = 1 - \frac{2}{\pi} \arctan t$. Graphs of the generator and its inverse function are depicted in Figure 6.13. The

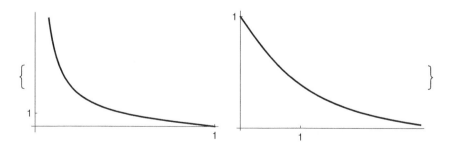

Figure 6.13 Graphs of generator $\phi(t) = \tan(\pi(1 - t)/2)$ and its inverse.

resulting copula is

$$C(u, v) = 1 - (2/\pi) \arctan(\tan(\pi(1 - u)/2) + \tan(\pi(1 - v)/2)).$$

Three one-parameter families of copulas from the Archimedean class are used in applications more frequently than the others: Clayton copula, Frank copula, and Gumbel–Hougaard copula. Now we write down the expressions for these copulas, their generators, and their densities.

6.5.3 Clayton Copula

The generator

$$\phi(t) = \frac{1}{\alpha}(t^{-\alpha} - 1)$$

with the pseudo-inverse

$$\phi^{[-1]}(s) = \max\{(1 + \alpha s)^{-1/\alpha}, 0\}$$

is used to define Clayton's class of copulas:

$$C_\alpha(u, v) = \max\{(u^{-\alpha} + v^{-\alpha} - 1)^{-1/\alpha}, 0\}, \ \alpha \in [-1; 0) \cup (0; \infty). \tag{6.21}$$

This copula is typically applied for $\alpha > 0$, then

$$C_\alpha(u, v) = (u^{-\alpha} + v^{-\alpha} - 1)^{-1/\alpha}, \ \alpha > 0. \tag{6.22}$$

Two of the basic examples considered above correspond to particular cases of Clayton's class: Example 6.5.2 to $\alpha = -1$ and Example 6.5.3 to $\alpha = 1$.

In this situation the copula density is given by the formula

$$c_\alpha(u, v) = \frac{(\alpha + 1)(uv)^\alpha}{(u^\alpha + v^\alpha - (uv)^\alpha)^{\frac{1}{\alpha}+2}}, \ \alpha > 0. \tag{6.23}$$

Figure 6.14 demonstrates a three-dimensional graph and a contour plot of the joint density $h(x, y)$ of the distribution $H(x, y)$ defined by Clayton copula and Weibull marginals corresponding to the joint life example. The value of copula parameter is chosen as $\alpha = 1.28$.

Sample numerical results yield $H(65, 65) = 0.0338$ and $H(55, 55) = 0.0069$. Left tail is visibly seen on the graphs, especially in the contour plot, which explains values higher than for Gaussian copulas. If we want to be conservative in evaluation of insurance risks, we usually prefer overestimation of the left tail probabilities to their underestimation. Comparing (6.22) to (5.19) we can see that Clayton's copulas also arise in noncopula context as shared frailty models. Indeed, historically model (6.22) based on shared frailty predates the general theory of copulas.

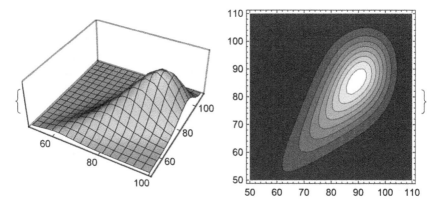

Figure 6.14 Clayton copula with Weibull margins (6.12) and (6.13).

6.5.4 Frank Copula

The generator

$$\phi(t) = -\ln \frac{e^{-\alpha t} - 1}{e^{-\alpha} - 1}$$

with the pseudo-inverse

$$\phi^{[-1]}(s) = -\frac{1}{\alpha} \ln[1 + e^{-s}(e^{-\alpha} - 1)]$$

brings about Frank copula

$$C_\alpha(u, v) = -\frac{1}{\alpha} \ln \left(1 + \frac{(e^{-\alpha u} - 1)(e^{-\alpha v} - 1)}{e^{-\alpha} - 1} \right), \quad \alpha \neq 0. \tag{6.24}$$

The density of this copula is

$$c_\alpha(u, v) = \frac{\alpha(1 - e^{-\alpha})e^{-\alpha(u+v)}}{(e^{-\alpha} - 1 + (e^{-\alpha u} - 1)(e^{-\alpha v} - 1))^2}. \tag{6.25}$$

Figure 6.15 demonstrates a three-dimensional graph and a contour plot of the joint density $h(x, y)$ of the distribution $H(x, y)$ defined by Frank copula and the same Weibull marginals as in the previous example.

Sample numerical results yield tail values $H(65, 65) = 0.0039$ and $H(55, 55) = 0.0022$ for Frank copula. Tails are not strongly expressed on the contour plot, which corresponds to relatively low tail values. The following copula is also frequently called Gumbel copula, but we will use two names in its title to avoid possible confusion. There exist different copulas, also named after Gumbel, though they will not be discussed in this book.

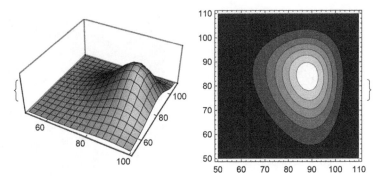

Figure 6.15 Frank copula with Weibull margins (6.12) and (6.13).

6.5.5 Gumbel–Hougaard Copula

The generator

$$\phi(t) = (-\ln t)^\alpha$$

with pseudo-inverse

$$\phi^{[-1]}(s) = e^{-s^{1/\alpha}}$$

defines

$$C_\alpha(u, v) = \exp\left(-((-\ln u)^\alpha + (-\ln v)^\alpha)^{1/\alpha}\right), \ \alpha \geq 1. \tag{6.26}$$

The density of this copula is

$$c_\alpha(u, v) = (uv)^{-1}(\ln u \cdot \ln v)^{\alpha-1}(w^{2/\alpha-2} + (\alpha - 1)w^{1/\alpha-2})C_\alpha(u, v),$$
$$w = (-\ln u)^\alpha + (-\ln v)^\alpha. \tag{6.27}$$

Figure 6.16 demonstrates a three-dimensional graph and a contour plot of the joint density $h(x, y)$ of the distribution $H(x, y)$ defined by Gumbel–Hougaard copula and the same Weibull marginals as in the two previous examples.

The survival Gumbel–Hougaard copula is also very popular in applications, especially in combination with exponential or Weibull margins, in which case its algebraic expression is fairly simple. This copula is determined by the formula

$$\bar{C}_\alpha(u, v) = u + v - 1 + \exp\left(-((-\ln(1 - u))^\alpha + (-\ln(1 - v))^\alpha)^{1/\alpha}\right). \tag{6.28}$$

Sample numerical results yield $H(65, 65) = 0.0154$ and $H(55, 55) = 0.0017$ for Gumbel–Hougaard copula, which has a visible right tail (values concentrated in the upper right corner of the graph). However the survival copula rotates the picture by 180 degrees bringing about an expressed lower tail: $H(65, 65) = 0.0295$ and $H(55, 55) = 0.0060$ for survival version. That makes survival Gumbel–Hougaard a good model for capturing the left tail behavior.

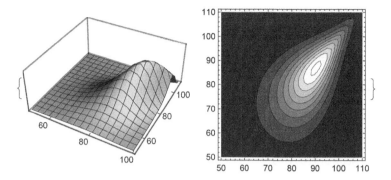

Figure 6.16 Gumbel–Hougaard copula with Weibull margins (6.12) and (6.13).

Comparing (6.26) to (5.20) we can see that this copula along with Clayton copula may arise in noncopula context as a shared frailty model.

These three copulas are widely used because they allow for modeling various types of non-linear dependence, especially the tail dependence, and are flexible enough for many diverse applications. The Clayton copula is used in situations where the dependence between low values of u and v (left tail) is stronger than the dependence between values close to 1. The Frank copula is used when the strength dependence is relatively similar for all values of u and v. The Gumbel–Hougaard copula is used when a stronger dependence is observed for values of u and v close to 1.

To illustrate these points without having to take into account the effect of the marginals, we also build the graphs of the diagonal cross-sections of the densities $\delta_\alpha(u) = c_\alpha(u, u)$. We present these graphs in Figure 6.17 for the following values of the copula parameters: for Clayton copula (solid line) and Gumbel–Hougaard copula (dotted line) $\alpha = 2$, for Frank copula (dashed line) $\alpha = 5.737$. As we will see later, all these values characterize the same overall strength of dependence, corresponding to the value of the Kendall's concordance $\tau = 0.5$.

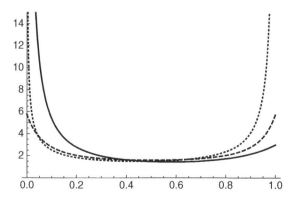

Figure 6.17 Diagonal cross-sections for Clayton, Frank, and Gumbel–Hougaard copulas.

Table 6.1 Copula results for joint life distribution

Copula	H(55, 55)	H(65, 65)
Independent	0.0003	0.0045
Gaussian	0.0016	0.0139
Student's t	0.0058	0.0290
Clayton's	0.0069	0.0338
Frank's	0.0022	0.0039
Gumbel–Hougaard	0.0017	0.0154
Gumbel–Hougaard (survival)	0.0060	0.0295

Before moving on, let us summarize the information obtained from using several different copulas for the joint life example. This is done in Table 6.1. The values of copula parameters were estimated by MLE as discussed in detail in Chapter 7. The main issue we want to point out is a huge variety in final numerical results depending on the particular choice of copula family.

6.5.6 Two-Parameter Copulas

Let us suppose that there exists the second derivative $\phi''(t)$. Then $\phi(t)$ is a generator if $\phi'(x) < 0$ and $\phi''(x) > 0$. In order to construct an Archimedean copula, we just need to specify a generator. We will consider some simple examples. The following theorem provides new methods to construct wider subclasses of Archimedean copulas based on the simplest generators.

Theorem 6.4 *If $\phi(t)$ is a generator and $\phi''(t)$ exists, then $\phi^\beta(t^\alpha)$ is also a generator if $0 < \alpha \le 1$ and $\beta \ge 1$.*

Proof: We will check the statement of the theorem using the existence of the second derivative. We denote $\psi(t) = \phi(t^\alpha)$. Then for $\alpha > 0$,

$$\psi'(t) = \alpha t^{\alpha-1}\phi'(t^\alpha) < 0,$$

and for $\alpha \le 1$,

$$\psi''(t) = \alpha^2 t^{2\alpha-2}\phi''(t^\alpha) + \alpha(\alpha-1)t^{\alpha-2}\phi'(t^\alpha) > 0.$$

We denote $\eta(t) = \phi^\beta(t)$ and then for $\beta > 0$

$$\eta'(t) = \beta\phi^{\beta-1}(t)\phi'(t) < 0$$

and for $\beta \ge 1$

$$\eta''(t) = \beta(\beta-1)\phi^{\beta-2}(t)(\phi'(t))^2 + \beta\phi^{\beta-1}(t)\phi''(t) > 0.$$

∎

Thus we obtain a two-parameter family of generators starting from a single generator ϕ. This opens a way to construct two-parameter subclass of Archimedean copulas.

Example 6.5.5 Using Theorem 6.4 we can construct a generator $\psi(t) = \phi^\beta(t^\alpha) = (t^{-\alpha} - 1)^\beta$ and $\psi^{[-1]}(s) = (1 + s^{1/\beta})^{-1/\alpha}$. So starting with the copula from Example 6.5.2, we obtain a two-parameter family with "inner power" $0 < \alpha \leq 1$ and "outer power" $\beta \geq 1$.

$$C_{\alpha,\beta}(u, v) = (1 + ((u^{-\alpha} - 1)^\beta + (v^{-\alpha} - 1)^\beta)^{1/\beta})^{-1/\alpha}. \tag{6.29}$$

This copula was introduced by Joe [17] and is known as **BB1 copula**. We will further use this name. Notice that BB1 copula is a natural extension of two one-parametric families: When $\beta = 1$, we obtain Clayton's copula, and when $\alpha \to 0$, BB1 copulas converge to Gumbel–Hougaard family (see Exercises).

Example 6.5.6 Using Theorem 6.4 we construct generators $\psi(t) = \phi^\beta(t^\alpha) = (-\log(t^\alpha))^\beta = (-\alpha \log t)^\beta$, then $\psi^{-1} = \exp\left(-\frac{1}{\alpha}t^{1/\beta}\right)$. Thus we obtain a two-parameter family of copulas

$$C_{\alpha,\beta}(u, v) = \exp\left(-\frac{1}{\alpha}((-\alpha \log u)^\beta + (-\alpha \log v)^\beta)^{1/\beta}\right). \tag{6.30}$$

This copula includes Gumbel–Hougaard class as a particular case for $\alpha = 1$. It is possible to build many interesting two-parameter families of copulas using the construction of Theorem 6.4, see for instance [18] and [26]. Two-parametric copulas allow for more flexibility in modeling tails and other important features of joint distributions.

6.6 Simulation of Joint Distributions

All discussions regarding the statistics of copulas, namely: how to select an appropriate copula family, how to estimate copula parameters, and how to validate the models will be postponed until Chapter 7. In this section we will only discuss the issue of simulation in copula models, or more precisely: how to simulate dependent random variables whose dependence structure is defined by a copula. We will consider only the simplest examples concentrating on two most important families of copulas: elliptical and Archimedean. We will also restrict ourselves to the treatment of bivariate (pair) copulas. For further reading [10], [14], and [22] can be recommended. Also see [35] for computer supported procedures.

6.6.1 Bivariate Elliptical Distributions

Simulation of elliptical copulas has a very important practical purpose. As we saw in Section 6.4, the joint c.d.f. of two variables whose dependence is expressed in terms of an elliptical copula is defined through integrals which are often intractable. Thus the only way to specify probabilities defined by the joint distribution would be Monte Carlo integration. However the simulation itself is not hard, handily utilizing the techniques of Chapters 3 and 5.

Let us begin with Gaussian copulas. Suppose that our goal is to estimate probability $H(x_0, y_0) = P(X \leq x_0, Y \leq y_0)$, where $X \sim F(x)$ and $Y \sim G(y)$ and $H(x, y) = C_\rho(F(x), G(y))$ defined in (6.9). Then we implement the following procedure with the first three steps coinciding with the simulation of bivariate normal distribution described in Section 5.6, and the particular techniques of the fourth step depend on the specifics of the marginal $F(x)$ and $G(y)$ and go along the lines of Section 3.2.

Sampling from Gaussian Copula
1. Generate independently two standard normal variables $z_1, z_2 \sim N(0, 1)$.
2. Define correlated standard normal variables as $w_1 = z_1$ and $w_2 = \rho z_1 + \sqrt{1 - \rho^2} z_2$.
3. Set $u = \Phi(w_1), v = \Phi(w_2)$.
4. Set $x = F^{-1}(u), y = G^{-1}(v)$. Exact implementation of this step depends on the distributions $F(x)$ and $G(y)$.

Repeat n times to obtain sample $(x_i, y_i), i = 1, \ldots, n \sim C_\rho(F(x), G(y))$. The proportion of the sample elements satisfying conditions $x_i \leq x_0, y_i \leq y_0$ estimates the probability in question.

Example 6.6.1 Let us consider generating a small sample size 3, which will illustrate the procedure above. Our simulation will correspond to Weibull marginals (6.12) and (6.13). If we use Gaussian copula with $\rho = 0.575$ (estimated by the paired data considered in the study), we obtain Table 6.2. Following, for instance, the last row corresponding to $i = 3$, we generate $z_1 = -1.23$

Table 6.2 Gaussian copula with Weibull margins

i	z_1	z_2	w_1	w_2	u	v	x	y
1	0.45	0.18	0.45	0.34	0.67	0.63	90.53	86.05
2	−1.61	−0.08	−1.61	−0.70	0.05	0.24	66.89	72.63
3	−1.23	0.34	−1.23	−0.16	0.11	0.44	72.15	79.94

and $z_2 = 0.34$ independently from standard normal distribution according to step 1, and then set $w_1 = z_1$ and w_2 according to step 2. We apply standard normal distribution on step 3: $\Phi(-1.23) = 0.11$ and $\Phi(-0.16) = 0.44$. Finally, apply inverse transform for Weibull distribution:

$$u = 1 - \exp\{-(x/\theta)^{1/\beta}\} \Leftrightarrow x = \theta(-\log(1-u))^{1/\beta},$$

plugging in the values of shape and scale parameters, and get the values of x and y reflecting the required dependence structure.

For Student t-copula with $H(x, y) = C_{\nu\rho}(F(x), G(y))$ defined in (6.14) we will need a slight modification requiring additional simulation from chi-square distribution, which conveniently belongs to the class of Gamma distributions thoroughly treated in Section 3.2. Here we will use the facts mentioned in Chapter 1, combining (1.13) and (1.22), that if $Z \sim N(0, 1)$ and $S \sim \chi^2(\nu)$ are respectively a standard normal variable and a chi-square variable with ν degrees of freedom, then $T = Z/\sqrt{S/\nu}$ has t-distribution with ν degrees of freedom, see also [7].

Sampling from t-Copula

1. Generate independently two standard normal variables $z_1, z_2 \sim N(0, 1)$.
2. Generate a random variable s from $\chi^2(\nu)$ independent from z_1, z_2.
3. Define correlated standard normal variables as $w_1 = z_1$ and $w_2 = \rho z_1 + \sqrt{1 - \rho^2} z_2$.
4. Set $t_1 = w_1/\sqrt{s/\nu}, t_2 = w_2/\sqrt{s/\nu}$.
5. Set $u = T_\nu(w_1), v = T_\nu(w_2)$.
6. Set $x = F^{-1}(u), y = G^{-1}(v)$. Exact implementation of this step depends on the distributions $F(x)$ and $G(y)$.

Repeat n times to obtain sample $(x_i, y_i), i = 1, \ldots, n \sim C_{\nu\rho}(F(x), G(y))$.

6.6.2 Bivariate Archimedean Copulas

In the case of Archimedean copulas there usually exist convenient formulas for direct calculation of probabilities $H(x_0, y_0) = P(X \leq x_0, Y \leq y_0)$. That makes the problem of sampling from Archimedean copulas somewhat less urgent. Besides, sampling procedures for Archimedean copulas require additional understanding of Archimedean copula structure. However, simulation from Archimedean copulas is important enough to consider at least two different basic methods. The first method of sampling from Archimedean copulas as suggested by McNeil [22] (see also [9]) utilizes the concept of Kendalls' distribution.

If we consider random variable $S = \phi(U)/(\phi(U) + \phi(V))$ for U and V uniform on $[0, 1]$, we can prove that it is also uniformly distributed on $[0, 1]$. Consider also random variable $T = C(U, V)$ and following [13] define *Kendall's distribution function* $K_C(t) = P(C(U, V) \leq t)$. It can be proven (see [9]) that S and T are independent, thus the joint distribution of S and T is the product $H(s, t) = sK_C(t)$. This factorization allows for the following procedure, which is facilitated by the fact that for Archimedean copulas there exists a simple representation of Kendall's distribution function through its generator as

$$K_C(t) = t - \frac{\phi(t)}{\phi'(t)}. \tag{6.31}$$

Assume that the marginal distributions are $U = F(x)$ and $V = G(y)$ and the generator ϕ_α corresponds to an Archimedean copula with the association parameter value α.

Sampling via Kendall's Distribution
1. Generate independently two variables s and w, uniform on $[0, 1]$.
2. Solve for $t = K_C^{-1}(w)$.
3. Set $u = \phi_\alpha^{[-1]}(s\phi_\alpha(t))$, $v = \phi_\alpha^{[-1]}((1 - s)\phi_\alpha(t))$.
4. Set $x = F^{-1}(u)$, $y = G^{-1}(v)$.

Repeat n times to obtain sample (x_i, y_i), $i = 1, \ldots, n \sim C_\alpha(F(x), G(y))$.

Example 6.6.2 If we want to apply Kendall's distribution to simulate values from the joint distribution with the same Weibull margins (6.12) and (6.13) and Clayton's copula (we choose the association parameter value $\alpha = 1.2$), we start with independent generation of two uniform variables s and w, the latter being transformed into $t = K_C^{-1}(w)$. This transformation can be achieved by solving for t a polynomial equation for Kendall's distribution for Clayton's copula:

$$K_C(t) = t - \frac{1}{\alpha}(t - t^{\alpha+1}) = \frac{(\alpha + 1)t - t^{\alpha+1}}{\alpha},$$

and then applying generator transforms on step 3

$$u = \left(1 + \frac{s}{\alpha}(t^{-\alpha} - 1)\right)^{-1/\alpha}, \quad v = \left(1 + \frac{(1 - s)}{\alpha}(t^{-\alpha} - 1)\right)^{-1/\alpha},$$

and transforming marginals on step 4 as shown in Table 6.3

$$x = \theta_x(-\ln(1 - u))^{1/\tau_x}, \quad x = \theta_y(-\ln(1 - v))^{1/\tau_y}.$$

An alternative procedure introduced earlier by Marshall and Olkin [21] utilizes the theorem stating that for many useful Archimedean copulas with generator $\phi_\alpha(t)$ there exists a nonnegative random variable W such that $\phi_\alpha(t)$ is

Table 6.3 Clayton Copula with Weibull Margins

i	s	w	t	u	v	x	y
1	0.06	0.03	0.01	0.15	0.02	74.68	50.71
2	0.79	0.71	0.47	0.57	0.83	87.98	92.70
3	0.93	0.27	0.16	0.19	0.72	76.57	88.80

the inverse of its moment generating function $M(t) = E(e^{-tW})$, also known as its *Laplace transform*. This construction is also handy in shared frailty models discussed in Chapter 5. Marshall–Olkin's construction based on this representation suggests that if S and T are two independent variables uniformly distributed on $[0, 1]$,

$$U = M\left(-\frac{\ln S}{W}\right), \ V = M\left(-\frac{\ln T}{W}\right) \tag{6.32}$$

are also uniform on $[0, 1]$, and their joint distribution is an Archimedean copula with generator $\phi_\alpha(t)$. Neither the theorem nor the properties of the construction (6.32) will be proven here. However, it brings about some nice pairings of positive variables W with Archimedean copulas: for instance, Gamma distribution with Clayton's family, discrete logarithmic with Frank's, and stable with Gumbel–Hougaard's. If the distribution of W is relatively easy to sample, then the following algorithm can be recommended.

Sampling via Marshall–Olkin Construction
1. Generate a copy of random variable w.
2. Draw s and t independently from $Unif[0, 1]$.
3. Set $u = \phi_\alpha^{[-1]}(-\ln(s)/w), v = \phi_\alpha^{[-1]}(-\ln(t)/w)$.
4. Set $x = F^{-1}(u), y = G^{-1}(v)$.

Repeat n times to obtain sample $(x_i, y_i), i = 1, \ldots, n \sim C_\alpha(F(x), G(y))$.

Example 6.6.3 To apply Marshall–Olkin's method to simulate values from the joint distribution with the same Weibull margins (6.12) and (6.13) and Gumbel–Hougaard's survival copula with association parameter value $\alpha = 1.64$ obtained in [29], begin with the new step: generating a value w from stable distribution. This can be done in many ways [14], for example by using Chambers–Mallows–Stuck algorithm [3]. For our purposes, we can use the version suggested by Weron [34]. Then we draw two independent uniform variables s and t and use generator transform to obtain, on step 3,

$$u = \exp\{-(\ln s/w)^{1/\alpha}\}, \ v = \exp\{-(\ln t/w)^{1/\alpha}\},$$

Table 6.4 Gumbel–Hougaard copula with Weibull margins

i	w	s	t	u	v	x	y
1	2.08	0.25	0.43	0.46	0.56	87.39	80.08
2	1.68	0.59	0.03	0.61	0.21	83.30	91.22
3	1.63	0.06	0.19	0.25	0.37	92.45	86.02

and on step 4 as demonstrated in Table 6.4

$$x = \theta_x(-\ln u)^{1/\tau_x}, \ x = \theta_y(-\ln v)^{1/\tau_y}.$$

Steps 3 and 4 can be naturally merged into one formula

$$x = \theta_x \left(\frac{\ln s}{w}\right)^{1/\alpha\tau_x}, \ y = \theta_y \left(\frac{\ln t}{w}\right)^{1/\alpha\tau_y}.$$

6.7 Multidimensional Copulas

All discussions in Chapter 6 so far were restricted to bivariate distributions which could be modeled by pair copulas. This restriction is quite serious since most interesting applications deal with ensembles of more than two dependent random variables. More than two in this context may mean dozens or even hundreds of variables. Constructions of elliptical and Archimedean copulas which we discussed for dimension two can be naturally extended to higher dimensions. Let us begin with the definition of elliptical copulas.

Elliptical Copulas of Order $d > 2$

Elliptical distribution $Q_{d,R}$ of a random vector $t = (t_1, \ldots, t_d)$ can be defined by its joint density function

$$|\Sigma|^{-1/2} k((t - \mu)^T \Sigma^{-1}(t - \mu)),$$

where μ is a $d \times 1$ vector of means, Σ is a positively defined $d \times d$ covariance matrix, and $k(t)$ is some nonnegative function of one variable integrable over entire real line. Matrix R with elements $R_{ij} = \Sigma_{ij}/\sqrt{\Sigma_{ii} \times \Sigma_{jj}}$ is the correlation matrix determining all pairwise associations between the components of the random vector t. Define also by $Q_i(t_i)$ the marginal distribution of t_i. Then we can define an elliptical copula

$$C_R(u_1, \ldots, u_d) = Q_{d,R}(Q_1^{-1}(u_1), \ldots, Q_d^{-1}(u_d)). \tag{6.33}$$

The most popular elliptical copulas are Gaussian copulas which when combined with marginal distributions $u_i = F_i(x_i)$ for the components of the data vector $X = (X_1, \ldots, X_n)$, define the joint distribution $H(x)$ of vector X as

$$H(x) = C_R(u_1, \ldots, u_d) = \Phi_{d,R}(\Phi^{-1}(F_1(x)), \ldots, \Phi^{-1}(F_d(x))), \qquad (6.34)$$

where $\Phi(x)$ is standard normal distribution and $\Phi_{d,R}$ is d-variate normal with zero mean, unit variances, and correlation matrix R. Off-diagonal elements of matrix R describe pairwise associations, so the strength of association may differ for different pairs of components of vector X.

Archimedean Copulas of Order $d > 2$

A formal extension of Archimedean copula construction to dimension $d > 2$ is straightforward:

$$C_\alpha(u_1, \ldots, u_d) = \phi^{[-1]}(\phi_\alpha(u_1) + \cdots + \phi_\alpha(u_d))$$

is a legitimate copula with generator ϕ_α, satisfying the additional condition of complete monotonicity [25]. The only problem here is the exchangeability or symmetry requirement, which suggests that one parameter α describes all pairwise associations, thus all these associations have to be of equal strength. This is a substantial limitation of the modeling process. Therefore, to address nonexchangeable situations, this construction has to be modified. Three more sophisticated constructions suggested below are not necessarily limited to Archimedean copulas, but they are very effective for this class relaxing the exchangeability requirement. In order to avoid technicalities, we will demonstrate these constructions for simpler cases $d = 3$ and $d = 4$, keeping in mind that working in higher dimensions is also possible.

Vine Copulas

This construction, described in detail in Aas et al. [1], became very popular lately. Let us begin with $d = 3$ and consider the problem of modeling joint distribution $P(X \leq x, Y \leq y, Z \leq z)$ using marginals $u = F(x), v = G(y), w = H(z)$ and pair copulas. Let us designate one out of three variables (say, Y) as the central variable, whose associations with both X and Z are most important. Modeling the association between X and Z will have a lower priority. This "hierarchy" of dependence structure is inevitable, and can be established either from context, or by preliminary estimation of the strength of pairwise associations. Two equivalent ways to graphically illustrate the hierarchy of pairwise associations are suggested in a centered vine diagram in Figure 6.18 and an equivalent linear vine diagram in Figure 6.19. In both vines, primary links between the variables are indicated by solid lines, and the secondary links by dashed lines.

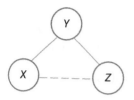

Figure 6.18 Vine structure, centered, $d = 3$.

Let us denote by $f(x), g(y)$, and $h(z)$ respective marginal densities of X, Y, and Z. By $g_{Y|X}(y|x), h_{Z|X}(z|x), h_{Z|Y}(z|y)$, and $h_{Z|X,Y}(z|x,y)$ we will denote corresponding conditional densities, and by $fg(x,y), gh(y,z)$, and $fgh(x,y,z)$ the double and triple joint densities.

Combining expressions for double and triple joint densities

$$fg(x,y) = g_{Y|X}(y|x)f(x)$$

and

$$fgh(x,y,z) = h_{Z|X,Y}(z|x,y)fg(x,y),$$

obtain the joint density of all three variables as a chain of conditionals:

$$fgh(x,y,z) = h_{Z|X,Y}(z|x,y)g_{Y|X}(y|x)f(x). \tag{6.35}$$

We will model primary associations for pairs (X, Y) and (Y, Z) using two pair copulas:

$$C_1(u, v) = C_1(F(x), G(y))$$

and

$$C_2(v, w) = C_2(G(y), H(z))$$

with respective densities $c_1(F, G)$ and $c_2(G, H)$. Double joint densities can be represented through these copula densities

$$fg(x,y) = c_1(F, G)f(x)g(y), \quad gh(y,z) = c_2(G, H)g(y)h(z),$$

therefore corresponding conditionals also can be expressed as

$$g_{Y|X}(y|x) = c_1(F, G)g(y), \quad h_{Z|Y}(z|y) = c_2(G, H)h(z). \tag{6.36}$$

Combining (6.36) with (6.35), we arrive at the expression

$$fgh(x,y,z) = \frac{h_{Z|X,Y}(z|x,y)}{h_{Z|Y}(z|y)}c_1(F, G)c_2(G, H)f(x)g(y)h(z).$$

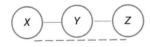

Figure 6.19 Vine structure, linear, $d = 3$.

The first term on the right-hand side requires a special treatment. If the value of Y were known (conditioning on Y), one would be able to introduce one more copula linking X and Z similar to (6.36):

$$h_{Z|X}(z|x) = c_3(F,H)h(z), \quad c_3(F,H) = \frac{h_{Z|X}(z|x)}{h(z)}.$$

However, if Y varies, this copula will be also dependent on the value of Y. This problem can be resolved if we assume that the same copula structure can be applied to link two conditional distributions rather than two marginals and thus introduce the conditional copula $C_3(F_{X|Y}, H_{Z|Y})$ with the density

$$c_3(F_{X|Y}, H_{Z|Y}) = \frac{h_{Z|X,Y}(z|x,y)}{h_{Z|Y}(z|y)}.$$

Therefore, we can express the triple joint density in terms of marginal densities, two pair copulas for (X, Y) and (Y, Z), and one additional copula, which is defined on conditional distributions rather than on marginals. This construction is known as a ***vine copula*** or pair copula construction (PCC). The following diagram describes the two-level tree of associations, where at the first level we need two pair copulas, and at the second just one conditional pair copula which uses conditional distributions instead of marginals as its arguments.

$$(X, Y) \, (Y, Z)$$
$$(X, Z|Y)$$

For dimension $d = 4$, we can use the vine structure above but also have to cope with an additional variable W, which should be linked to the dependence diagram for three variables. We will have to add a new link. The new link connects W either with the central variable Y or with a noncentral Z (or X)—whatever association is more important. This distinction brings about a classification of vine diagrams into two popular types: C-vines and D-vines. This importance hierarchy can be determined by contextual meaning of the variables or by rough estimation of the strength of their association. The diagram in Figure 6.20 corresponds to a C-vine, where Y plays the role of the central variable, primary links are shown by solid lines (star symbol indicates the

Figure 6.20 C-vine for $d = 4$.

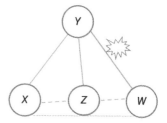

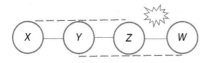

Figure 6.21 D-vine for $d = 4$.

new link), secondary links—by dashed line, and the third-level links by a dotted line. Figure 6.21 shows a so-called D-vine, where the variables are linked in a straight chain.

Three-level trees of the process for a C-vine and D-vine respectively are:

$$(X, Y) \, (Y, Z) \, \mathbf{(Y, W)}$$
$$(X, Z|Y) \, \mathbf{(Z, W|Y)}$$
$$(X, W|Y, Z)$$

and

$$(X, Y) \, (Y, Z) \, \mathbf{(Z, W)}$$
$$(X, Z|Y) \, \mathbf{(Y, W|Z)}$$
$$(X, W|Y, Z)$$

with the highlighted differences between C-vine and D-vine caused by the star-designated link in Figures 6.20 and 6.21 [1]. A comprehensive guide to vine copulas is presented in [20], many statistical aspects such as estimation and model selection are developed by Czado et al. [5] and [6], and some computational aspects are discussed in [19].

Hierarchical Archimedean Copulas

Let us suppose that $u = F(x), v = G(y), w = H(z)$ are three marginal distributions in case of $d = 3$. Hierarchical copula construction

$$C(u, v, w) = C_2[C_1(u, v), w] \tag{6.37}$$

uses two different generators ϕ_1 for the inner (*cluster*) copula C_1 and ϕ_2 for the outer (*nesting*) copula C_2. Not only their association values α_1 and α_2, but also the subclasses of Archimedean copulas (e.g., Clayton, Frank, Gumbel–Hougaard's families) could be different for C_1 and C_2. The suggested hierarchical clustering/nesting diagram is shown in Figure 6.22.

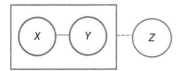

Figure 6.22 Hierarchical Archimedean copulas, $d = 3$.

Figure 6.23 Hierarchical
Archimedean copulas, fully nested.

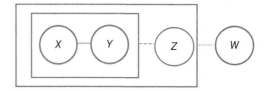

Using generator representation, since $C_1(u,v) = \phi_1^{[-1]}(\phi_1(u) + \phi_1(v))$ and $C_2(C_1,w) = \phi_2^{[-1]}(\phi_2(C_1) + \phi_2(w))$,

$$C(u,v,w) = \phi_2^{[-1]}[\phi_2 \circ \phi_1^{[-1]}(\phi_1(u) + \phi_1(v)) + \phi_2(w)]. \tag{6.38}$$

Additional conditions are required for the key element of this construction, the superposition $\psi(t) = \phi_1 \circ \phi_2^{[-1]}(t)$ to guarantee that $C(u,v,w)$ is a legitimate copula function. Here is a version of these conditions:

(i) $\psi(0) = 0$.
(ii) $\psi(0+) = 0$.
(iii) ψ is completely monotone ($\psi' < 0, \psi'' > 0$, etc.).

Example 6.7.1 Let C_1 and C_2 be two Gumbel–Hougaard copulas with respective parameters α_1 and α_2. Then $\phi_i(u) = (-\log u)^{\alpha_i}$ and $\phi_1^{[-1]}(t) = \exp(-t^{1/\alpha_i})$. Function $\psi(t) = t^{\alpha_2/\alpha_1}$ satisfies conditions (1)–(3) above if $\alpha_2 < \alpha_1$. This assumption determines the hierarchy of the dependence structure. The link between X and Y should be stronger than the link between the cluster (X,Y) and Z.

For this example we can combine two stages of hierarchical modeling into one formula, which can be directly used for estimation of probabilities $P(X \le x, Y \le y, Z \le z)$ using copula $C(F,G,H)$:

$$C(F,G,H) = \exp(-([(-\log F)^{\alpha_1} + (-\log G)^{\alpha_1}]^{\alpha_2/\alpha_1} + (-\log H)^{\alpha_2})^{(1/\alpha_2)}. \tag{6.39}$$

In case of four variables one may consider the following two possibilities of initial clustering depending on the importance of pairwise associations. In Figure 6.23 the hierarchy diagram corresponds to a fully nested model, and in Figure 6.24 to a non-nested model. For more detailed treatment, simulation

Figure 6.24 Hierarchical Archimedean
copulas, non-nested.

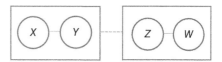

with nested Archimedean copulas, and applications in higher dimensions see, for example, Hofert [14], [15], [16], and McNeil [22].

Hierarchical Kendall Copulas

This construction introduced in [2] utilizes Kendall's distribution $K_C(t)$ defined in the previous section. For a simple illustration let us suppose that a triple copula $C(u, v, w)$ is built in two steps:

1. Choose a cluster (inner) copula $C_1(u, v)$, estimate its association parameter, and calculate $K_{C_1}(t)$.
2. Choose a nesting (outer) copula C_2 and use $C(u, v, w) = C_2(K_{C_1}(t), w)$.

This construction can be used not only for Archimedean copulas, but in the latter case the calculation of Kendall's distribution can be done straightforwardly through its generator. Differences between these methods become more dramatic in higher dimensions.

Vine copulas offer much flexibility in combining various types of copulas in one model and provide a huge variety of dependence hierarchies (compare C-vines and D-vines for different component ordering). This flexibility might also be a problem for model selection and parametric estimation, because it brings about a variety of possible models and a very large number of parameters. One possible way out of trouble is ignoring some weak associations and assuming conditional independence of certain pairs of variables. The most important technical problem encountered while working with vine copulas is the difficulty of obtaining expressions for joint distributions, while expressions for densities are easily available.

On the other hand, hierarchical Archimedean copulas have a potential advantage of logical cluster organization thus reducing the total number of parameters. The principal drawback of these models is the necessity to check nontrivial conditions (1)–(3). Also, a technical difficulty encountered with HAC deals with necessity to derive and calculate formulas for copula density functions involving higher order derivatives of the generators, while the copulas *per se* are typically easy to find. The third option, hierarchical Kendall copulas promise to become a powerful modeling tool which has yet to prove its value in practical applications.

References

1 Aas, K., Czado, C., Frigessi, F., and Bakken, H. (2009). Pair-copula constructions of multiple dependence. *Insurance: Mathematics and Economics*, 44, 182–198.
2 Brechmann, E. C. (2014). Hierarchical Kendall copulas: properties and inference. *The Canadian Journal of Statistics*, 42(1), 78–108.

3 Chambers, J. M., Mallows, C. L., and Stuck, B. W. (1976). A method for simulating stable random variables. *Journal of the American Statistical Association*, 71(354), 340–344.

4 Cherubini, U., Luciano, E., and Vecchiato, W. (2004). *Copula Methods in Finance*. London-New York: John Wiley & Sons, Inc.

5 Czado, C., Schepsmeier, U., and Min, A. (2012). Maximum likelihood estimation of mixed C-vines with application to exchange rates. *Statistical Modelling*, 12(3), 229–255.

6 Czado, E., Brechmann, C., and Gruber, L. (2013). Selection of vine copulas. In: P. Jaworski, F. Durante, and W. Haerdle, (editors). *Copulae in Mathematical and Quantitative Finance*, Lecture Notes in Statistics, Proceedings. Springer, Berlin. *Quantitative Finance*, 16, 775–787.

7 Demarta, S., and McNeil, A. (2005). The t copulas and related copulas. *International Statistical Review*, 73, 111–129.

8 Durante, F., and Sempi, C. (2016). *Principles of Copula Theory*, CRC, Chapman & Hall, Taylor & Francis Group.

9 Embrechts, P., Lindskog, F., and McNeil, A.J. (2003). Modelling dependence with copulas and applications to risk management. In: S Rachev (editor). *Handbook of Heavy Tailed Distributions in Finance*, 329–384.

10 Embrechts, P., McNeil, A., and Straumann, D. (2003). Correlation and dependency in risk management: properties and pitfalls. In: *Risk Management: Value at Risk and Beyond*. Cambridge University Press, 1S-223.

11 Genest, C., and J. MacKay. (1986a). The joy of copulas: bivariate distributions with uniform marginals. *American Statistician*, 40, 280–283.

12 Genest, C., and J.R. MacKay. (1986b). Copules Archimédiennes et familles de lois bidimensionnelles dont les marges sont donnees. *The Canadian Journal of Statistics*, 14, 145–159.

13 Genest, C., and Rivest, L.-P. (1993). Statistical inference procedures for bivariate Archimedean copulas. *Journal of the American Statistical Association*, 88, 1034—1043.

14 Hofert, M. (2008). Sampling Archimedean copulas. *Computational Statistics and Data Analysis*, 52, 5163–5174.

15 Hofert, M., and Maechler, M. (2011). Nested Archimedean copulas meet R: the nacopula package. *Journal of Statistical Software*, 39, 1–20.

16 Hofert, M., and Scherer, M. (2011). CDO pricing with nested Archimedean copulas. *Quantitative Finance*, 11(5), 775–787.

17 Joe, H. (1997). *Multivariate Models and Dependence Concepts*. London: Chapman & Hall.

18 Joe, H. (2014). *Dependence Modeling with Copulas*. London: Chapman & Hall/CRC.

19 Kojadinovic, I., and Yan, J. (2010). Modeling multivariate distributions with continuous margins using the copula R Package. *Journal of Statistical Software*, 34(9), 1–20.

20 Kurowicka, D., and Joe, H. (2011). *Dependence Modeling: Vine Copula Handbook*. Singapore: World Scientific Publishing Co.

21 Marshall, A. W., and Olkin, I. (1988). Families of multivariate distributions. *Journal of the American Statistical Association*, 83, 834–841.

22 McNeil, A. J. (2008). Sampling nested Archimedean copulas. *Journal of Statistical Computation and Simulation*, 78, 567–581.

23 Meucci, A. (2011). A new breed of copulas for risk and portfolio management, *Risk*, 24(9), 122–126.

24 Mikosch, T. (2006). Copulas: tales and facts. *Extremes*, 9, 3–20.

25 Nelsen, R. B. (2006). *An Introduction to Copulas*. 2nd ed. New York: Springer Science and Business Media Inc.

26 Nicoloutsopoulos, D. (2005). Parametric and Bayesian non-parametric estimation of copulas, Ph. D. Thesis, University College London.

27 Schweizer, B., and Sklar, A. (1983). *Probabilistic Metric Spaces*. New York: North Holland.

28 Schweizer, B., and Wolff, E. (1981). On non-parametric measures of dependence for random variables. *Annals of Statistics*, 9, 879–885.

29 Shemyakin, A., and Youn, H. (2006). Copula models of joint last survivor insurance. *Applied Stochastic Models of Business and Industry*, 22(2), 211–224.

30 Sklar, A. (1959). Fonctions de ráepartition à n dimensions et leurs marges. *Publications de l'Institut de Statistique de l'Universitée de Paris*, 8, 229–231.

31 Sklar, A. (1996). Random variables, distribution functions, and copulas - a personal look backward and forward. In: L. Ruschendorf, B. Schweizer, and M. D. Taylor (editors), *Distributions with Fixed Marginals and Related Topics*, pp. 1–14. Hayward, CA: Institute of Mathematical Statistics.

32 Sungur, E., and Orth, J. M. (2012). On modeling directional dependence by using copulas. *Model Assisted Statistics and Applications*, 7(4), 305–313.

33 Trivedi, K. S. (2008). *Probability and Statistics with Reliability, Queueing and Computer Science Applications*. London/New York: John Wiley & Sons, Inc.

34 Weron, R. (1996). On the Chambers-Mallows-Stuck method for simulating skewed stable random variables. *Statistics and Probability Letters*, 28, 165–171.

35 Yan, J. (2007). Enjoy the joy of copulas: with a package *copula*. *Journal of Statistical Software*, 21(4), 1–21.

Exercises

6.1 Check that the minimum copula $W(u, v) = \max\{u + v - 1, 0\}$ satisfies all copula properties.

6.2 Check that the product copula $P(u, v) = uv$ satisfies all copula properties.

6.3 Prove that FGM copula (6.6) does not satisfy some of the copula properties for $\alpha > 1$. What properties are violated?

6.4 What are the values $C(u, v)$ of the following copulas at the point $(u, v) =$
(0.5, 0.5)?
(a) Maximum copula $M(u, v)$?
(b) Minimum copula $W(u, v)$?
(c) Gaussian copula with $\rho = 0$?
(d) Gaussian copula with $\rho = 0.5$?
(e) Gaussian copula with $\rho = 0.6$?
(use numerical approximation whenever necessary)?

6.5 Apply Clayton's and Gumbel–Hougaard's copulas to estimate the joint
probability of two executive stock options being exercised in 5 years if
individual times-to-exercise for both ESOs with vesting period of 4 years
are modeled by Pareto distribution with c.d.f.

$$F(x) = 1 - \left(\frac{\gamma}{x}\right)^\tau, \ x \geq 4,$$

with parameters $\gamma = 4$, $\tau = 3$, and the association parameters of the copulas are given as
(a) $\alpha = 0.7$ (Clayton's);
(b) $\alpha = 0.8$ (Clayton's);
(c) $\alpha = 1.35$ (Gumbel–Hougaard's);
(d) $\alpha = 1.4$ (Gumbel–Hougaard's);

6.6 Thirty-year mortgage contracts are subject to refinancing and/or delin-
quency and default. Therefore their lifetimes could be much shorter than
30 years. The first-to-default (FTD, F2D) swap is a simple form of struc-
tured credit protection with the pay-off and contract termination after
the credit event which corresponds to the first default in the basket. Consider a basket consisting of two identical 30-year mortgage contracts.
The investor commits to cover the loss if any one of the two mortgage
holders defaults on their debt. If that happens, the FTD terminates, and
the investor no longer has exposure to the other name. Calculate probability of the FTD termination within the first 12 years using the following model for the time-to-default: Gaussian copula with $\rho = 0.3$ and the
marginal distribution being
(a) exponential with the mean (scale parameter) $\theta = 10$
(b) Weibull with scale parameter $\theta = 10$ and shape $\tau = 2$.
(c) Pareto with parameters $\gamma = 4$, $\tau = 3$ (see Exercise 6.3).
Repeat (a)–(c) for Clayton's copula with $\alpha = 1.2$. Use any available software to obtain exact results.

6.7 Apply simulation procedures described in Section 6.6 to confirm the results of the previous problem.

6.8 Calculate the values of joint life distribution $H(65, 65)$ and $H(55, 55)$, corresponding to marginals (6.12) and (6.13) and BB1 copula (6.29) with $\alpha = 1$, $\beta = 2$.

6.9 Prove that BB1 copula defined in (6.29) for $\alpha \to 0$ can be reduced to Gumbel–Hougaard's copula with the copula parameter β.

6.10 What kind of Archimedean copula corresponds to generator $\phi(t) = t^{-\alpha} - 1$?

7

Statistics of Copulas

7.1 The Formula that Killed Wall Street

This is how copulas became infamous. Very few mathematical formulas ever achieved such a huge popular recognition, even with clearly negative connotations, as the title of this section may suggest. Formulas rarely kill. Gaussian copulas (or, rather, applications of Gaussian copulas to pricing credit derivatives) have earned this distinction. As Felix Salmon [38] tells the story, it all started in the year 2000 when David X. Li, a former actuary turned financial mathematician, suggested a smart way to estimate default correlations using historical credit swap rates [29].

The approach suggested by Li was indeed very attractive. Being able to use correlation structure to define dependence, as was discussed in Chapter 5, really simplifies the problem of building a model for joint distribution. As we have mentioned, this is how it works for multivariate normal distribution: all dependence that may exist between the components of a Gaussian random vector can be expressed in terms of the correlation matrix. In the bivariate case the correlation matrix degenerates into just one number: the correlation coefficient.

By that time (the turn of the new millennium) it was already well known that Gaussian approximation was not working very well for many financial variables. Empirical observations of the long-term behavior of daily stock price returns, market indexes, and currency exchange rates had exposed some strange patterns hardly consistent with the assumption of normality. Their distributions were characterized by fat tails and skewness, which gave evidence that large deviations of these variables, especially on the down side, were much more likely to occur in reality than it would be predicted by Gaussian model. By that time the Black–Scholes formula, the crown achievement of mathematical finance (pricing European options based on the assumption of normality) already ceased to satisfy financial practitioners, and alternative distribution models were being investigated making full use of the Monte Carlo approach.

Introduction to Bayesian Estimation and Copula Models of Dependence, First Edition.
Arkady Shemyakin and Alexander Kniazev.
© 2017 John Wiley & Sons, Inc. Published 2017 by John Wiley & Sons, Inc.
Companion Website: http://www.wiley.com/go/shemyakin/bayesian_estimation

However, the Gaussian copulas looked like a very different tool distinctive from regular multivariate normal models. They did not require any assumptions of normality. On the contrary, arbitrary random variables, X and Y, characterized by fat tails, skewness, and/or other indecent behavioral patterns, in short, variables behaving as wildly as they could with whatever exotic distribution functions F and G, would be transformed into uniform $U = F(X)$ and $V = G(Y)$ and then again individually transformed into normal variables $S = \Phi^{-1}(U)$ and $T = \Phi^{-1}(V)$ and sent to live and interact in the normal world, where their dependence patterns were to be studied. And as we know, in the world of Gaussian copulas $C_\rho(u, v) = \Phi_\rho(\Phi^{-1}(u), \Phi^{-1}(v))$ all the dependence between the marginals may be expressed in terms of correlation ρ. Then the backward transform would send our variables back into the wild, but with their dependence patterns well recorded. Temptation was obviously there and a nice modeling algorithm was open for the takers: The study of the marginal distributions was isolated from the study of the dependence structure, and these two could be done separately, with the latter carried out in a surgically clean world of Gaussian variables. That is how the default correlations were estimated and pricing of credit derivatives became technically possible. The result was a spectacular growth in the industry of credit derivatives, especially, CDOs, which became extremely popular as evidenced by McKenzie et al. in [33].

CDOs or collateralized debt obligations are special credit derivative securities packaging and repackaging debt, buying and selling risks in exchange for positive cash flows. A simplified example of a CDO, three loan portfolio, where we priced the separate tranches, for example, obligations to pay back specially allocated portions of the mortgage debt in case of defaults was discussed in Section 5.7. Much more complex CDOs were created, priced, and sold in the real world, with their senior tranches deemed virtually riskless and thus being assigned very high credit ratings based on Gaussian copula models. Credit agencies trusted the models and also added to the common illusion of safety.

On the top of CDO popularity, this is what an analytic paper [8] indicated as their main benefits (2004 Outlook):

- Investment diversification
- Customized risk–reward
- Attractive yields
- Leveraged returns (optional redemption)
- Transparency

Evidently, things looked good in 2004. Then bad things started happening. With the US housing market going down in 2008, all the estimates of probabilities of joint defaults on mortgages went down in flames. The simplest reasons for that were suggested in Section 5.7: First, it was extremely hard to estimate correlation using historical data. Second, even if we knew everything about correlations, this still did not give us enough information to estimate probabilities of joint defaults.

One possible solution was to throw out all copula techniques as compromised by one of the worst crises of the Wall Street and therefore one of the worst moments in the history of financial mathematics. The authors of this book cannot recommend this solution with two motivations: First, we have been indeed heavily invested in copula modeling for the last fifteen years. This however would be an insufficient motivation to recommend the copula approach to anyone, if not for the other reason: in applications, we have encountered many examples when copulas do their job perfectly well, providing good estimation for actuarial, financial, and other problems. We certainly acknowledge the deep and serious critiques of copula models expertly done by Mikosch in [32] and also many other sources. But we do not find enough evidence to totally deny copula methods their rightful place among the other statistical techniques.

Another possible solution of which we can approve: always look for an adequate copula model. Do not fall for the first or the easiest choice. As we will see very soon, attractive Gaussian copulas obtain one property (or rather, lack one feature), which makes them often unfit for risk management purposes. This lacking feature is known as tail dependence.

Sklar's theorem tells us that for any joint distribution there always exists a copula representation, which means that the true model of dependence can be found inside the class of copula functions. Unfortunately, no one has provided an algorithm yet, which can efficiently search for the true solution over the entire set of copula models. Mathematically speaking, we can just regret the absence of a general parametrization for the entire set of copula functions. Such a parametrization would reduce the search for a good copula model to a routine problem of parametric estimation. What statisticians have learned to do very well, is to search for the best solution in a given specific parametric subclass of copulas such as any of those described in Chapter 6: Gaussian, FGM, Clayton, etc.

We can think of this problem as similar to the one we may encounter while looking for a treasure chest hidden in a huge office building. If we know that this treasure chest exists, but do not have keys to all the offices in the building, we might be restricted to the confines of the offices which either are already open, or to which the keys are readily available. Nobody guarantees the existence of true distribution or even a good approximation in these open offices. Actually, the optimal solution for a given office might be a quarter or even a penny we find in a dusty corner, and it is worth much less than the treasure chest which is still sitting behind a closed door in a different room.

Thus we can consider an entirely different problem: We wish to develop the methods which will allow us to compare the relative wealth of open offices so that we could select the copula model which were at least the best available for our application. Following George Box's idea that "essentially, all models are wrong, but some are useful" [5], we would like to suggest a useful model inside the subclass which we can properly analyze.

Section 7.2 will contain the general discussion of the most logical criteria of quality of statistical models helpful in choosing one copula model over another one. We will follow the empirical principle: the best model is the one that fits our data the best. However, keeping the best fit in mind, we should not forget about the predictive quality of the model, which may often suffer as the result of overfitting. We will consider standard goodness-of-fit criteria which will typically provide some measure of the overall fit of a statistical model. Then we will review information criteria such as AIC and its later versions, which will equip us with certain measures against overfitting. Two criteria of fit which are more specific for copulas address the tail dependence and such measures of overall dependence as Kendall's concordance and Spearman's rank correlation.

Given we know the office we want to search, how to find a strategy to find the wealth it contains? Which methods can we use to find the optimal solution for a given parametric subclass of copula functions? This is the problem of parametric estimation which we will consider in Section 7.3. First of all, the structure of copula models allows for two options: we either can estimate all parameters at once, including parameters of the marginals and the parameters of association. If this one-step approach is too cumbersome, we may opt for a two-step approach: estimate the marginals first, and then estimate the parameters of association given marginals. This can be done either parametrically at both steps or semiparametrically (using nonparametric estimates of the marginals and parametric estimates of the association). We will not consider fully nonparametric estimation, which will stay out of the limits of this discussion.

We will also make use of the Bayesian approach. The difference between Bayesian and classical methods consists in more than just the difference between two point estimators: say, maximum likelihood estimator versus the posterior mean incorporating some prior information. The most important aspect of Bayesian estimation is that the inference gives us entire posterior distribution rather than one point estimate. To choose the best representative of a given parametric class, we will utilize some of the the criteria discussed in Section 7.2 and will compare the results of parametric and semiparametric estimation with these criteria in mind. Among the methods we will consider, the most important role will belong to a version of the method of moments, minimum distance (least squares) approach, method of maximum likelihood, and, naturally, Bayesian estimation methods.

Section 7.4 will be dedicated to the selection of the office (sorry, the subclass of copulas), which is the best of those being open. To make sure that a good enough choice exists, we will discuss goodness-of-fit tests which can address the availability of feasible solutions in these classes. One logical way would be to designate the best solution for each open office (subclass of copulas), and compare these solutions using the criteria discussed in Section 7.2. Another possible solution is to apply Bayesian model selection procedures which will base the comparison between the subclasses not on the sole representatives of each, but rather on the integral performance of models from these subclasses.

Thus Bayesian statistics plays two separate roles in this chapter: We apply Bayesian estimation to find the best solutions in the subclasses of copulas, and also Bayesian hypothesis testing for the purpose of model selection between these subclasses.

In order to illustrate our point, we will consider two applications which utilize some of these approaches in actuarial and financial contexts mentioned in Chapter 5. Section 7.5 considers Bayesian copula estimation for a joint survival problem and Section 7.6 touches on Bayesian model selection for the related failures of vehicle components. Chapter 8 will contain a study of international financial markets which will combine most of the elements of statistical analysis considered in Chapter 7.

7.2 Criteria of Model Comparison

Suppose we use a finite sample $X = (X_1, \ldots, X_n)$ to build a distribution model, which will both explain the data we obtain and provide a valid prediction for future, hopefully supported by the data yet to come. If a distribution model $F_0(x)$ is suggested, what makes it a viable candidate? How can we choose a better model out of two candidates? Most of the criteria providing comparison for copula models are based on the closeness of the empirical distribution or its specific characteristics to those of the candidate distribution. In parametric case a special role is played by the likelihood function and its modifications.

7.2.1 Goodness-of-Fit Tests

Traditional goodness-of-fit approach is about testing the null hypothesis $H_0 : X \sim F_0(x)$ versus $H_1 : otherwise$. In general setting, we define a nonnegative "distance" (which may or may not be a proper mathematical distance or metric, satisfying conditions of identity of indiscernibles, symmetry, and triangle inequality) $d(F_n^*(x), F_0(x))$ between the theoretical c.d.f. $F_0(x)$ suggested by the model and the empirical c.d.f.

$$F_n^*(x) = \frac{1}{n+1} \sum_{i=1}^{n} I_{(-\infty, x)}(X_i), \tag{7.1}$$

of the data sample $X = (X_1, \ldots, X_n)$, where $I_A(x)$ is the standard indicator function of the set A. Notice the slightly nonstandard use of the denominator $n + 1$

in (7.1) instead of more common n, in which we follow Genest and Rivest [19] and others, who find it convenient that the empirical c.d.f. never takes the value 1. We will briefly discuss two common measures of goodness-of-fit, which are most popular in application to copula structures. These measures are **Kolmogorov–Smirnov distance**

$$d_{KS} = d_{KS}\big(F_n^*(x), F_0(x)\big) = \sqrt{n} \max_x \big|F_n^*(x) - F_0(x)\big|, \tag{7.2}$$

and **Cramer–von Mises distance**

$$\omega^2 = \omega^2\big(F_n^*(x), F_0(x)\big) = \int_x \big(F_n^*(x) - F_0(x)\big)^2 dx, \tag{7.3}$$

defining two corresponding criteria of closeness.

The tests based on these two distances with rejection regions of the form $d_{KS} > c$ and $\omega^2 > c$ are convenient because the distributions of the test statistics and thus the p-values of the tested data are available under the assumptions of H_0 being true, at least for one-dimensional case. In order to determine d_{KS} one has to compare the vertical distance between two functions (empirical and theoretical), only at the sample points, which represent the jumps of empirical c.d.f., so that the search for the maximum is simplified. Similarly, instead of integral ω^2 the following test statistic can be calculated:

$$T = \frac{1}{12n} + \sum_{i=1}^{n} \left(\frac{2i-1}{n} - F_0(x_{(i)})\right)^2,$$

where $x_{(i)}$ is the ith order statistic.

One way both of these measures can be applied beyond testing the null hypothesis is to compare several theoretical candidate models, selecting those with the lower values of d_{KS} or ω^2.

Let us notice that while goodness-of-fit tests are essentially non-Bayesian, both Kolmogorov–Smirnov and Cramer–von Mises distances may be used in a Bayesian context as we will see later on. We should also be aware that in case of copula models $F_0(x)$ is a joint distribution of vector X with two or more components, which makes practical application of goodness-of-fit tests much more technically difficult. In particular, distributions of test statistics are no longer distribution free and finding the maximum for Kolmogorov–Smirnov statistic is not as trivial as in one dimension. That is why the determination of even approximate p-values of the tests requires a substantial amount of work such as Monte Carlo generation of multiple samples from the theoretical distribution [18]. However, this is not an issue for model comparison purposes.

The real issue is: using goodness-of-fit criteria for model selection, we do not put any barriers against **overfitting**. Which means that the theoretical model with a c.d.f. close to the empirical c.d.f. is always preferable. It is always easy to imagine a theoretical model with way too many parameters (overparame-terized), which will provide a really good fit for the empirical c.d.f. However this model will take all observed features of the sample for granted, even those "features" which have completely random origin. At the same time, it will not allow for the features not observed. It is like looking at a picture of a human face which has a wart on the nose, an ink stain in the middle of the forehead, and one ear cut-off. Should we take all other human faces to have warts and ink stains in appropriate places, but just one ear? Such a model will provide a good description of a particular portrait, but a poor prediction of what most humans look like.

7.2.2 Posterior Predictive *p*-values

Before moving on to the measures of model fit, which will address overfitting, let us introduce a fully Bayesian concept which may serve the purpose of testing goodness-of-fit hypotheses without applying Bayesian factors as was done in Chapter 2. To be more precise, we will follow the idea of Bayesian predictive checking introduced by Rubin [37].

Let us review a parametric version of the general goodness-of-fit test and the frequentist concept of *p*-value, which is consistently criticized by Bayesian statisticians and lately (along with confidence intervals) became a sore spot of applied statistics as one of the most intuitively misunderstood and misused tra-ditional statistical techniques. This critique was epitomized in an editorial by Trafimov and Marks [44] in a popular applied journal *Basic and Applied Social Psychology* and further developed by Wasserstein and Lazar [47].

For simplicity, let us consider only the continuous case treating all data as one vector and parameter as another. Let us denote by X the data (it could be a finite i.i.d. sample or a different object taking values x on the sample space S, $X = x \in S$) with the distribution density (likelihood) $f(x; \theta)$ depending on parameter θ. Consider a null hypothesis $H_0 : \theta = \theta_0$ to be tested against the composite alternative $H_1 : \theta \neq \theta_0$. For test statistic $T(X)$ and critical region of the form $R_T(x) = \{x \in S : T(x) \geq c\}$, define the test:

"***Reject*** H_0" if $T(x) \geq c$ or "***Do not reject*** H_0" if $T(x) < c$.

Significance level approach suggests determining first the ***size*** or significance level of the test as the tolerable level of the Type I Error $\alpha = P(T(X) \geq c|\theta_0)$, so that the critical value $c(\alpha)$ depends on α. However a more popular approach is to determine the ***p-value*** of the data, $p = P(T(Y) \geq T(x)|x, \theta_0)$, which may be interpreted as the tail probability of statistic $T(Y)$ for the given x and θ_0. We also may rewrite it as

$$p(x; \theta_0) = \int_{T(y) \geq T(x)} f(y; \theta_0) dy. \tag{7.4}$$

Notice that under the null hypothesis p-value as a function of x is uniformly distributed on $[0, 1]$, so that exactly 5% of all data from $S(\theta_0)$ will have $p \leq 0.05$. General recommendation is to reject the null hypothesis for data with low p-values.

If the choice of statistic $T(X)$ is smart, this test will have high power defined as rejection probability under the alternative: $k = P(T(X) \geq c | H_1)$. Famous Lindley's paradox [30] guarantees that for large enough sample sizes, virtually all values $\theta \neq \theta_0$ will be eventually rejected. However high rejection probability is not the goal of Bayesian hypothesis testing. If the model is false (and all models **are** false, according to George Box), but fits our data well, we do not want to reject it. This is the motivation to introduce a Bayesian concept, which will measure the feasibility not of a single value of parameter θ_0, but rather of the entire posterior distribution, see also Meng [31].

In order to formulate a Bayesian alternative to the p-value approach, we need to introduce Bayesian predictive distributions, which we ignored so far being focused on estimation. In estimation framework we have considered prior $\pi(\theta)$ and posterior $\pi(\theta|x)$ densities. Now let us consider a new sample, $y \in S$. What is the distribution of y given our information on θ, provided both by the prior and the data x? We can define **Bayesian predictive density** as the integral over posterior

$$g(y|x) = \int f(y; \theta)\pi(\theta|x)d\theta = \frac{\int f(y; \theta)\pi(\theta)f(x; \theta)d\theta}{\int \pi(\theta)f(x; \theta)d\theta}. \tag{7.5}$$

It is also possible to define **prior predictive density** as an extreme case with no data available (with prior replacing posterior):

$$g(y) = \int f(y; \theta)\pi(\theta)d\theta.$$

It is easy to see that the prior predictive density $g(y)$ coincides with the marginal density $m(y)$ in the denominator of the formula (2.26) playing the central role in Bayesian estimation.

Let us use the posterior predictive density to define the **posterior predictive** **p-value** similar to (7.4) as

$$pp(x) = \int_{T(y) \geq T(x)} g(y|x)dy. \tag{7.6}$$

It is important that the latter does not depend on a particular value of θ_0, but integrates the information from the posterior. The role of $pp(x)$ is similar to traditional p-values: reject null when $pp(x)$ is small; let us say, if it does not exceed usual benchmark of 0.1, 0.05, or 0.01.

In order to illustrate the difference between (7.4) and (7.6), let us consider a slightly paraphrased example introduced by Gelman [15].

Example 7.2.1 Let us test a hypothesis regarding the normal mean using a single data point: $H_0 : \theta = 0$ versus a two-sided alternative $H_1 : \theta \neq 0$. Assume $f(x; \theta) \sim N(\theta, 1)$ and $\pi(\theta) \sim N(0, \tau^2)$. It is easy to see (compare to Exercise 2.8) that

$$\pi(\theta|x) \sim N\left(\frac{\tau^2}{\tau^2 + 1}x, \frac{\tau^2}{\tau^2 + 1}\right),$$

$$g(y|x) \sim N\left(\frac{\tau^2}{\tau^2 + 1}x, 1 + \frac{\tau^2}{\tau^2 + 1}\right), \ g(y) \sim N(0, \tau^2 + 1).$$

Defining $T(X) = |X|$ and taking into account that one tail of the test statistic translates into two symmetric tails of the standard normal distribution $\Phi(u)$, we obtain from (7.4) and (7.6) the following expressions:

$$p(x; 0) = 2(1 - \Phi(x)),$$

$$pp(x) = 2\left(1 - \Phi\left(\frac{x - \frac{\tau^2}{\tau^2+1}x}{\sqrt{1 + \frac{\tau^2}{\tau^2+1}}}\right)\right) = 2\left(1 - \Phi\left(\frac{x}{\sqrt{(\tau^2 + 1)(2\tau^2 + 1)}}\right)\right).$$

What can we say about particular data values which are extreme from the point of view of the likelihood? Consider, for instance, $x = 3$, three standard deviations away from the mean. It gives ample evidence against H_0 from frequentist point of view, bringing about p-value of 0.0027. However, if we consider a vague prior with $\tau = 10$, $pp(3) = 0.9832$ and H_0 should hardly be rejected from a Bayesian standpoint. Even if $x = 30$, which is three standard deviations of prior predictive distribution away from the mean, $p(30, 0) \approx 0$, but $pp(30) = 0.8333$ and hypothesis H_0 stays highly feasible and should not be rejected.

But why do we not want our hypothesis to be rejected when the observation $x = 30$ seriously contradicts the prior? Yes, in this case the prior is far away from the likelihood, but somehow the posterior is O'K: $\pi(\theta|30) \sim N(29.7, 0.99)$, which means that the Bayesian model works too well to be discarded. The difference between the classical decision to reject H_0 and the Bayesian decision not to reject corresponds to the difference between the way we treat the null hypothesis: in classical setting it characterizes the point θ_0 of the parametric space, while in Bayesian setting it corresponds to the entire posterior distribution.

7.2.3 Information Criteria

In parametric setting, a special role when the theoretical c.d.f. is defined as $F_0(x) = F_{\theta_0}(x)$ is played by the likelihood function $L(x; \theta)$ defined in (2.13). Let us define **deviance** of a parametric model M with the k-dimensional parameter value θ as

$$D(\theta) = -2 \ln L(x; \theta). \tag{7.7}$$

The smaller values of the deviance correspond to the higher likelihood. The maximum likelihood estimator $\hat{\theta}$ will provide the smallest possible deviance. We will define **Akaike Information Criterion** or AIC [1] as

$$AIC(M) = D(\hat{\theta}) + 2k. \tag{7.8}$$

Information criterion represents the relative loss of information which occurs when we use model M instead of the true model [2]. The lower values of AIC correspond to the better quality of the model. Most of it is determined by likelihood, but additional term $2k$ represents a penalty for overfitting: larger models with more parameters (higher dimension of vector θ) will be penalized, and smaller *parsimonious* models will be preferred.

Another information criterion was suggested by Schwarz [39]. It penalizes overparameterized models even more substantially. It is called **Bayesian Information Criterion** or BIC and is defined as

$$BIC(M) = D(\hat{\theta}) + k \ln n. \tag{7.9}$$

The names might be misleading, and there is nothing particularly Bayesian in BIC. Both AIC and BIC are based on maximizing the likelihood function, and only the sizes of the penalty for the parametric dimension are different, for BIC increasing with the sample size. The main advantage of both AIC and BIC and their later versions [2] becomes evident when we compare two or more parametric models of different parametric dimensions or even different parametrizations with regard to overfitting. Let us say, there exist two models M_1 with parameter θ_1 of dimension k_1 and M_2 with parameter θ_2 of dimension k_2. Then we say that a model M_1 is better than M_2 if $AIC(M_1) < AIC(M_2)$ or $BIC(M_1) < BIC(M_2)$, depending on your preference of AIC or BIC. Models M_1 and M_2 may be non-nested (e.g., Clayton copula with Weibull margins vs. Gumbel–Hougaard copula with Gompertz margins).

None of these criteria are especially convenient for comparison of models within Bayesian framework. First, they are based on the value of MLE, which might not be directly available in Bayesian estimation. Second, they do not utilize the entire form of the posterior distribution. **Deviance Information Criterion** introduced by Speielgelhalter et al. [43] resolves these problems. For a model M with parameter θ it may be defined as

$$DIC(M) = D(\bar{\theta}) + 2(\bar{D} - D(\bar{\theta})) = 2\bar{D} - D(\bar{\theta}), \tag{7.10}$$

where $D(\bar{\theta})$ is the deviance at the posterior mean and $\bar{D}$ is the posterior mean of the deviance $D(\theta)$. In this presentation the first term characterizes the fit of the model, and the second term implicitly penalizes overparameterization. Another version of DIC suggested by Gelman et al. in Reference [16]

$$DIC(M) = D(\bar{\theta}) + \hat{V}ar(D(\theta)) \tag{7.11}$$

is based on the posterior variance. The most attractive feature of DIC is its straightforward application to MCMC procedures of Bayesian estimation, where all three characteristics $D(\bar{\theta})$, $\bar{D}$, and $\hat{V}ar(D(\theta))$ can be estimated directly from the generated sample. As with other information criteria, the lower values of DIC indicate the better model.

7.2.4 Concordance Measures

What is the most important feature of pair copula models? It is the way they make it possible to model joint distribution of two variables $P(X \leq x, Y \leq y) = C_\alpha(F(x), G(y))$, and also to separate the estimation of dependence from the estimation of marginal distributions $F(x)$ and $G(y)$. A better copula model should accurately estimate the strength of dependence whatever the marginals are. The strength of dependence suggested by model C_α within a given copula family is determined by the association parameter(s). The strength of dependence between components of the random vector (X, Y) can be also expressed in terms of Spearman's coefficient of rank correlation ρ^* or Kendall's concordance τ defined in (5.4). It is important that both the rank correlation and concordance are invariant with respect to marginals F and G and are fully determined by the copula function $C_\alpha(u, v)$. Relationships between the copula function and its concordance measures are well established:

$$\rho^*(X, Y) = 12 \int\int_{[0,1]^2} C_\alpha(u, v) du dv - 3 \tag{7.12}$$

and

$$\tau(X, Y) = 4 \int\int_{[0,1]^2} C_\alpha(u, v) dC_\alpha(u, v) - 1. \tag{7.13}$$

In particular, for Archimedean copula with generator $\varphi_\alpha(t)$ it holds that

$$\tau = 1 + 4 \int_0^1 \frac{\varphi_\alpha(t)}{\varphi'_\alpha(t)} dt. \tag{7.14}$$

For instance, see [20], for Clayton's family

$$\tau = \frac{\alpha}{\alpha + 2}, \tag{7.15}$$

for Gumbel–Hougaard's family (also, its survival copula version)

$$\tau = \frac{\alpha - 1}{\alpha}. \tag{7.16}$$

For Frank's family the formula is not as simple, though we still can express concordance

$$\tau = 1 + \frac{4(D(\alpha) - 1)}{\alpha} \tag{7.17}$$

through Debye's integral

$$D(\alpha) = \frac{1}{\alpha} \int_0^{\alpha} \frac{t}{e^t - 1} dt \tag{7.18}$$

and approximate it numerically. Also, for FGM family

$$\tau = \frac{2\alpha}{9},$$

and for all elliptical copulas (including Gaussian and t-copulas, see [9])

$$\tau = \frac{2}{\pi} \arcsin \alpha. \tag{7.19}$$

For two-parametric BB1 copulas (6.19) concordance may be expressed through both association parameters:

$$\tau = 1 - \frac{2}{\beta(\alpha + 2)}.$$

Using the fact that Kendall's tau may be estimated with the help of its empirical value $\hat{\tau}$ according to (5.5), we can suggest two options for one-parametric copulas. First, $\hat{\tau}$ may be used to directly obtain an association parameter estimate within a given family of copulas. Second, if several models with different values of association are considered, the difference between the values of τ induced by models via the formulae above and the empirical value $\hat{\tau}$ may be treated as a concordance-based criterion of model comparison.

7.2.5 Tail Dependence

In applications of copula models to risk management, the attention is often concentrated on the tails of the joint distributions indicating extreme co-movements of two or more variables of interest. These tails correspond to catastrophic events. That is why it makes sense to construct the models with especially good fit at the tails. Coefficients of **lower tail dependence**

$$\lambda_L = \lim_{u \downarrow 0} \frac{C(u, u)}{u} \tag{7.20}$$

and **upper tail dependence**

$$\lambda_U = \lim_{u \uparrow 1} \frac{\bar{C}(u, u)}{1 - u}, \tag{7.21}$$

where $\bar{C}(u, u)$ is the survival copula defined in (6.4), characterize respectively the behavior of the lower left and upper right tails of the copula function. If tail dependence coefficient is zero, we can talk about "no tail dependence." As well as the concordance measures from the previous subsection, tail dependence is invariant with respect to marginal distributions.

For popular elliptical and Archimedean copulas tail dependence can be expressed in terms of association parameter, namely, for Clayton's family

$$\lambda_L = 2^{-1/\alpha}, \lambda_U = 0, \tag{7.22}$$

for Gumbel–Hougaard copula

$$\lambda_L = 0, \lambda_U = 2 - 2^{-1/\alpha}, \tag{7.23}$$

and for its survival version

$$\lambda_L = 2 - 2^{-1/\alpha}, \lambda_U = 0. \tag{7.24}$$

Frank and Gaussian copula have no tail dependence, while t-copula has both lower and upper ones:

$$\lambda_L = \lambda_U = 2T_{\eta+1}(t), \ t = \left(-\sqrt{\eta+1}\sqrt{\frac{1-\alpha}{1+\alpha}}\right), \tag{7.25}$$

where $T_\nu(t)$ is the c.d.f. of t-distribution with ν degrees of freedom [9]. Two-parametric BB1 copulas allow for two different tail dependence coefficients:

$$\lambda_L = 2^{-1/\alpha\beta}, \lambda_U = 2 - 2^{-1/\beta}. \tag{7.26}$$

A better copula model should give a good correspondence of the model-induced tail dependence with the empirical tail dependence. The only problem is caused by a certain ambiguity in estimating empirical tail dependence.

7.3 Parametric Estimation

In this chapter we concentrate on pair copula analysis, restricting ourselves to two marginal distributions tied together with a pair copula. Suppose we have chosen a particular family of copulas (say, FGM or Clayton's). How can we specify the values of parameters (both marginal parameters and ***copula parameter***, which is also called the ***association parameter***)? Typically, we will have to work in the parametric set of three to six dimensions (one or two per marginal and one or two for the association).

First of all, we have to determine whether we want to estimate all parameters at once ***(one-step estimation)*** or we prefer to use an important property of copula models, which allows us to estimate the marginal parameters first, and then estimate the association given the marginals. This constitutes ***two-step***

estimation also known as IFM: ***inference functions for margins***. As shown by Joe [22], see also [23], both procedures are consistent. The first one seems to be more theoretically appealing, while the second is certainly more practical, reducing the number of parameters to be estimated on one step, therefore reducing the dimensionality of the parametric set.

Example 7.3.1   For a numerical example, let us consider a sample of size $n = 10$ consisting of pairs (x_i, y_i), $i = 1, \ldots, n$, where variables X and Y are distributed exponentially with scale parameters θ_1 and θ_2, so that $F(x) = 1 - e^{-x/\theta_1}$ and $G(y) = 1 - e^{-y/\theta_2}$. Suppose that their association can be described by survival Gumbel–Hougaard copula with copula parameter α, so that their joint survival function is

$$P(X > x, Y > y) = S(x, y) = \exp\left\{-\left[\left(\frac{x}{\theta_1}\right)^{\alpha} + \left(\frac{y}{\theta_2}\right)^{\alpha}\right]^{1/\alpha}\right\}.$$

Actually, this is exactly how they are distributed, because for the construction of this mini-sample we used the simulation procedure from Example 6.6.3, which can be easily adjusted for exponential marginals. The choice of parameters was: $\theta_1 = 3$, $\theta_2 = 5$, and $\alpha = 2$. This choice of association parameter corresponds to $\tau = 0.5$ and also simplifies simulation in Marshall–Olkin construction, because for this case the stable distribution reduces to Levy distribution and its values can be generated as $W = \frac{1}{2}\left(\Phi^{-1}(1 - U/2)\right)^{-2}$, where $U \sim Unif[0, 1]$.

Using simulated data for illustration purposes is a very common practice. It allows for some meaningful comparisons, even for such a small dataset. Here is the sample, and respective ranks of the sample values listed in Table 7.1. Further in this section we will use it to demonstrate results of application of the most common techniques of parametric estimation and obtain numerical estimates for θ_1, θ_2, and α.

Aside from practicality, there are some additional justifications for the two-step approach. One of them is evident when we analyze multiple pair copulas connecting many possible pairs on the same set of marginal distributions. In this case it will be counter-intuitive to allow different parametric models for the same marginal distributions when they are coupled into different pairs. This is the setting of Section 7.6, all cases in Chapter 8, and many other popular applications.

7.3.1   Parametric, Semiparametric, or Nonparametric?

If we choose the two-step IFM approach, it allows for one more choice. In general, one may choose nonparametric approach for both steps: estimating marginal parameters and building a nonparametric copula, see, for example,

Table 7.1 Simulated exponential/Gumbel–Hougaard sample

i	1	2	3	4	5	6	7	8	9	10
x	1.95	5.53	0.65	0.19	2.34	3.22	0.48	2.59	5.73	4.60
y	1.86	5.81	1.53	3.57	4.32	1.13	0.16	6.23	6.86	8.14
Rank (x)	4	9	3	1	5	7	2	6	10	8
Rank (y)	4	7	3	5	6	2	1	8	9	10

Genest and Rivest [19] and Nicoloutsopoulos [34]. In the book dedicated to parametric estimation we will choose not to consider a fully nonparametric approach and concentrate on parametric copulas. However it still allows for both parametric and nonparametric estimation of the marginals thus bringing about either fully parametric estimation (for both marginal and association parameters) or semiparametric estimation (nonparametric marginal and parametric copula model).

Example 7.3.2　In order to apply IFM approach to the data in Table 7.1, we have to estimate the margins first. We will start with parametric estimation. Using standard procedure for exponential distribution, we can use either MLE or the method of moments with the same results: $\hat{\theta}_1 = \bar{x} = 2.73$ and $\hat{\theta}_2 = \bar{y} = 3.96$. Corresponding values of survival functions for sample data are listed in the first and the second rows of Table 7.2. Similarly, we can obtain empirical estimates for the same survival functions using empirical c.d.f. as defined in (7.1), see the third and the fourth rows of Table 7.2.

In general, fully parametric models will be less data dependent and will be expected to have higher predictive quality. On the other hand, semiparametric estimation is fully theoretically justifiable as discussed by Genest et al. [17] and may be more appropriate for the comparison of copula models and copula model selection, eliminating the effect of possible misspecification of the marginal distributions discussed by Kim et al. [24]. From this point on we will concentrate our attention on the copula parameter of association, estimating marginal distributions parametrically or nonparametrically, separately

Table 7.2 Parametric and nonparametric estimates of marginals

i	1	2	3	4	5	6	7	8	9	10
$S_1(x; 2.73)$	0.52	0.16	0.81	0.94	0.46	0.34	0.85	0.42	0.15	0.22
$S_2(y; 3.96)$	0.62	0.23	0.68	0.40	0.33	0.75	0.96	0.21	0.18	0.13
$S_1^*(x)$	0.64	0.18	0.73	0.91	0.55	0.36	0.82	0.45	0.09	0.27
$S_2^*(y)$	0.64	0.36	0.73	0.55	0.45	0.82	0.91	0.27	0.18	0.09

or simultaneously with the copula parameter as dictated by the context and convenience.

7.3.2 Method of Moments

Generalized method of moments or plug-in approach suggests (a) finding functional relationships between the parameters of interest and some distribution characteristics, such as moments; (b) using sample moments to estimate distribution moments; and (c) estimating the parameters by plugging in sample moments instead of distribution moments into the formulas established in (a). In case of copula parameters the role of moments may be played by concordance measures: Kendall's and Spearman's rank correlation. Formulas introduced in Subsection 7.2.4 provide functional relationships for (a) and can be used to estimate association for one-parameter Archimedean copulas and correlation for elliptical copulas. It is less obvious what moment-like measures should be used in case of two-parameter families.

Example 7.3.3 It is easy to estimate Kendall's concordance τ for the initial sample from Table 7.2 or even from ranks in Table 7.1. Kendall's concordance is based on ranks, so parametric (rows 1 and 2) and nonparametric (rows 3 and 4) estimates of the margins in Table 7.2 will bring about the same result $\hat{\tau} = 0.511$. Solving Equation (7.16) for α, we obtain the method of moment estimate

$$\hat{\alpha}_{MM} = \frac{1}{1 - \hat{\tau}} = 2.045, \tag{7.27}$$

which is pretty close to the true value. This degree of precision cannot be guaranteed for such a small sample, but confirms that the method of moments provides a "fast and dirty" procedure for the estimation of copula parameters.

7.3.3 Minimum Distance

This method suggests minimizing certain distance between the empirical distribution and distributions in the chosen family of copulas. The distribution from the family minimizing the distance is supposed to be the best choice. Difficulties with this approach may be related both to the proper choice of a distance and the technical task of finding its minimum. For certain classes of Archimedean copulas this approach may utilize the concept of Kendall's distribution defined in Section 6.6 and was proven to be very productive by Nicoloutsopoulos [34].

To illustrate this point, let us define **pseudo-observations** or **empirical copula** for $i = 1, \ldots, n$ as

$$Z_i = \frac{1}{n-1} \# \{(X_j, Y_j) : X_j < X_i, Y_j < Y_i, j = 1, \ldots, n\}, \tag{7.28}$$

Table 7.3 Ranks, pseudo-observations, and empirical Kendall's function

i	1	2	3	4	5	6	7	8	9	10
Rank (x)	4	9	3	1	5	7	2	6	10	8
Rank (y)	4	7	3	5	6	2	1	8	9	10
Z_i	2/9	6/9	1/9	0	4/9	1/9	0	5/9	8/9	7/9
$K_n(Z_i)$	0.46	0.73	0.36	0.18	0.55	0.36	0.18	0.64	0.91	0.82

the number of sample points both strictly to the left and below of (X_i, Y_i). Slightly unusual normalizing factor $n-1$ may be explained by convenience of the following representation of sample concordance in terms of pseudo-observations: $\hat{\tau} = 4\bar{Z} - 1$. Then define for any $t \in [0, 1]$ the empirical Kendall's function

$$K_n(t) = \frac{1}{n+1} \sum_{j=1}^{n} I\{Z_j \leq t\}, \tag{7.29}$$

which can serve as an empirical analog (and estimator) of the theoretical Kendall's function $K_\alpha(t) = P(C_\alpha(U, V) \leq t)$ corresponding to the true model. The values of Z_i and $K_n(Z_i)$ calculated by the ranks are summarized in Table 7.3.

Then we can compare the distances

$$d_K = \left(\sum_{j=1}^{n} (K_n(Z_j) - K_\alpha(Z_j))^2 \right)^{1/2} \tag{7.30}$$

for different values of association. The value $\hat{\alpha}_{MD}$ minimizing d_K may serve as an estimate of α.

Example 7.3.4 In the case of Gumbel–Hougaard family, according to (6.31), Kendall's function may be represented as

$$K_\alpha(t) = t \left(1 + \frac{\ln t}{\alpha} \right) \tag{7.31}$$

and the numerical minimization of d_K brings about $\hat{\alpha}_{MD} = 2.032$.

7.3.4 MLE and MPLE

The most popular method used in applications is probably still the method of maximum likelihood estimation (MLE). In the IFM setting [22], when on the second step we form pseudo-likelihood function by plugging in estimates of the marginals instead of their values and maximize this function to obtain the estimate for the association, it is more appropriate to talk about **maximum pseudo-likelihood estimation** (MPLE). In case of empirical estimates of the marginals, this method was discussed and justified by Genest et al. [18] and is also known as **canonical maximum likelihood** (CMLE).

Example 7.3.5   We will use both MPLE in IFM setting and full MLE to obtain copula parameter estimate for the data in Table 7.1. Pseudo-likelihood function for IFM can be obtained as

$$L(u, v; \alpha) = \prod_{i=1}^{10} c_\alpha(\hat{u}_i, \hat{v}_i), \tag{7.32}$$

where $c_\alpha(u, v)$ is the copula density defined in (6.27); $\hat{u}_i = S_1(x_i; 2.73)$ and $\hat{v}_i = S_2(y_i; 3.96)$ for parametric estimation and $\hat{u}_i = S_1^*(x_i)$ and $\hat{u}_i = S_2^*(y_i)$ for semiparametric estimation. Maximizing in (7.32) brings about the estimates for the association parameter: $\hat{\alpha}_P = 1.69$ for parametric and $\hat{\alpha}_{SP} = 2.03$ for semiparametric estimation (CMLE).

Full likelihood function for MLE can be expressed as

$$L(u, v; \theta_1, \theta_2, \alpha) = \prod_{i=1}^{10} c_\alpha(S_1(x; \theta_1), S_2(y, \theta_2)) \frac{\partial S_1}{\partial \theta_1} \frac{\partial S_2}{\partial \theta_2}, \tag{7.33}$$

and maximized with respect to all three parameters at a time. The numerical results yield: $\hat{\theta}_1 = 2.64$, $\hat{\theta}_2 = 3.70$, and $\hat{\alpha} = 1.69$.

It will be presumptuous to make conclusions based on a small sample numerical example. However, it seems to be true in general that the method of moments using Kendall's tau is the simplest and fastest way to obtain numerical value of association. Maximum likelihood approach is preferable for large samples because of its good large sample properties (asymptotic optimality). Using either nonparametric or parametric margins in IFM two-step context is possible, but numerical results may differ. The role of marginal estimation and possible marginal misspecification may be significant as our examples demonstrate. Full MLE and MPLE in parametric version typically bring about close numerical results, but it has to be decided whether the change of the marginal estimates from one-dimensional to two-dimensional case is an advantage or a drawback, and whether it is appropriate at all in the context of complex research projects with many pairs of variables being analyzed.

7.3.5   Bayesian Estimation

In Bayesian framework, we can certainly utilize both one-step and two-step approaches, and also use nonparametric estimates for the marginal. The difference between Bayesian approach and such classical methods as MLE, MPLE, or the method of moments as it was stated in Chapter 2 is more than just the choice of the best fit for our data within the given family of copulas. The benefit, as usual, is the use of entire posterior distribution instead of one representative of a family, providing more stable procedures and fully utilizing the benefits of integration versus optimization. The problem, as usual, is the choice of priors. In the presence of prior information we can elicit prior distributions using subjective considerations or use relevant data to implement empirical Bayes approach. In the absence of prior information we use noninformative priors. Examples and case studies in Sections 7.5 and 7.6 and Chapter 8 demonstrate a variety of ways in which Bayesian approach can be implemented to copula estimation. Here we will provide one particular implementation of Bayesian estimation via Metropolis algorithm using the familiar 10-point example and assuming a lack of reliable prior information.

Example 7.3.6 Let us consider a two-step IFM scheme, where both marginals are estimated first. This approach will allow us to avoid the problem of defining a joint prior for all three parameters and to assign priors independently. Assume very weak priors for the inverses of marginal parameters $\lambda_i = 1/\theta_i$, $i = 1, 2$, $\pi(\lambda_1) \sim Gamma(\varepsilon, \varepsilon)$ and $\pi(\lambda_2) \sim Gamma(\varepsilon, \varepsilon)$, where ε is chosen as small as possible in order not to sacrifice computational stability. This way, the prior means are equal to 1, and prior variances are both equal to $1/\varepsilon$, which makes priors as nonintrusive as possible. The resulting chains with 10,000 iterations produce the left and central trace plots in Figure 7.1 with Bayes estimates of the posterior mean $\tilde{\theta}_1 = 2.72$ and $\tilde{\theta}_2 = 3.98$.

At the second stage, using the estimated marginals, assume a conservative prior $\pi(\alpha) \propto \frac{1}{\alpha}, \alpha \in (1, \infty)$, penalizing higher values of association. The resulting trace plot is shown in the bottom part of Figure 7.1 and the estimate of posterior mean is $\tilde{\alpha} = 1.733$, which as can be expected for weak priors is similar to MLE.

Using the same independent priors for one-step analysis with simultaneous block update of all three parameters, we can expect a lower acceptance rate. Corresponding calculations bring about numerical estimates of $\tilde{\theta}_1 = 2.65$, $\tilde{\theta}_2 = 3.65$, and $\alpha = 1.747$. The corresponding trace plots are demonstrated in Figure 7.2.

In this example we cannot see any distinctive advantage of using Bayesian approach. First of all, we use no prior information, and therefore the final numerical estimates are not too far from MLE or method of moments estimates. Second, there is no technical problem with appying MLE for such a simple setting, therefore Bayesian approach is not necessary. However, it is definitely applicable, and leads to reasonable numerical results.

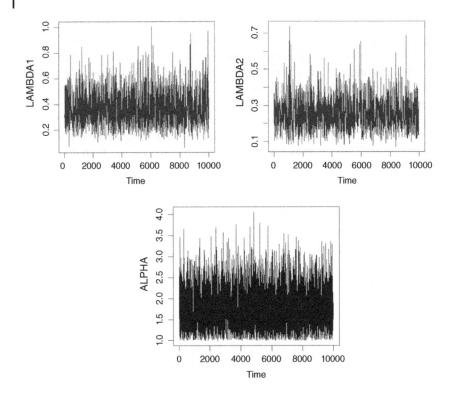

Figure 7.1 Trace plot for two-step Bayesian estimation.

7.4 Model Selection

In the previous section we have discussed parametric estimation within a chosen family of copulas. Here we will consider a different problem. How to make this choice of a family? Why should we prefer Gaussian to Frank's or Clayton's to Gumbel–Hougaard's? We will consider several principles which can help us to make this choice.

Let us say we want to compare two families of copulas and determine which of the two provides the best fit of the data. We can use a method like MLE, MPLE, method of moments, or Bayesian estimation to determine the best parametric fit in each family separately. Then we will select the family whose representative provides a better fit.

7.4.1 Hybrid Approach

This approach will combine two methods of estimation of the copula parameter: say, method of moments and MPLE. Suppose we consider two different

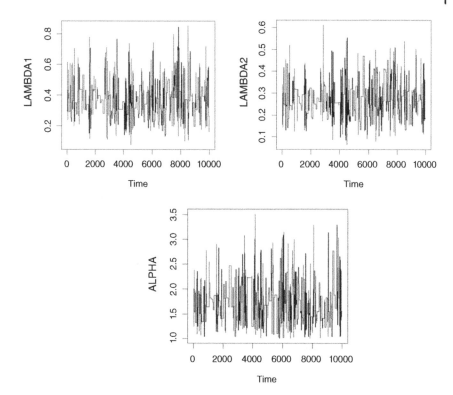

Figure 7.2 Trace plots for one-step Bayesian estimation.

families of copulas and want to choose the better model. To reduce the role of marginal specification, we will use the same estimate for the marginal distributions in both cases. Let us determine MPLE for the parameter of association for each of the two families. Then we can use the idea of generalized moments or concordances. Calculate concordance values implied by MPLEs in Subsection 7.2.3 for each family. Then compare these values with empirical concordance, and the smaller deviation will determine the best fit. For instance, model-induced values of τ obtained from estimated parameter values via 7.15 and 7.16 can be compared with the empirical value $\hat{\tau}$, and a smaller discrepancy will indicate a better model fit.

We can also use the hybrid between the minimum distance method and MLE or MPLE. In this case we choose two best representatives of two different copula families, and then compare them from the point of view of a distance from the empirical measure. The smaller distance wins. This approach is similar to the minimum distance method from the previous section, but now the problem is much simplified by the need to compare two numbers instead of running a minimization procedure over entire family. For instance, we can calculate the values of distances d_K defined in (7.30) for two best representatives of two families and choose the one granting the smaller value. The role of such distances in hybrid approach can be also played by goodness-of-fit measures discussed in Subsection 7.2.1. However there exists some danger of choosing an overfitted model if we have to choose between two families of different parametric complexity.

7.4.2 Information Criteria

Information criteria defined in Subsection 7.2.4 penalize model complexity and overfitting, and their application makes more sense for comparison of models with a different number of parameters. These criteria are based on the values of likelihood function, so they cannot be used "across the board" and we have to compare separately two-step to two-step, IFM to IFM, or CMLE to CMLE estimates obtained for two families. These criteria are not necessarily good for nonlikelihood-based methods of estimation.

Example 7.4.1 Let us extend the example with data from Table 7.1 to a different family of copulas. Clayton's family is often applied in the same context as survival Gumbel–Hougaard, when the modeling of the lower tails is of primary importance. Let us use Clayton copula to obtain MPLE estimates for α in parametric and nonparametric setting and also full MLE. Table 7.4 will summarize all information on numerical estimates obtained by full one-step MLE, parametric IFM (MPLE), and semiparametric method CMLE. We also calculate the model-induced values of tau and Kendall's distance d_K to implement the hybrid

Table 7.4 Comparison of survial Gumbel–Hougaard and Clayton copulas

Copula	Gumbel–Hougaard			Clayton		
method	MLE	IFM	CMLE	MLE	IFM	CMLE
$\hat{\theta}_1$	2.64	2.73	*	2.72	2.73	*
$\hat{\theta}_2$	3.70	3.96	*	3.85	3.96	*
$\hat{\alpha}$	1.69	1.69	2.03	0.86	0.86	1.48
τ	0.41	0.41	0.51	0.30	0.30	0.43
d_K	0.37	0.37	0.36	0.47	0.47	0.46
Log like	−42.06	1.72	2.45	−42.51	1.48	1.98
AIC	90.12	−1.44	−2.90	91.02	−0.96	−1.96
BIC	91.03	−1.14	−2.60	91.93	−0.66	−1.66
Tail	0.49	0.49	0.59	0.45	0.45	0.63

approach to model selection. In order to calculate the distance in (7.30) for the Clayton's family we can use the following formula obtained from (6.31):

$$K_\alpha(t) = \frac{1}{\alpha}((\alpha + 1)t - t^{\alpha+1}), \tag{7.34}$$

log-likelihood function, information criteria values, and tail dependence according to Subsection 7.2.5. We will leave the derivation of Bayes estimates for Clayton's family and comparison of two Bayes estimates for the end-of-the-chapter exercises.

It is clear that no parameter estimates are obtained for marginals when semi-parametric estimation is applied. Also, we can see slight discrepancies between the values of marginal parameters obtained by full one-step MLE and MPLE with IFM. Comparing values of α between Gumbel–Hougaard and Clayton model is in general not practical, because the meaning of these parameters is quite different. In this particular case $\tau = 0.5$ corresponds to $\alpha = 2$ for both families, but this is a very rare occasion.

However, according to the hybrid approach to estimation, comparison of corresponding values of τ is always informative. The model providing concordance estimates closer to the empirical value $\hat{\tau} = 0.511$ is more adequate. We should notice though that $\hat{\tau}$ is not a very reliable estimate of the theoretical τ, especially for small samples. For instance, in our example data were generated using Gumbel–Hougaard copula with association parameter $\alpha = 2$, corresponding by (7.16) to theoretical concordance value of $\tau = 0.5$. In this case survival Gumbel–Hougaard model performs somewhat better. This is to be expected, because the data were generated using this model, and Clayton copula should not necessarily provide a good fit.

From row 7, corresponding to d_K, it also looks like Gumbel–Hougaard family provides a better fit than Clayton: values of d_K are smaller for all pairs of estimates: MLE versus MLE, IFM versus IFM, and semiparametric CMLE versus CMLE.

Comparison of two families using AIC and BIC based on log-likelihood function should also be done pairwise. Full MLE will bring about the largest values of information criteria, which reflects the more complex structure of the model. Semiparametric model by definition will provide a better fit than the parametric IFM. We can compare though the pairs provided by similar settings for different copula choices. In all the three cases (MLE, IFM, CMLE) the Gumbel–Hougaard model ends up slightly on top. It would be interesting to compare the tail behavior of two copula families, but it is much harder to do so for the lack of reliable empirical measures of tail dependence, especially for small samples.

Bayesian estimation for Clayton family and comparison of Bayesian estimates for two families of copulas using information criterion DIC specific for Bayesian estimation is left for the end-of-chapter exercises in Chapter 8.

7.4.3 Bayesian Model Selection

Following Bretthorst [7] and Huard et al. [20], we suggest to compare the data fit provided by several pair copula models not at a single value of association parameter(s) obtained by MPLE, but rather over the entire range of possible association values. This can be accomplished by specifying a prior distribution for association parameter(s) and integrating the likelihood with respect to the prior distribution. The problem is the difference of meaning and ranges of association parameters for different copula classes. If we want to compare several classes of copulas in Bayesian framework, we need to establish the common basis of comparison. For that purpose we need to suggest a universal parameter, which can be evaluated for all classes of copulas under consideration.

One such universal parameter is Kendall's concordance τ, which as we have seen can be expressed in terms of association for many copula families. Sample concordance $\hat{\tau}$ is a reasonable nonparametric estimator of τ [19]. Using formulas expressing concordance through association parameters, for example, (7.15), (7.16), and (7.19), we can calculate values of τ induced by MPLE for parameters of elliptic or Archimedean copulas and compare them to the sample values of $\hat{\tau}$, which is done in Table 7.4. Proximity of model-induced values of τ to the sample value $\hat{\tau}$ may serve as a measure of the model fit and help to compare the model performance as in hybrid approach to model selection discussed above in Example 7.4.1. However, this comparison is still using single values to represent entire families.

We will assume that the classes of copulas we choose correspond to exhaustive and mutually exclusive hypotheses $H_1, H_2, \ldots, H_m$. Posterior probabilities of hypotheses $H_k, k = 1, \ldots, m$, for data D may be rewritten as

$$P(H_k \mid D) = \int P(H_k, \tau \mid D) d\tau = \frac{\int P(D \mid H_k, \tau) P(H_k \mid \tau) \pi(\tau) d\tau}{P(D)}, \quad (7.35)$$

where we will consider all m hypotheses a priori equally likely. If the dependence between variables is positive for all hypotheses, we can assume $\tau \geq 0$. In this case the natural choice of prior for τ is beta distribution, and the choice of parameters for the prior can be subjective or noninformative objective. However, in the presence of relevant additional data, similar in nature, the prior might be suggested by sample concordance for this data consistently with empirical Bayes approach: $P(D \mid H_k, \tau) = L_k(D \mid \alpha(\tau))$, $P(H_k \mid \tau) = P(H_k) = \frac{1}{m}$, $\pi(\tau) \sim Beta(\hat{a}, \hat{b})$. If we have multiple pairs of variables included in the study, estimates of parameters of the beta distribution for empirical Bayes can be obtained from all pairs of components.

We will not need to calculate the denominator of the posterior in (7.35). It suffices to calculate the weights

$$W_k = \int_0^1 L_k(D \mid \alpha(\tau))\pi(\tau)d\tau = \int_0^1 \Pi_{i=1}^n c_k(\hat{u}_i, \hat{v}_i \mid \alpha(\tau))\pi(\tau)d\tau, \quad (7.36)$$

or, using Monte Carlo approach and drawing samples from the Beta prior, evaluate

$$\hat{W}_k = \frac{1}{N} \sum_{j=1}^N \Pi_{i=1}^n c_k(\hat{u}_i, \hat{v}_i \mid \alpha(\tau_j)). \quad (7.37)$$

Then we select the class with the highest weight and obtain the Bayes estimate of the association parameter using MCMC.

7.5 Copula Models of Joint Survival

This study has been mentioned in the introduction to Chapter 5 and some of its results were used to numerically illustrate copula models in Chapter 6. Now we will consider it in more detail. The data we use consists of 11,457 joint last survivor annuity contracts of a large Canadian insurer in payoff status over a 5-year observation period 1989–1993. Most of the sample represented heterosexual married couples. We will construct copula models for bivariate joint distributions of the couples. As discussed in Section 5.1, there exist multiple factors of dependence between two lives in a couple such as common shock, common lifestyle, or broken heart syndrome. To address all of these factors, it would be beneficial to model entire joint distribution, and copulas provide convenient tools for that.

Frees et al. [14] were the first to suggest copula models of joint survival in this context, and they used this dataset for their research. Shemyakin and Youn applied the same dataset for illustration of different estimation techniques in [40], [41], and [42], and we will use some results of these papers. However, we will restrict ourselves to joint first life analysis (time to the first death in a couple) and choose not to discuss more complicated joint last survivor models (time to the second death in a couple), which are built, for example, in [42]. The main objective of this example is to compare the results of maximum likelihood estimation to the results of Bayesian analysis with informative priors.

Estimating the joint first life survival function of a married couple requires observing a pair of associated lives until the first death. Each observation of a pair of associated lives (L_{i1}, L_{i2}) in a sample $y = (y_1, y_2, ..., y_n)$ represents a vector

$y_i = (a_{i1}, a_{i2}, t_i, c_i)$, where a_{ij}, $j = 1, 2$ are entry ages of lives L_{ij}, t_i is the time till termination, and c_i is the censoring indicator:

$$c_i = \begin{cases} 0, & t_i = T_i \ (\textit{no death during observation period}) \\ 1, & t_i < T_i \ (\textit{first death occurred during observation period}) \end{cases},$$

where T_i is the duration of observation period for y_i. This setting represents right censoring (not all lives terminate during the observation period) and left truncation (represented by entry ages).

Conventional copula modeling based on two marginal distribution for male and female lives requires a very careful treatment because of the necessity to address both censoring and truncation. Therefore in this setting the idea of using **conditional copula** is very attractive. We will apply a copula function to conditional survival functions instead of the marginals so that the **joint first life survival function**

$$p_{FL}(t; a_1, a_2)$$
$$= P\left(\min\{L_{i1} - a_1, L_{i2} - a_2\} > t \mid \min\{L_{i1} - a_1, L_{i2} - a_2\} > 0\right)$$

can be represented through conditional survival functions

$$p_{FL}(t; a_1, a_2) = C_\alpha \left(\frac{S_1(a_1 + t)}{S_1(a_1)}, \frac{S_2(a_2 + t)}{S_2(a_2)} \right),$$

where the marginal survival functions S_j and the copula function C_α are left to be specified. Using Weibull marginals with scale parameters θ_j and shapes τ_j with a survival Gumbel–Hougaard copula brings about expression

$$p_{FL}(t; a_1, a_2) = exp\left\{ -\left(w_1^\alpha + w_2^\alpha\right)^{1/\alpha} \right\},$$

where for $j = 1, 2$,

$$w_j = w(t, a_j, \tau_j, \theta_j), \quad w(t, a, \tau, \theta) = \left(\frac{a + t}{\theta}\right)^\tau - \left(\frac{a}{\theta}\right)^\tau,$$

and the likelihood function

$$l(\alpha, \tau_1, \tau_2, \theta_1, \theta_2 | y_1, \ldots, y_n) = exp\left\{ -\sum_{i=1}^{n} \left(\sum_{j=1}^{2} w_{ij}^\alpha \right)^{1/\alpha} \right\}$$

$$\times \prod_{i=1}^{r} \left[\sum_{j=1}^{2} \frac{\tau_j w_{ij}^{\alpha-1}}{\theta_j^{\tau_j}} \left(a_{ij} + t_i\right)^{\tau_j - 1} \cdot \left(\sum_{j=1}^{2} w_{ij}^\alpha \right)^{(1/\alpha)-1} \right] \tag{7.38}$$

with $c_i = 1$ for $i = 1, \ldots, r$ and $c_i = 0$ for $i = r + 1, \ldots, n$.

Table 7.5 Parameter estimates (with standard errors)

Parameters		M_1 Independence	M_2 MLE	M_3 Bayes
Female	θ_1	92.62	89.51	89.59
		(0.59)	(0.48)	(0.6332)
	τ_1	9.98	9.96	8.614
		(0.38)	(0.34)	(0.655)
Male	θ_2	86.32	85.98	87.06
		(0.37)	(0.40)	(0.6324)
	τ_2	7.94	7.65	7.202
		(0.26)	(0.26)	(0.68)
Association	α	1	1.64	1.812
			(0.49)	(0.729)

Frees et al. [14] used a different model specification, not requiring conditional copulas. They also considered Frank's copula along with Gumbel–Hougaard family, and Gompertz model for the marginals along with Weibull model. Prior to their work, most related actuarial studies of joint life assumed independence of the marginal distributions and in pricing joint life products applied special corrections compensating for underestimation of risks caused by this assumption. Following the guidelines of [14], Shemyakin and Youn applied one-step maximum likelihood estimation to Weibull marginals and survival Gumbel–Hougaard copula in [40]. Some of the results of estimation and standard errors are presented in Table 7.5. Model M_1 assumes independence of the marginals, which can be treated as Gumbel–Hougaard copula with $\alpha = 1$. Parameters in Model M_2 were estimated by one-step full MLE [40]. The estimates from M_2 were used in joint life examples in Chapter 6.

Discrepancies between the marginal estimates for models M_1 and M_2 can be explained by construction of MLE, and more general model M_2 should be preferable. However, with massive information available on male and female mortality (mortality tables), it seems reasonable to use Bayesian approach. Bayesian models, based on the conditional copula approach with Weibull marginals and stable copula, were first applied to joint mortality in [41]. The structure of priors suggested normal or lognormal for the shape and scale parameters, and diffuse or noninformative priors on the parameter of association. Hyperparameters for the shape and scale distributions can be chosen to fit the existing male and female ultimate mortality tables. This choice reflects the belief that a substantial amount of prior information on individual male and female mortality can be incorporated in a copula model improving the estimation of marginals.

Model M_3 in Table 7.5 is based on applying copula structure to conditional survival functions. It requires Bayesian estimation of five parameters of the conditional copula function with likelihood (7.38) and the following priors: $\theta_j \sim N(\mu_{\theta_j}, \sigma_{\theta_j})$, $\tau_j \sim N(\mu_{\tau_j}, \sigma_{\tau_j})$, $\pi(\alpha) \propto \alpha^{-1}$. The hyperparameters are determined from the US male and female mortality tables as: $\mu_{\tau_1} = 8.535$, $\sigma_{\tau_1} = 0.454$, $\mu_{\theta_1} = 89.62$, $\sigma_{\theta_1} = 0.412$, $\mu_{\tau_2} = 7.097$, $\sigma_{\tau_2} = 0.485$, $\mu_{\theta_2} = 86.96$, $\sigma_{\theta_2} = 0.422$.

Bayesian computation is performed using independent Metropolis–Hastings algorithm in special software package WinBUGs [35] with chain length of 10,000. In order to compare performance of the models, we calculate Bayesian Information Criterion (BIC) based on the likelihood (7.38) and using Model 1 as the baseline. In order to take into account the effect of censoring, we use the methodology of [46], suggesting

$$BIC(M_k) = -2(L_k - L_1) + (m_k - m_1)\ln(r),$$

where L_k and m_k, $k = 2, 3$, are, respectively, maximum log-likelihoods and number of parameters for models M_k, r is the number of noncensored observations. We arrive at the lowest value of BIC for Model M_3.

Table 7.5 demonstrates that the results of Bayesian estimation with informative priors may substantially differ from MLE, especially when a one-step full MLE procedure is applied. Bayesian approach in this case allows for the use of mortality tables representing substantial prior information on the marginals.

7.6 Related Failures of Vehicle Components

We will review the example from Section 3.6, where Markov chains were applied to model-related failures of vehicle components. In this section we will take a different approach: we will analyze time-to-failure (TTF) variables for different car parts using the methods of survival analysis similar to those applied to time-to-default (TTD) variables in Section 4.5. Now we will use copulas to model dependence between TTF variables for different vehicle components, providing a valid alternative to Markov chain analysis of related failures. Actually, copula approach has an advantage of modeling entire joint distribution of components, directly evaluating all relevant conditional probabilities and assessing the chances of a new failure given the history of vehicle repairs. In this example based on research paper [27], we will restrict ourselves to pair copulas and will not address data censoring and selection bias.

The main purpose of this example is to illustrate the principles of copula model selection considered in Section 7.4, choosing a family of copulas based on multiple pairs of variables. We will consider hybrid approach, information criteria, and finally, Bayesian model selection.

Let us recall the problem from Section 3.6. Five main components chosen out of 60 detected in Hyundai Accent warranty claim records are: the spark plug assembly (A), ignition coil assembly (B), computer assembly (C), crankshaft position sensor (D), and oxygen heated sensor (E). Data were recorded for the cars that had at least one of the five main components fail within the warranty period. We analyze TTF of each component measured in days from the car sale (or the previous repair) to the next repair. The ultimate goal is to suggest a pair copula model for the joint distribution of all 10 possible pairs of TTF variables of five engine assembly components. If one pair copula provides a better fit for most of the pairs, it can be used in further multidimensional analysis.

In the choice of copula families, the factor of tail dependence plays a very important role. Preliminary analysis indicated the presence of two tails of bivariate distributions corresponding to joint or related failures of two parts occurring either very early in the car history, or very late (close to the end of the warranty period). It is evident that both tails represent a special interest for car manufacturers and insurers. Therefore we will consider six different copula families: Clayton (H1) and survival Gumbel–Hougaard (H4), representing the left tail dependence, Gumbel–Hougaard (H2) and survival Clayton (H3), capturing the right tail behavior, and also two two-parameter copulas allowing for both: BB1 copulas (H5) defined in (6.29) and t-copulas (H6) defined in (6.14).

For each of these families we can define copula density as

$$c_k(u, v|\alpha) = \frac{\partial^2 C_k(u, v|\alpha)}{\partial u \partial v}, \quad k = 1, ..., 6.$$

Consider all 10 possible pairs (X, Y) of TTF for components (A)–(E). For a matched i.i.d. sample $(x_i, y_i), i = 1, ..., n$, we use copulas to model the joint distribution of the underlying TTF variables X and Y. First, estimate the marginal distributions of X and Y as $\hat{u} = \hat{F}(x)$ and $\hat{v} = \hat{G}(y)$ to obtain the sample $D = (\hat{u}_i, \hat{v}_i), i = 1, ..., n$ and then estimate the association parameter α for the copula $C(\hat{u}, \hat{v} \mid \alpha)$.

IFM approach suggests using a sensible parametric model for marginals and then estimating the association. Properties of the estimates obtained by this approach were fully investigated in [21]. Weibull distribution often provides a good fit for individual parts' TTFs, see Baik [3], Lawless [28], and also Wu [48]. However, here we concentrate on the identification of a good copula model of association. Therefore in order to minimize the role of possible misspecification of the marginal distributions, we apply semiparametric CMLE approach: use corresponding empirical distribution functions as the estimates $\hat{u}$ and $\hat{v}$ as suggested by Genest and Rivest [19] and followed by many other authors.

7.6.1 Estimation of Association Parameters

The pseudo-likelihood function for the copula estimation can be written as

$$L_k(D|\alpha) = \prod_{i=1}^{n} c_k(\hat{u}_i, \hat{v}_i|\alpha),$$

Table 7.6 Estimates of association for six classes of copulas

	H1 $\hat{\alpha}$	H2 $\hat{\alpha}$	H3 $\hat{\alpha}$	H4 $\hat{\alpha}$	H5 $\hat{\alpha}$	H5 $\hat{\beta}$	H6 $\hat{\rho}$	H6 $\hat{\eta}$
AB	3.02	3.12	2.95	3.14	0.62	2.48	0.92	2.00
AC	0.58	1.36	0.57	1.37	0.20	1.26	0.42	2.44
AD	1.05	1.61	0.93	1.64	0.52	1.33	0.54	2.00
AE	0.62	1.44	0.71	1.42	0.15	1.36	0.48	2.34
BC	0.74	1.41	0.58	1.44	0.44	1.19	0.46	2.37
BD	0.93	1.64	0.95	1.63	0.26	1.48	0.67	2.00
BE	0.81	1.57	0.87	1.54	0.27	1.41	0.57	3.46
CD	1.12	1.74	1.10	1.75	0.42	1.48	0.64	2.98
CE	0.62	1.46	0.73	1.42	0.15	1.37	0.49	2.51
DE	0.43	1.21	0.30	1.24	0.32	1.07	0.28	3.43

where $\hat{u}_i$ and $\hat{v}_i$ are marginal empirical distribution functions. Using a numerical implementation of the maximum likelihood method [6, 25], obtain for each class of copluas H1–H5 the CMLE $\hat{\alpha}$ (also $\hat{\beta}$ for BB1 and $\hat{\rho}$ and $\hat{\eta}$ for t-copula). We present these values for each of the 10 pairs of five components in Table 7.6.

Notice that the number of degrees of freedom η was set to be at least 2, and this limit was attained for AB, AD, and BD.

The choice of a specific copula class from the set H1–H6 is important. Let us consider the example of early failure of both components A (spark plug assembly) and B (ignition coil assembly) in the first year of the vehicle use. There was a total of K failures for the first 365 days out of N vehicles on the record, corresponding to the relative frequency of the first-year failure of $\hat{p} = K/N = 0.398$. If we use the parameter values from Table 7.6, Clayton's copula H1 would suggest the probability of failure $p_1 = 0.365$, copula H2 would yield $p_2 = 0.371$, BB1 copula H5 would yield $p_5 = 0.373$, and t-copula gives $p_6 = 0.387$.

7.6.2 Comparison of Copula Classes

We will first use information criteria to determine which of the six copula classes H1–H6 provides the best fit. Applying Akaike information criterion (AIC) and Bayes information criterion (BIC) brings about the results in Tables 7.7 and 7.8. The larger negative values correspond to the better models. The best values of the criteria are boldfaced. Two-parametric class H6 provides the best fit in most of the cases.

The Kolmogorov–Smirnov distance defined in (7.2) measures the maximum distance between a cumulative distribution function (c.d.f.) and its empirical cumulative distribution function (e.c.d.f.), and is applicable to joint distributions as well. Table 7.9 summarizes the Kolmogorov–Smirnov distances

Table 7.7 AIC values for six classes of copulas

	H1	H2	H3	H4	H5	H6
AB	−521.8	−633.4	−518.9	−637.7	−661.9	**−809.5**
AC	−57.8	−79.3	−64.6	−81.4	−82.1	**−103.7**
AD	−50.5	−54.1	−40.8	−58.8	−60.2	**−73.5**
AE	−46.6	−72.9	−57.0	−61.8	−72.7	**−89.0**
BC	−55.6	−51.6	−34.3	−63.6	−61.6	**−70.5**
BD	−41.3	−54.9	−39.9	−54.3	−55.0	**−81.9**
BE	−65.0	−83.2	−62.4	−77.5	−85.4	**−88.8**
CD	−106.4	−127.8	−102.0	−129.9	−135.6	**−143.3**
CE	−114.6	−179.5	−141.1	−150.7	−182.2	**−208.8**
DE	−24.4	−17.1	−10.8	−28.1	−23.9	**−28.8**

Table 7.8 BIC values for six classes of copulas

	H1	H2	H3	H4	H5	H6
AB	−517.7	−629.2	−514.7	−633.5	−653.6	**−801.2**
AC	−53.7	−75.2	−60.5	−77.3	−74.0	**−95.5**
AD	−47.5	−51.1	−37.8	−55.7	−54.2	**−67.4**
AE	−42.8	−69.2	−53.3	−58.0	−65.2	**−81.5**
BC	−52.1	−48.1	−30.7	−60.0	−54.5	**−63.4**
BD	−38.3	−51.8	−36.8	−51.2	−48.8	**−75.7**
BE	−61.5	−79.6	−58.9	−73.9	−78.3	**−81.7**
CD	−102.8	−124.2	−98.4	−126.3	−128.4	**−136.1**
CE	−110.1	−174.9	−136.5	−146.2	−173.2	**−199.8**
DE	−20.8	−13.5	−7.2	**−24.5**	−16.7	−21.7

Table 7.9 d_{KS} distances for six classes of copulas

	H1	H2	H3	H4	H5	H6
AB	0.245	0.167	0.179	0.161	0.0317	**0.0301**
AC	0.379	0.115	0.115	0.111	0.0283	**0.0274**
AD	0.215	0.106	0.112	0.102	0.0600	**0.0533**
AE	0.144	0.215	0.166	0.132	0.0347	**0.0314**
BC	0.242	0.146	0.141	0.146	0.0368	**0.0312**
BD	0.414	0.157	0.173	0.158	**0.0304**	0.0344
BE	0.266	0.144	0.133	0.164	**0.0287**	0.0310
CD	0.324	0.178	0.175	0.206	0.0382	**0.0368**
CE	0.176	0.083	0.065	0.096	0.0256	**0.0225**
DE	0.498	0.264	0.230	0.228	0.0392	**0.0341**

between the joint e.c.d.f.s and model c.d.f.s corresponding to the MPLE of parameters for each copula class.

From the computed Kolmogorov–Smirnov distance values, H5 and H6 are two best choices by far when compared to the other hypotheses. The distance values for H5 and H6 range from 0.0225 up to 0.06, whereas the other four combined range from 0.0653 up to 0.498. Each of H1–H4 misrepresents the e.c.d.f.s at some point, but H5 and H6 accurately represent the entirety of the data. H5 and H6 can be concluded as the obvious selections when comparing Kolmogorov–Smirnov distance values.

Tail dependence describes extreme co-movements between a pair of random variables in the tails of the distributions. The tail dependence coefficients for copulas $C(u, v)$ was defined in (7.20) and (7.21). Certain copulas, such as the Gaussian copula, do not admit tail dependence (their tail dependence coefficients are equal to 0). Some copulas (e.g., H1–H4) exhibit only either upper or lower tail dependence, and the t-copula and BB1 family exhibit both. Thus, when selecting an accurate copula model for data it is important to consider whether the data displays upper and/or lower tail dependence. Formulas (7.22)–(7.26), from Section 7.2 can be used to calculate the lower and/or upper tail dependences of the selected five copula models: H1: $\lambda_l = 2^{-1/\alpha}$, H2: $\lambda_u = 2 - 2^{1/\alpha}$, H3: $\lambda_u = 2^{-1/\alpha}$, H4: $\lambda_l = 2 - 2^{1/\alpha}$, H5 reflects both lower and upper tail dependences, and they are allowed to be asymmetric: $\lambda_l = 2^{-1/\alpha\beta}$, and $\lambda_u = 2 - 2^{1/\beta}$, H6: $\lambda_l = \lambda_u = 2t_{\eta+1}(-\sqrt{\eta + 1}\sqrt{\frac{1-\alpha}{1+\alpha}})$.

All 10 pairs in Table 7.10 exhibit both upper and lower tail dependence. Since H5 and H6 are the only two classes that allow for both, they demonstrate the best fit, though the symmetry of the tails is forced for H6. Lower values for H5, especially at the lower tails, may indicate some problems.

Table 7.10 Tail dependence induced by the models

	H1 (L)	H2 (U)	H3 (U)	H4 (L)	H5 (L)	H5 (U)	H6 (L=U)
AB	0.795	0.751	0.790	0.753	0.639	0.678	0.741
AC	0.300	0.334	0.297	0.340	0.063	0.268	0.313
AD	0.515	0.460	0.475	0.472	0.367	0.318	0.415
AE	0.324	0.384	0.376	0.365	0.032	0.336	0.351
BC	0.394	0.364	0.305	0.380	0.270	0.214	0.339
BD	0.475	0.473	0.483	0.471	0.161	0.404	0.501
BE	0.427	0.446	0.449	0.431	0.158	0.365	0.322
CD	0.540	0.511	0.532	0.515	0.325	0.405	0.404
CE	0.327	0.392	0.388	0.369	0.036	0.342	0.341
DE	0.200	0.228	0.099	0.250	0.136	0.094	0.184

All of these approaches (using AIC, BIC, Kolmogorov–Smirnov statistic, or tail dependence) can help us to determine the best class of pair copula models. However they share the same weakness: the analysis is restricted to the comparison of the single representatives of each class obtained by CMLE. The following section discusses a possibility to make conclusions based on multiple representatives of six classes H1–H6.

7.6.3 Bayesian Model Selection

We suggest Kendall's concordance τ as a universal parameter, which can be evaluated for all six classes H1–H6 defined earlier. Sample concordance $\hat{\tau}$ is a reasonable nonparametric estimator of τ. Using parametric estimates from Table 7.5 and formulas expressing concordance through association parameters (see definitions of classes H1 and H2, (7.15) and (7.16)), we can calculate the values of τ induced by CMLE for parameters of copula classes H1–H6, and compare them to the sample values of $\hat{\tau}$, which is done in Table 7.11.

Proximity of model-induced values of τ to the sample value $\hat{\tau}$ may serve as a measure of the model fit and help to compare the model performance. However, this comparison is still using single values representing entire families.

Following Bretthorst [7] and Huard [20], in Section 7.4 we suggested to compare the data fit provided by models H1–H6 not at a single value of association parameter(s) obtained by MLE, but rather over the entire range of possible association values. This can be accomplished by specifying a prior distribution for association parameter(s) and integrating the likelihood with respect to the prior distribution. The problem is the difference of meaning and ranges of association parameters for different copula classes. However this problem is resolved by specifying a prior on τ.

Table 7.11 Sample $\hat{\tau}$ and model-induced values of τ

	$\hat{\tau}$	H1	H2	H3	H4	H5	H6
AB	0.707	0.601	0.679	0.596	0.682	0.692	0.737
AC	0.269	0.224	0.264	0.222	0.269	0.278	0.278
AD	0.375	0.343	0.377	0.318	0.389	0.403	0.364
AE	0.314	0.235	0.307	0.262	0.291	0.316	0.320
BC	0.307	0.271	0.290	0.226	0.304	0.311	0.306
BD	0.400	0.318	0.389	0.323	0.387	0.402	0.471
BE	0.386	0.289	0.364	0.302	0.350	0.375	0.382
CD	0.452	0.360	0.425	0.354	0.429	0.442	0.443
CE	0.318	0.237	0.315	0.268	0.295	0.321	0.324
DE	0.188	0.177	0.174	0.130	0.193	0.194	0.182

There exist convenient formulas expressing association as a function of τ for all four one-parametric copulas H1–H4, inverting (7.15) and (7.16):

$$\alpha = \frac{2\tau}{(1 - \tau)}$$

for H1 and H3, and

$$\alpha = \frac{1}{1 - \tau}$$

for H2 and H4.

In the case of the two-parametric class H5 we will use the conditional relationship between α and τ treating β as the nuisance parameter and using its MPLE estimate $\hat{\beta}$:

$$\alpha = \frac{2}{\hat{\beta}(1 - \tau)} - 2 \Leftrightarrow \tau = 1 - \frac{2}{\hat{\beta}(\alpha + 2)}.$$

This approach can be justified by the higher risks of early related failures corresponding to the lower tails of the joint distribution represented by α, which becomes the critical parameter for the class H5. Notice that in the two-parametric class H6, there is an additional parameter η (degrees of freedom), which is not directly related to τ. We will apply (7.19) and use CMLE $\hat{\eta}$ in the further analysis.

We will assume that six classes H1–H6 represent exhaustive and mutually exclusive hypotheses. Posterior probabilities of these hypotheses may be rewritten as

$$P(H_k \mid D) = \int P(H_k, \tau \mid D)d\tau = \frac{\int P(D \mid H_k, \tau)P(H_k \mid \tau)\pi(\tau)d\tau}{P(D)}, \quad (7.39)$$

where we will consider all six hypotheses a priori equally likely, the dependence between variables being positive which suggests $\tau \geq 0$. In this case the natural choice of prior for τ is Beta distribution, and the choice of parameters for the prior is suggested by sample concordance for the entire dataset consistently with empirical Bayes approach: $P(D \mid H_k, \tau) = L_k(D \mid \alpha(\tau))$, $P(H_k \mid \tau) = P(H_k) = \frac{1}{6}$, $\pi(\tau) \sim Beta(\hat{a}, \hat{b})$. Estimates of parameters of the Beta distribution are obtained from 10 pairs of components as $\hat{a} = 4.095$ and $\hat{b} = 6.920$.

We will not need to calculate the denominator of the posterior in (7.39). It suffices to calculate the weights

$$W_k = \int_0^1 L_k(D \mid \alpha(\tau))\pi(\tau)d\tau = \int_0^1 \Pi_{i=1}^n c_k(\hat{u}_i, \hat{v}_i \mid \alpha(\tau))\pi(\tau)d\tau, \quad (7.40)$$

Table 7.12 Logs of posterior weights for H1–H6

	H1	H2	H3	H4	H5	H6
AB	258.41	313.16	256.99	315.24	327.88	**401.31**
AC	27.84	38.83	31.18	39.88	41.14	**52.22**
AD	24.83	26.73	20.03	29.03	30.53	**37.58**
AE	22.40	35.87	27.71	30.29	36.60	**45.09**
BC	27.11	25.29	16.32	31.28	31.13	**35.90**
BD	20.29	27.12	19.59	26.82	27.98	**41.48**
BE	31.80	41.06	30.58	38.23	42.94	**44.97**
CD	52.49	63.13	50.29	64.16	67.75	**71.96**
CE	56.03	88.76	69.37	74.38	90.92	**104.61**
DE	10.96	7.46	3.61	13.12	11.88	**14.52**

or, using Monte Carlo approach and drawing samples from the Beta prior, evaluate

$$\hat{W}_k = \frac{1}{N} \sum_{j=1}^{N} \Pi_{i=1}^{n} c_k(\hat{u}_i, \hat{v}_i \mid \alpha(\tau_j)). \tag{7.41}$$

Then we choose the class with the highest weight and obtain the Bayes estimate of the association parameter using MCMC.

Table 7.12 gives the values of the natural logarithms of weights $\ln(\hat{W}_k)$ calculated by $N = 20,000$ runs of Monte Carlo sampling from the Beta prior. The highest values for each pair of components are boldfaced.

7.6.4 Conclusions

As we can see from the warranty claim data, a substantial dependence is observed between the TTFs for automotive components related to engine subassembly of Hyundai Accent vehicles. This dependence cannot be ignored, because the assumption of independence would lead to grossly underestimated related failure risks and warranty costs. However it can be addressed using pair copula models. Open source software packages in R environment [6, 25] allow for a straightforward implementation of parametric and semiparametric estimation for different classes of copulas: direct and survival Archimedean copulas (Clayton and Gumbel–Hougaard classes), two-parameter Archimedean copulas (BB1), and elliptical copulas (Student t-class).

The problem of model selection a.k.a. an adequate choice of a copula class stays central, because misspecification of the class of copulas can be critical. For model comparison, one can consider using information criteria or Kolmogorov–Smirnov statistic as a good measure of overall fit, or studying tail dependence to assess the model quality. A convenient tool of model selection is provided by Bayesian framework using Kendall's tau as the common parameter for different classes of copulas.

For TTFs of automotive components most of the tools of model selection indicate t-copulas being superior to four one-parameter Archimedean copulas and even to BB1. This fact can be explained by t-copulas exhibiting tail dependence in both lower and upper tails of the joint TTF distribution, while direct or survival Archimedean copulas concentrate at one of the tails or, as seems to be the case with BB1, underestimates tail dependence. However, it is not clear whether the assumption of symmetry of the tails of t-copulas is plausible and not too restrictive. One possible suggestion for future work is to use more complex hybrid or mixed Archimedean copula models as an alternative to t-copulas (see also Reference [26]). Another alternative would be to consider skewed or asymmetric t-copulas [9] or other more complex extensions of elliptical copula models.

References

1 Akaike, H. (1974). A new look at the statistical model identification. *IEEE Transactions on Automatic Control*, 19(6), 716–723.

2 Akaike, H. (1985), Prediction and entropy. In: A. C. Atkinson, and S. E. Fienberg (editors), *A Celebration of Statistics*, 1–24. Springer.

3 Baik, J. (2010). Warranty analysis on engine: case study. Technical Report, Korea Open National University, Seoul, 1–25.

4 Berg, D. (2007). Copula goodness-of-fit testing: an overview and power comparison. Statistical Research Report No. 5, University of Oslo, Oslo.

5 Box, G. E. P., and Draper, N. R. (1987). *Empirical Model-Building and Response Surfaces*, John Wiley & Sons, Inc.

6 Brechmann, E. C., and Schepsmeier, U. (2013). Modeling dependence with C- and D-vince copulas: the R package CDVine. *Journal of Statistical Software*, 52(3), 1–27.

7 Bretthorst, G.L. (1996). An introduction to model selection using probability theory as logic. In: G. Heidbreger (editor,) *Maximum Entropy and Bayesian Methods*, 1–42. Kluwer Academic Publishers.

8 CDO Primer: The Bond Market Association (2004).

9 Demarta, S., and McNeil, A. (2005). The t copulas and related copulas. *International Statistical Review*, 73, 111–129.

10 Embrechts, P., and Hofert, M. (2014). Statistics and quantitative risk management for banking and insurance. *Annual Review of Statistics and its Application*, 1, 492–514.

11 Embrechts, P., McNeil, A., and Straumann, D. (2003). Correlation and dependency in risk management: properties and pitfalls. In: M. Dempster, and

H. K. Moffat (editors) *Risk Management: Value at Risk and Beyond.* Cambridge University Press, 176–223.

12 Fermanian, J-D., and Scaillet, O. (2005). Some statistical pitfalls in copula modelling for financial applications. In: E. Klein (editor), *Capital Formation, Governance and Banking*, 59–74. New York: Nova Science Publication.

13 Fermanian, J-D., Charpentier, A., and Scaillet, O. (2006). The estimation of copulas: theory and practice. In: J. Rank (editor), *Copulas, from Theory to Application in Finance.*, New York: Risk Books.

14 Frees, E. W., Carriere, J. F., and Valdez, E. (1996), Annuity valuation with dependence mortality. *Journal of Risk and Insurance*, 63, 2, 229–261.

15 Gelman, A. (2013). Two simple examples for understanding posterior p-values whose distributions are far from unform. *Electronic Journal of Statistics*, 7, 2595–2602.

16 Gelman, A., Carlin, J. B., Stern, H.S., and Rubin, D. B. (2004). *Bayesian Data Analysis*, 2nd ed. Texts in Statistical Science. CRC Press.

17 Genest, C., Ghoudi, K., and Rivest, L. P. (1995). A semiparametric estimation procedure of dependence parameters in multivariate families of distributions. *Biometrika*, 82(3), 543–552.

18 Genest, C., Remillard, B., and Beaudoin, D. (2009). Goodness-of-fit tests for copulas: a review and a power study. *Insurance: Mathematics and Economics*, 44, 199–213.

19 Genest, C., and Rivest, L.-P. (1993). Statistical inference procedures for bivariate Archimedean copulas. *Journal of the American Statistical Association*, 88, 1034–1043.

20 Huard, D., Evin, G., and Favre, A.-C. (2006). Bayesian copula selection. *Computational Statistics and Data Analysis*, 51(2), 809–822.

21 Joe, H. (1997). *Multivariate Models and Dependence Concepts*. London: Chapman & Hall.

22 Joe, H. (2005). Asymptotic efficiency of the two-stage estimation method for copula-based models. *Journal of Multivariate Analysis*, 94, 401–419.

23 Joe, H. (2014). *Dependence Modeling with Copulas*. London: Chapman & Hall/CRC.

24 Kim, G., Silvapulle, M., and Silvapulle, P. (2007). Comparison of semiparametric and parametric methods for estimating copulas. *Computational Statistics and Data Analysis*, 51(6), 2836–2850.

25 Kojadinovic, I., and Yan, J. (2010). Modeling multivariate distributions with continuous margins using the copula R package. *Journal of Statistical Software*, 34(9), 1–20.

26 Komornikova, M., and Komornik, J. (2010). A copula based approach to the analysis of returns of exchange rates to EUR of the Visegrad countries. *Acta Polytechnica Humgarica*, 7(3), 79–91.

27 Kumerow, J., Lenz, N., Sargent, K., Shemyakin, A., and Wifvat, K. (2014). Modelling related failures of vehicle components via Bayesian copulas, *ISBA-2014, Cancun, Mexico*, § 307, 195.

28 Lawless, J. F., Hu, J., and Cao, J. (1995). Methods for estimation of failure distribution and rates from automobile warranty data. *Lifetime Data Analysis*, 1, 227–240.

29 Li, D.X. (2000) On default correlation: a copula function approach. *Journal of Fixed Income*, 9, 43–52.

30 Lindley, D.V. (1957). A statistical paradox. *Biometrika*, 44, 187–192.

31 Meng, X. L. (1994). Posterior predictive *p*-values. *Annals of Statistics*, 22(1), 1142–1160.

32 McKenzie, D., and Spears, T. (2014). The formula that killed Wall Street: the Gaussian copula and modelling practices in investment banking. *Social Studies of Science*, 44, 393–417.

33 Mikosch, T. (2006). Copulas: tales and facts. *Extremes*, 9, 3–20.

34 Nicoloutsopoulos, D. (2005). Parametric and Bayesian non-parametric estimation of copulas, Ph.D. Thesis.

35 Ntsoufras, I. (2009). *Bayesian Modeling Using WinBUGs*. London - New York: John Wiley & Sons, Inc.

36 Okhrin, O., and Ristig, A. (2014). Hierarchical Archimedean copulae: the HAC package. *Journal of Statistical Software*, 58, 4.

37 Rubin, D. B. (1984). Bayesianly justifiable and relevant frequency calculations for the applied statistician. *Annals of Statistics*, 12, 1151–1172.

38 Salmon, F. (2012) The formula that killed Wall Street. *Significance*, 9(1), 16–20.

39 Schwarz, G. E. (1978). Estimating the dimension of a model. *Annals of Statistics*, 6(2) 461–464.

40 Shemyakin, A., and Youn, H. (2000). Statistical aspects of joint-life insurance pricing. *1999 Proceedings of American Statisical Association*, 34–38.

41 Shemyakin, A., and Youn, H. (2001). Bayesian estimation of joint survival functions in life insurance. In: *Bayesian Methods with Applications to Science, Policy and Official Statistics*, 489–496. European Communities.

42 Shemyakin, A., and Youn, H. (2006) Copula models of joint last survivor insurance. *Applied Stochastic Models of Business and Industry*, 22(2), 211–224.

43 Spiegelhalter, D. J., Best, N. G., Carlin, B. P., and van der Linde, A. (2002). Bayesian measures of model complexity and fit (with discussion). *Journal of the Royal Statistical Society, Series B*, 64(4), 583–639.

44 Trafimow, D. and Marks, M. (2015). Editorial. *Basic and Applied Social Psychology*, 37(1), 1–2.

45 Trivedi, K. S. (2008). *Probability and Statistics with Reliability, Queueing and Computer Science Applications*. London/New York: John Wiley & Sons, Inc.

46 Volinsky, C.T., and Raftery, A.E. (2000). Bayesian information criterion for censored survival models. *Biometrics*, 56, 256–262.

47 Wasserstein, R. L., and Lazar, N.A. (2016). The ASA's statement on p-values: context, process, and purpose. *American Statistician*, 20(2), 129–133.

48 Wu, J., McHenry, S., and Quandt, J. (2000). An application of Weibull analysis to determine failure rates in automotive components, *NHTSA*, U.S. Department for Transportation, Paper No.13-0027.

Exercises

7.1 Calculate the values of association parameter corresponding to Kendall's $\tau = 0.5$ for Clayton's, Frank's, and Gumbel–Hougaard's families. Use it to calculate values $C_\alpha(u = 0.1, v = 0.1)$ and $C_\alpha(u = 0.9, v = 0.9)$ for all three classes. Compare.

7.2 Verify all results obtained in Table 7.4.

7.3 Develop parametric models for the adjusted residual returns for daily index values of IGBM Madrid Stock Exchange and JSE Africa stock indexes in the file *IGBMJSE.xlsx* or use empirical distributions to estimate for both separately the probabilities of one-day drop by one or more standard deviations.

7.4 Based on the models of the marginals from the previous problem, estimate copula parameters for (a) Gaussian copula model; (b) Clayton copula model. Use any of the statistical methods discussed in Chapter 7. Explain your choice of the method, and the choice of the copula between (a) and (b).

7.5 Using results of the previous problem, estimate probability of simultaneous one-day drop of both indexes by one or more standard deviations.

7.6 Suggest and implement a Bayesian model for data in Table 7.1, based on Clayton copula and non-informative priors.

8

International Markets

8.1 Introduction

One of the important fields of research in modern financial mathematics and risk management deals with stock indexes. Global, regional, and national stock indexes and index futures contracts serve as instruments for hedging and diversification in the international markets. Statistical modeling of the joint behavior of stock indexes has been of special interest recently because such models can be used directly to hedge complex multinational investment portfolios, see Sharma and Seth [34].

Portfolio diversification can be attained through taking positions in futures, which are indexed through geographically or economically remote markets. In the classical Markowitz model this remoteness is modeled using an insignificant or even negative correlation between the national indexes. However, considering modern financial data, correlation analysis often proves to be insufficient because linear correlation, which fits perfectly dependence in multivariate normal models, poorly describes nonnormal joint distributions, since they allow for such deviations as asymmetry, heavy tails, and nonlinear dependence of distribution components, see [9], [10], or [30]. That is why we are especially interested in the analysis of tails of joint distributions, see Fortin and Kuzmics [12], as it helps us to assess the probabilities of several stock indexes plummeting simultaneously, which could cause global markets to collapse and inflict great losses on the multinational investment portfolios, see Hodrick and Xiaoyan [17] or Su [37].

Copula modeling makes it possible to release the assumption of normality and thus to avoid the limitations of the correlation analysis. Using copula models allows us to consider more realistic models of asymmetric marginal distributions with heavy tails as well as more adequate models of nonlinear dependence between the components of the multivariate distribution.

Practical risk management challenges require the construction of high-dimensional copula models. This leads to certain fundamental and technical problems described by Embrechts and Hofert in [8]. These problems are

Introduction to Bayesian Estimation and Copula Models of Dependence, First Edition.
Arkady Shemyakin and Alexander Kniazev.
© 2017 John Wiley & Sons, Inc. Published 2017 by John Wiley & Sons, Inc.
Companion Website: http://www.wiley.com/go/shemyakin/bayesian_estimation

associated with theoretical aspects as well as with numerical implementation of multivariate copula models. Building multivariate copula models requires, first and foremost, the definition of the hierarchy structure. As of today, three multivariate constructions introduced in Section 6.7 have gained most popularity: vine copulas described by Aas et al. [1], see also [6] and [7]; hierarchical Archimedean copulas (HAC) discussed in detail by Hofert et al. in [18] and [19], see also [32]; and hierarchical Kendall copulas (HKC) recently introduced by Brechmann [5]. All these multivariate models are based to some extent on pair copula constructions [14]. The most important question to address in building multivariate copula models is the choice of parametric class of pair copulas and the methods of parametric estimation.

The process of construction and validation of multivariate parametric dependence models for stock index data may be divided in four stages:

- preliminary data cleaning and transformation,
- selection of parametric models for marginals and estimation of marginal parameters,
- selection of parametric pair copula models and estimation of association parameters,
- selection of hierarchical structure for multivariate model based on pair copulas and estimation of hierarchical parameters.

We will discuss the first three stages of this construction.

Throughout the chapter, we will be driven by the principle that one family of pair copulas should be ideally chosen for the entire model. In the present situation it is certainly not a technical requirement, because most hierarchies and vines discussed in Section 6.7 can handle various types of copulas within one structure. However, this principle reflects the authors' belief that one may expect certain copula families to provide overall better fit for specific applications. If this assumption tested on a massive dataset turns out to be feasible, it certainly would simplify the further studies in related fields. This approach explains the setting of the study, where instead of starting directly with one multidimensional problem, we prefer to consider all possible pairs of variables in search of a universally applicable pairwise dependence structure. It is left to the readers to decide whether this effort is successful.

In Chapter 8 we will use the methods of model selection and parametric estimation covered in Chapter 7. We will also make a substantial use of the code written in the R environment, parts of which are embedded in the text directly, and the rest can be found online on the companion website (see page xxv) at ***http://www.wiley.com/go/shemyakin/bayesian_estimation*** in **Appendices** for Chapter 8 folder. We are far from asserting that the suggested implementation of the basic algorithms is in any sense optimal. There exist some excellent R packages, for example, see [38] and [18], providing effective solutions to a wide variety of copula modeling problems. It is not our goal to provide the most efficient programming solutions. Also, teaching our readers to work in R is not our goal either. The algorithms supporting the study are rather meant to prompt readers to use their own creativity. The book companion website also contains the data illustrating the chapter material and can be used in the exercises. Some of the results used in this chapter were introduced in Kangina et al. [26, 27], Kniazev et al. [28], and Shemyakin et al. [35] and summarized in **results.xlsx**.

The data used in the study are daily settlement values of stock indexes representing 27 national markets (the index abbreviations are given in parentheses): Argentina (MERVAL), Australia (ASX), Austria (ATX), Brasil (BUSP), Canada (TSE 300 Comp), Chile (IPSA), China (SSEC), Czech Republic (PX 50), France (CAC), Germany (DAX), Hong Kong (HSI), Hungary (BUX), India (BSE 30), Indonesia (JKSE), Japan (NIKKEI 225), Mexico (IPC), the Netherlands (AEX), Singapore (STI), USA (S&P 500, SPX), Spain (IGBM), Switzerland (SSMI), Turkey (XU 100), Malaysia (KLSE), Great Britain (FTSU 100), the Republic of South Africa (JSE), Ukraine (PFTS), and two indexes for Russia (MICEX and RTSI). The data covers the period from January 1, 2009 to December 31, 2011. The source is Russian business consulting agency RBC (www.rbc.ru).

Let us describe the first stage of modeling. At this stage we need to clean-up the errors and handle the missing data. Missing data can represent the nontrading days due to national holidays and other country-specific events. Therefore for different countries they may occur at different dates. Eliminating the dates of missing observations from all series brings about a significant reduction in the sample size. That is why imputation is inevitable. We will choose the standard way to address the nontrading days: the blanks will be filled up with the previous observations. Blanks in the beginning of the data will be filled with the first observation present. The series of index values with imputed missing data can be found on the companion website in **indexes.xlsx**.

Before we get to the construction of dependence models between the pairs of indexes, we will perform preliminary data processing, which becomes standard in dealing with stock market series. We will follow the scheme of [2] and [31], see also [24] and [11]. This scheme was briefly described in Section 1.6. Data filtration requires the followings steps:

- transforming daily index values S_t into scale-free logarithmic returns $R_t = \ln S_t - \ln S_{t-1}$,
- addressing autocorrelations via ARIMA, obtaining residuals ε_t,
- addressing heteroskedasticity via GARCH, estimating variance of residuals σ_t^2,
- using the standardized residuals $z_t = \frac{\varepsilon_t}{\sigma_t}$ for copula modeling.

This sequence of actions is justified by the assumption that while the autocorrelations and heteroskedasticity represent major effects in the series of national

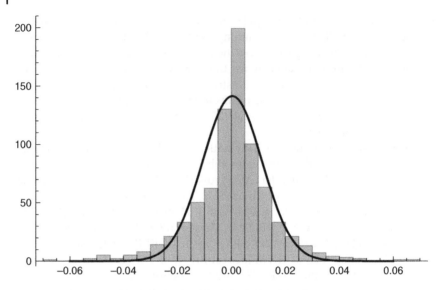

Figure 8.1 Histogram of standardized residuals for SPX and the Gaussian curve.

stock index prices, standardized residuals z_t (further dubbed residual series or z-series) for different countries, being free from scale factor and temporal structure, can reveal a finer effect of interdependence between the national markets.

8.2 Selection of Univariate Distribution Models

As might be suggested by the name z-series, it would be natural to assume normal (Gaussian) distribution for the standardized residuals. Indeed, historically this assumption has been extensively used. However, in the past couple of decades residual series were frequently noticed to obtain heavier tails than the normal model would allow, and also exhibit substantial asymmetry. These features are well illustrated in Figure 8.1.

To verify that normal distribution is not suitable to model residual series we can perform the Kolmogorov–Smirnov goodness-of-fit test. This test can be run for a normal distribution in R using **ks.test** command. The results of this test are provided in **Appendix 8.2**.

In order to model univariate distributions we will use Student's asymmetric t-distribution introduced in Section 1.3, see also [16]. The density of this distribution is defined by formula (1.12), where parameter η corresponds to degrees of freedom and λ determines asymmetry or skewness.

We are going to carry out parametric estimation for asymmetric t-distribution in three stages. First, we will obtain preliminary maximum likelihood estimates separately for all 28 residual series. Then we will use empirical

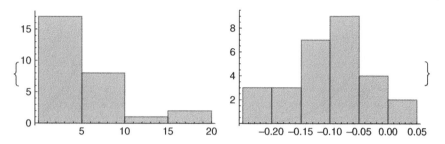

Figure 8.2 Histogram of η_1 and λ_1: transformed parameters of asymmetric *t*-distribution.

Bayes approach to elicit prior distributions for η and λ using the entire dataset. Finally, we will obtain Bayesian parameter estimates using these priors. Using this approach, we may expect some degree of "Bayesian shrinkage," where the most extreme parameter values obtained by MLE will shrink toward the prior means.

In order to obtain the MLE for η and λ, we need to define the likelihood function and perform restricted numerical optimization. Restriction $\eta \geq 2$ eliminates the numerically unstable cases of *t*-distribution, and by construction $\lambda \in (-1, 1)$. We suggest using the code in R provided in **Appendix 8.3**. Function **mlat** from this package can be used to obtain parameter estimates. You can also use the function **optimize**. Maximum likelihood estimates are provided in the fourth and the fifth column of **Appendix 8.2** and their values for all 28 indexes are graphically summarized in Figure 8.2, providing an insight into possible choices of the prior distributions.

In order to obtain more conventional priors, it will be convenient to transform both parameters. Let us consider a linear shift $\eta_1 = \eta - 2$. Judging by the histogram, we can naturally assume a gamma distribution. Let us calculate the sample mean and the sample variance: $\bar{\eta}_1 = 5.47, s^2_{\eta_1} = 17.46$. Based on these estimates we can suggest parameters for the prior $Gamma(\alpha, \lambda)$ distribution: $\alpha = 1.715, \lambda = 0.313$. Let us test the feasibility of gamma prior using the Kolmogorov–Smirnov test. The following code can be used to perform this test in R as an alternative to standard **ks.test** subroutine.

```
Fobb<-ecdf(eta1)
Fob<-Fobb(eta1)
Fm<-pgamma(eta1,shape,rate)
D<-max(abs(Fob-Fm))
s<-0
N<-28
for(k in 1:100){m=((-1)^k)*exp(-2*(k^2)*N*D^2); s=s+m}
P<-(-2)*s
print(P)
```

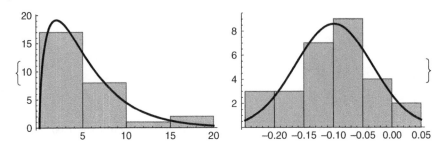

Figure 8.3 Histograms of η_1 and λ_1 with priors.

We obtain p-value $p = 0.70$, so the hypothesis is not rejected and gamma prior is viable.

Let us consider the standard transformation $\lambda_1 = \tan \frac{\pi \lambda}{2}$, which will expand the domain of the asymmetry parameter to $(-\infty, \infty)$ and allow for the use of normal prior. Judging by the histogram we can assume normality. Let us calculate the sample mean and the sample standard deviation: $\bar{\lambda}_1 = -0.102$, $s_{\lambda_1} = 0.065$. The p-value is $p = 0.9829$, and the normal distribution is not rejected, see Figure 8.3.

In order to use independent priors for two parameters, we need to establish the independence of parameters in Bayesian setting. To this end we will test the hypothesis of no correlation using sample Pearson's correlation (see verbatim inset from R):

```
Pearson's product-moment correlation
data:  eta1 and lambda1, t = -1.1048, df = 26,
p-value = 0.2794
alternative hypothesis:
true correlation is not equal to 0
95 percent confidence interval:  -0.5420091  0.1751643,
sample estimates:  cor = -0.2117505
```

Reviewing results of the test we can conclude that linear correlation between two parameters is not significant, see Figure 8.4. That is why we may consider the product of two distributions: $Gamma(1.715, 0.313)$ and $N(-0, 102, 0.065^2)$, a reasonable joint prior for η_1 and λ_1.

At the final stage of estimation we hope to improve preliminary MLE of the parameters of the marginal parameters using Bayes estimation with elicited priors. Due to intractability of the posterior, we will use random walk Metropolis algorithm (RWMA) discussed in Section 4.3. One could also use independent Metropolis algorithm, but its requirement to specify overdispersed proposals for two parameters makes the random walk algorithm more attractive. As a

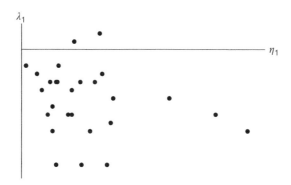

Figure 8.4 Scatterplot of MLE for η_1 and λ_1.

proposal distribution for RWMA we will take, as usual, a bivariate normal distribution with independent components and means equal to the previous values of the parameter. As we mentioned in Chapter 4, proper scaling (choosing the variance term) for the proposal is an important tool of algorithm calibration. The acceptance rate and therefore the rate of convergence depends on the scaling parameter. There are no firm rules dictating how to choose the scaling parameter, so it can be adjusted to achieve a reasonable acceptance rate. As a rule of thumb, one-half of the prior variance worked reasonably well for our application.

To implement this algorithm in R we need, first of all, to calculate the likelihood function, and then the acceptance ratio. Let us recall that the ratio used for updates in RWMHA is as follows:

$$R(\theta_t, \theta_{t-1}) = \frac{L(z, \theta_t)\pi(\theta_t)}{L(z, \theta_{t-1})\pi(\theta_{t-1})}, \tag{8.1}$$

where $\theta = (\eta_1, \lambda_1)$ is the vector of parameters updated on one step as a single block; θ_t is the proposed parameter value for step t, θ_{t-1} is the previous state of the chain, $L(z, \theta)$ is the likelihood function, and $\pi(\theta) \sim Gamma(1.715, 0.313) \times N(-0, 102, 0.065^2)$ is the prior density. While writing this algorithm we tried to minimize the number of operations. As a result, we have compiled the code in **Appendix 8.4**. The acceptance counter (ACRA) is built into the algorithm, which allows for in-process diagnostics. In Chapter 4 we have discussed the convergence issues for RWMA, therefore we apply a generous burn-in allowance, up to 50% of the resulting chain. The results of the algorithm's performance are summarized in columns 6–9 of **Appendix 8.2**. We can see that the acceptance rate is quite high for each series: from 26.7% to 59%. This indicates satisfactory mixing performance of the algorithm. According to Kolmogorov–Smirnov goodness-of-fit test applied to the posterior estimates, hypothesis of asymmetric t-distribution is not rejected for any series with the 0.01 level of

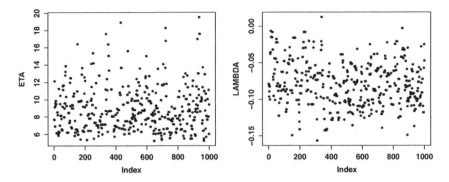

Figure 8.5 RWMHA chain for the residuals of FTSU100 (acceptance rate 57.6%).

significance. Corresponding p-values for the hypothesis of normality are much lower. Bayesian estimates turned out to be less scattered than MLE, which illustrates the shrinkage effect.

The behavior of the RWMHA chains is illustrated in Figure 8.5, where both parameters are transformed back to initial η and λ. We observe reasonably good mixing behavior of the chain. The trace plot indicates that the chains run across the entire state space and spend most of the time in the region of high posterior values.

8.3 Prior Elicitation for Pair Copula Parameter

We will now move on to the analysis of pairwise association between the standardized residuals (z series). In this section we are not going to provide the best or the most efficient solutions. On the contrary, our goal is to provide a wide variety of methods which can be used in practice for prior elicitation and to demonstrate the viability of these methods. The standard approach is to apply simple functional transformation to copula parameters, which would allow for the use of conventional priors (normal, Gamma, or Beta families). This approach facilitates the implementation of MCMC algorithms.

A total of 378 pairs can be formed from 28 index residuals. That should provide sufficient information to study regularities in the pairwise dependencies. Some of the scatterplots of residual pairs are presented in Figure 8.6.

In Figure 8.6 we do not observe substantial deviations from elliptical symmetry. Therefore we may expect elliptical models (Gaussian and t-copula) to be suitable. However, emphasis on the distribution tails suggests testing Archimedean copulas as an alternative. We will consider six models: Clayton, Frank, Gumbel–Hougaard, Gumbel–Hougaard survival copula, Gaussian, and Student's t-copulas, the first four from Archimedean family, and the last two

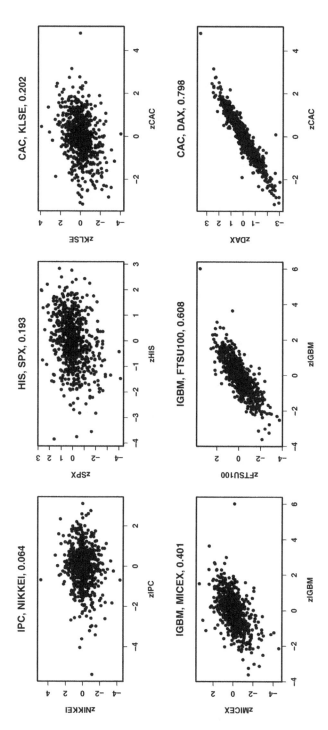

Figure 8.6 Scatterplots of z-series (with Kendall's $\hat{\tau}$).

from elliptical class. In Chapter 6 all of these copula functions were introduced, including formulas for their densities and other important properties.

In order to apply Bayesian approach to the estimation of copula parameters, the first step is to elicit prior distributions. Empirical Bayes method suggests estimating the copula parameter of association between standardized residuals separately for each pair of indexes, and then using these results to estimate the distribution of association parameter across the set of pairs. One may consider kernel density estimation to get nonparametric estimates of the prior, but we will prefer the convenience of parametric approach allowing for the standard choice of priors. With 378 different pairs we have the luxury of obtaining reasonable priors out of conventional families.

To obtain primary parametric estimates to use for the prior elicitation in case of Clayton, Frank, and Gumbel–Hougaard copulas, we can use the method of moments or, to be precise, the method of inversion of the sample Kendall's concordance $\hat{\tau}$. Values of $\hat{\tau}$ can be obtained in R via command **cor(x,y, method="kendall")**. Sample concordances for each pair are summarized in Appendix **results1.xlsx**. Let us point out that for all pairs of indexes these values are positive, which gives evidence of positive association of all residuals. This observation is going to be used in the further analysis. In particular, it lets us assume that Clayton copula is defined by the formula:

$$C_\alpha(u, v) = u^{-\alpha} + v^{-\alpha} - 1, \alpha \geq 0, \tag{8.2}$$

which does not involve the "maximum" operator and simplifies calculations relative to the general case (6.22).

The formulas that connect copula parameters and Kendalls correlation coefficients for Clayton, Gumbel–Hougaard, and Gaussian copulas reverting (7.15), (7.16), and (7.19) are very simple. Let us recall these formulas:

Clayton copula:

$$\alpha = \frac{2\tau}{1 - \tau}. \tag{8.3}$$

Gumbel–Hougaard copula:

$$\alpha = \frac{1}{1 - \tau}. \tag{8.4}$$

Gaussian copula:

$$\alpha = \rho = \sin\left(\frac{\pi\tau}{2}\right). \tag{8.5}$$

Let us point out that for Gaussian and t-copula the parameter of association $\alpha = \rho$ is also the Pearson's correlation coefficient.

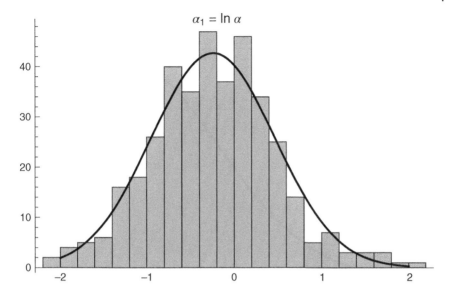

Figure 8.7 Clayton copula parameter $\alpha_1 = \ln \alpha$.

Let $\alpha_1 = \ln \alpha$ denote the natural logarithm of association α for Clayton copula. The histogram of this variable for the set of all pairs of index residuals, presented in Figure 8.7, was obtained by calculating $\hat{\tau}$ for each pair and then using (8.3) and logarithmic transform to obtain a sample of α_1. Judging by this graph, we can assume that α_1 has a normal distribution and further use the method of moments.

The sample mean of α_1 for the set of all 378 pairs is -0.241, the sample standard deviation of α_1 is 0.706. Let us test the hypothesis of α_1 being normally distributed.

```
One-sample Kolmogorov-Smirnov test
data:   alpha1
D = 0.0364, p-value = 0.6999
```

We can see that the hypothesis is not rejected and thus lognormal prior distribution $\ln \alpha \sim N(-0.241, 0.706)$ can be used for Clayton copula parameter.

The same approach will be used for other copulas. Let us consider transformation $\alpha_2 = \frac{1}{\alpha}$, where α is the parameter of Gumbel–Hougaard copula. With α having domain $[1, \infty)$, α_2 is defined on the finite interval $(0, 1]$. We will assume that this variable has beta distribution $Beta(a, b)$, see Figure 8.8. We have already discussed this distribution in Section 1.5 and it reappeared in Chapter 2 as a standard conventional prior. The sample characteristics of this

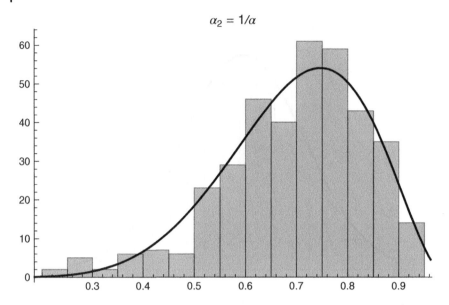

Figure 8.8 Gumbel–Hougaard copula parameter.

variable are *mean* = 0.699, *var* = 0.018, and the sample estimates of beta distribution parameters are $a = 7.322, b = 3.145$. Performing goodness-of-fit test, we obtain p-value $p = 0.667$. Thus, the prior $Beta(7.322, 3.145)$ can be used for Gumbel–Hougaard copula parameter. Let us point out that Gumbel–Hougaard survival copula can be assigned the same prior.

Beta distribution $Beta(a, b)$ is also suitable for Gaussian copula parameter. The parameters are $a = 3.055, b = 3.835$, the p-value of Kolmogorov–Smirnov test is $p = 0.78$, see Figure 8.9.

For Frank copula the dependence between τ and α is more complex. It is difficult to use directly (7.17) or its approximation introduced by Gordeev et al. [15]:

$$\tau = 1 + \frac{2\pi^2}{3\alpha^2} - \frac{4}{\alpha} + \frac{4}{\alpha} \ln\left(1 - e^{-\alpha}\right) - \frac{4}{\alpha^2} \sum_{k=1}^{\infty} \frac{e^{-\alpha k}}{k^2}. \tag{8.6}$$

It is less than straightforward how to express α from this formula through τ. For values of $\tau < 0.5$ we can use a rough linear approximation to transform $\ln \tau$ into $\ln \alpha$. Based on this approximation, the primary sample of α values can be generated from $\hat{\tau}$ values. Let us use the fact that all concordance values in our sample are positive. The histogram of this sample is shown in Figure 8.10.

Judging by Figure 8.10 we assume that α may be assigned a gamma distribution. We have considered this distribution in Section 1.3 and in

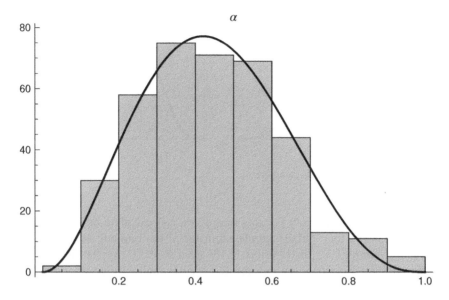

Figure 8.9 Gauss copula parameter.

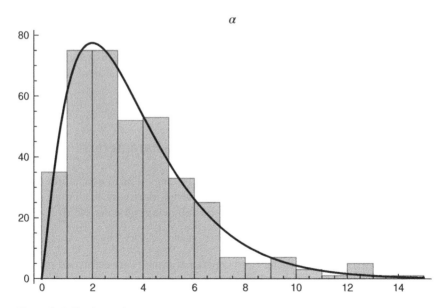

Figure 8.10 Frank copula parameter.

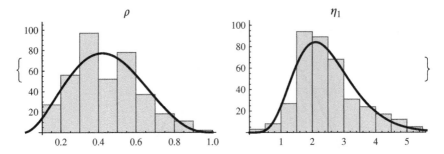

Figure 8.11 Student's *t*-copula parameters.

multiple examples later on. For this sample we can calculate *mean* = 3.655, *var* = 6.043, and then obtain sample estimates of the rate and scale parameters of *Gamma*(a, λ) distribution (1.18) as $a = 2.211$, $\lambda = 0.605$. Having performed the Kolmogorov–Smirnov test we get high *p*-value $p = 0.88$. So *Gamma*(2.211, 0.605) prior will be used for Frank copula parameter.

Student's *t*-copula has two parameters: correlation ρ and degrees of freedom η. However, ρ for *t*-copula has the same meaning as Gaussian copula parameter. For this parameter we use a beta prior *Beta*(3.204, 3.951). To obtain primary sample of the degrees of freedom we use the maximum likelihood method. Let us consider variable $\eta_1 = \ln(\eta - 2)$. The histogram of this variable is illustrated in Figure 8.11.

Let us check the hypothesis of this variable having gamma distribution with parameters $\alpha = 6.649$, $\lambda = 2.686$ (MLE). We obtain the *p*-value of 0.34, thus *Gamma*(6.649, 2.686) is a feasible prior for η_1.

8.4 Bayesian Estimation of Pair Copula Parameters

In Chapter 7 we have considered several different versions of estimation procedures for copula models. Most of them involved maximum likelihood estimation (MLE). We have distinguished the following versions of MLE (see [22] and [23]):

- Full MLE (marginal parameters and copula parameters are estimated simultaneously, in one step)
- IFM MPLE (inference from margins, pseudo-likelihood: parametric model for margins, then parametric estimation of copula parameters, in two steps)
- IFM CMLE (inference from margins, canonical MLE: empirical estimates for margins, then parametric estimation of copula parameters, in two steps)

The same three possibilities are open for Bayesian estimation. In Section 7.3 we considered an example similar to IFM MPLE: parametric estimates for margins and then parametric estimation for association. The difference is that on both steps Bayes estimates with noninformative priors are obtained instead of MLE. In Section 7.5 we considered a Bayesian analog of full MLE with informative priors for marginal parameters. In Section 7.6 we used IFM CMLE and applied Bayesian approach not to parametric estimation, but to model selection.

For the market index study we will use a two-step parametric IFM MPLE procedure, where empirical Bayes priors will be used both for marginal parameters and for the copula parameters. In the first case we use 28 indexes to elicit priors for asymmetric t-marginals, and in the second case we use 378 pairs of indexes to obtain priors for the association. For the sake of comparison we will also use semiparametric IFM CMLE with empirical estimates for marginal distributions.

To obtain Bayes estimates of pair copula parameters, we suggest using RWMHA. Our arguments for this are the same as in the previous section. Intractable posteriors do not allow for analytical solutions and require MCMC approach. Gibbs sampling is not universally applicable, because most of the conditional distributions obtained by fixing components of the vector parameter are also intractable. RWMHA algorithm allows us to adjust the acceptance rate by means of selecting the scaling parameter σ^2 (variance of the proposal distribution).

Let us look into the details of the random walk algorithm, as well as its realization in R environment using an example of Frank copula. First of all, the logarithm of this copula's density function must be written down.

```
ldfr<-function(u,v,a){
fa<-log(a-a*exp(-a))-a*(u+v)-log((exp(-a)-1+
(exp(-a*u)-1)*(exp(-a*v)-1))^2)
return(fa)
}
```

It is a function of two marginals: u and v and of the association parameter a.

All the further steps of the algorithm are merged in one function.

```
metrofr<-function(u,v,n,sigma,b0,shape,rate){
b<-b0
acra<-0
B<-0
for(k in 1:n){
r<-rnorm(1,log(b),sigma)
a<-exp(r)
LLH<-sum(ldfr(u,v,a)-ldfr(u,v,b))
```

```
LH<-exp(LLH)
G<-((a/b)^shape)*exp(rate*(b-a))
if(LH*G>runif(1)){
b<-a
acra=acra+1
}
B<-B+b
}
A<-B/n
Acra<-acra/n
res<-c(A,Acra)
return(res)
}
```

Let us describe the arguments of this function. Variables u and v correspond to the marginals. Parameter b_0 is the initial value of the chain. Preliminary method of moments estimates of the association parameter can be used as b_0. The *shape, rate* parameters are the parameters of the prior gamma distribution evaluated in the previous section. In our case they are the same for all pairs: rate is $\lambda = 0.605$, shape is $a = 2.211$. Parameter n is the number of steps of the algorithm (length of the chain). As the chain does not reach its stationary state quickly, its length has to be large, 100,000 and more. Finally, the scaling parameter *sigma* (the standard deviation of the normal proposal) is used to calibrate the chain to reach the desired acceptance rate which is usually between 25% and 50%.

At every step t function *metrofr* generates a random number a from normal distribution with the mean equal to the logarithm of the previous accepted value and standard deviation *sigma*. Then the ratio LH of the likelihood functions and the ratio G of the corresponding prior densities are calculated for the new state proposed for time t and the previous value from state $t - 1$. The product of these two ratios is compared to a random uniform number from the interval $[0, 1]$. If this ratio is larger, then the new proposed value is accepted. Otherwise, the previous value from step $t - 1$ is accepted as the chain value for step t. The values b of the generated chain are averaged (the value A) and the number of accepted values is counted (the *acra* counter). As a result, two numbers form the output: the estimate of the association parameter A (the mean value of the generated chain serving as an estimate of the posterior mean) and the acceptance rate *Acra*.

In order to illustrate the operation of the algorithm, let us apply it to estimation of Frank copula parameter for the residual series of two indexes: CAC (France) and HIS (Hong Kong). Let us use fully parametric approach, that is, according to Section 8.2, assume asymmetric t-distribution for the marginals

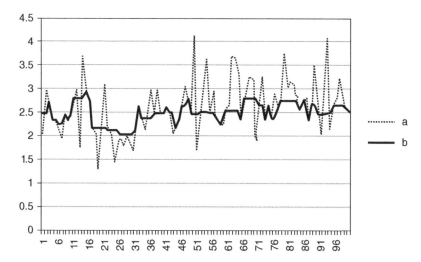

Figure 8.12 Proposed and accepted chain values. Sigma = 0.2, Acra = 0.49, A = 2.496

u and *v*. The marginal parameters for the selected series can be found in **Appendix 8.2**. We recommend using the **pat** function from **Appendix 8.3** to obtain the values of *u* and *v*. After that, we can import functions **ldfr** and **metrofr** into the workspace. Now we are ready to use the function **metrofr**. We will choose a relatively short chain in order to clearly illustrate the results. Evidently, for practical application we would need a much longer chain.

Let us first select the scaling parameter *sigma* = 0.2. The proposed values (dashed line) and the accepted values (solid line) are shown in Figure 8.12. The acceptance rate is 52%, the sample mean *A* estimating the posterior mean is 2.530, which gives us the Bayes estimate of the Frank copula parameter. In Figure 8.13, you can see the same series with calibration *sigma* = 1. In this case, the acceptance rate is 13% and the value of Bayes estimate *A* is 2.595. In Figure 8.13, the solid line consists almost entirely of horizontal segments, which might signify a bad mixing of values in the chain indicating low acceptance rate. This can affect the numerical estimate. In Figure 8.12, the chain does not stick at the same state for long periods of time, therefore the effect of autocorrelations slowing down the convergence is less manifested. Both chains however are fairly close to the method of moments estimate = 2.483, though for such a short chain it might well be a coincidence.

The algorithms in R used to calculate Bayesian parameter estimates of pair copula models are provided in **Appendices 8.5–8.9**. However, it is important to notice that these algorithms are adjusted to the particular problem we consider. Gaussian proposal or its parameter σ can be modified. The change of prior

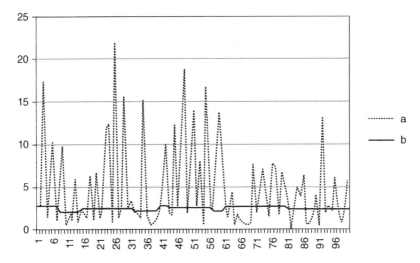

Figure 8.13 Proposed and accepted chain values. Sigma = 1, Acra = 0.12, A = 2.485.

denoted by *G* and likelihood *LH* should be made reflecting the specific problem. The algorithms given in the Appendices are already adjusted to apply to the end-of-the-chapter exercises.

The results of parameter estimation for all copulas and all pairs of indices are provided in **results1.xlsx**. They include the estimates obtained via both parametric and semiparametric (using no parametric model for marginals) methods. We performed 10,000 iterations for the Student's *t*-copula and 100,000 for all the other cases. This difference can be explained by *t*-copula having the most difficult density function of all the six models we used. Therefore, the generation of the chain might take more time than is needed for an illustration. A stable decrease in the acceptance rate from 90% to 40% is noticeable as the strength of dependence increases in all copulas with the exception of *t*-copula, for both parametric and semiparametric methods.

8.5 Selection of Pair Copula Model

There exist a variety of ways to compare the performance of pair copula models and select the best candidates from the list of viable alternatives [36]. According to the principles discussed in Chapter 7, we can use distances of the model characteristics from similar characteristics of empirical distribution or information criteria. In fully Bayesian setting one may argue for the use of DIC or Bayesian model selection. In this section we are going to present just a few from a wide range of options. We will consider three characteristics: model-induced Kendalls correlation coefficient, model distribution function (c.d.f.), and the model value of the tail parameter. For each of these characteristics we will calculate the empirical analog, and calculate the distances from the model to empirical value taken over the entire set of pairs and models. The sum of squared deviations of the model characteristic from the corresponding empirical characteristic is calculated:

$$RSS_\tau = \sum_{1 \leq i < j \leq 28} \left(\tau_{ij}^{(emp)} - \tau_{ij}^{(mod)} \right)^2, \tag{8.7}$$

$$RSS_{DF} = \sum_{1 \leq i < j \leq 28} \left(\sum_{k=1}^{780} \left(F_{ij}^{(emp)} \left(z_k^{(i)}, z_k^{(j)} \right) - F_{ij}^{(mod)} \left(z_k^{(i)}, z_k^{(j)} \right) \right)^2 \right), \tag{8.8}$$

$$RSS_\lambda = \sum_{1 \leq i < j \leq 28} \left(\lambda_{ij}^{(emp)} - \lambda_{ij}^{(mod)} \right)^2. \tag{8.9}$$

Let us begin with Kendall's τ, which is the easiest to evaluate. Model values of Kendall's concordance $\tau^{(mod)} = \tau(\alpha)$ are calculated using formulas (7.15), (7.16), (7.17), and (7.19) for Bayes estimates of association α, and sample concordance $\tau^{(emp)} = \hat{\tau}$ according to (5.5). Sample and model concordance for each pair of indexes and copula family are summarized in **results.xlsx**. The values of sample Kendalls correlation coefficients and of model coefficients for selected pairs of indexes are given in Table 8.1. For illustration purposes we chose five pairs of indexes representing five levels of association, from very weak (IPC

Table 8.1 Sample and model values of τ for selected pairs

Copulas	IPC NIKKEI	CAC KLSE	IGBM MICEX	IGBM FTSU100	CAC DAX
Sample tau	0.064	0.202	0.401	0.608	0.798
Semiparametric models					
Gumbel	0.070	0.178	0.360	0.566	0.776
Clayton	0.079	0.170	0.317	0.494	0.705
Frank	**0.064**	**0.203**	**0.403**	0.601	0.791
Gauss	0.100	0.157	0.255	0.393	0.562
Student *t*	0.081	0.252	0.392	**0.601**	**0.792**
Parametric models					
Gumbel	0.071	0.184	0.374	0.578	0.783
Clayton	0.077	0.170	0.319	0.494	0.715
Frank	**0.067**	**0.208**	0.412	**0.608**	**0.798**
Gauss	0.101	0.160	0.257	0.396	0.568
Student *t*	0.075	0.257	**0.401**	0.607	0.799

and NIKKEI), weak (CAC and KLSE), moderate (IGBM and MICEX) to strong (IGBM and FTSU 100) to very strong (CAC and DAX). The latter is the canonical example used in many other studies, see, for example [15].

All copula models tend to overestimate sample concordance for the pairs with weaker association and most of them underestimate sample concordance for the pairs with stronger association. Frank and Student copulas provide the best fit for this particular subset of the study. Notice that Gumbel–Hougaard and its survival version share the same model value of τ.

Another important indicator of the model's fit is the sum of squared deviations between the empirical and model distribution functions calculated at the sample points (RSS_{DF}). This indicator is subject to overfitting, but in our case all models have similar level of complexity within each of the two separate groups: semiparametric or parametric models. Comparison between the groups will indubitably give preference to semiparametric models.

This method of model comparison is technically more labor intensive. First, the values of bivariate empirical cumulative distribution function have to be calculated for each pair of indexes for each observation. For each observation $(z_j^{(1)}; z_j^{(2)})$ we have to calculate the number of such $(z_k^{(1)}; z_k^{(2)})$ that $z_k^{(1)} \leq z_j^{(1)}, z_k^{(2)} \leq z_j^{(2)}$. Then this number is to be divided by the total number of observations n or rather by $n + 1$ as we did in Chapter 7.

Model distribution functions are calculated for Archimedean copulas by direct u and v values for marginals into the expression for corresponding copula function. Depending on parametric or semiparametric approach, u and v are obtained either from the model asymmetric t-distribution or directly from marginal empirical distributions. Notice that the values of model distribution functions are different for Gumbel–Hougaard and survival Gumbel–Hougaard copulas.

More technical work is required to deal with the members of elliptical family: Gaussian copula and t-copula. These values have to be calculated from intractable integrals therefore numerical integration is in order. Monte Carlo methods provide a very logical avenue. The values of RSS_{DF} for all pairs of indices and for all the copulas are provided in **results1.xlsx**. The values of this quality indicator for selected pairs are summarized in Table 8.2.

Judging by the sum of squared deviations RSS_{DF}, the accuracy of model fit usually decreases as the dependence increases. Student copulas behave decently for all five selected pairs. However overall fit might not be the best indicator if special attention is paid (as often happens) to the distribution tails. In this case, the lower (left) tails may be of primary importance since they describe probabilities of a substantial simultaneous drop of two indexes represented by a pair copula model.

A rough empirical approximation of the tail dependence can be calculated in a very simple way. First, one needs to evaluate the number n_t of

Table 8.2 Squared deviations RSS_{DF} for selected pairs

Copulas	IPC NIKKEI	CAC KLSE	IGBM MICEX	IGBM FTSU100	CAC DAX
		Semiparametric models			
Gumbel	0.029	0.061	0.113	0.080	0.033
Survival Gumbel	**0.020**	0.025	0.046	0.061	0.046
Clayton	0.028	0.042	0.182	0.386	0.381
Frank	0.025	0.026	0.043	0.044	0.064
Gauss	0.170	0.162	0.484	0.917	1.232
Student t	0.032	**0.021**	**0.019**	**0.007**	**0.015**
		Parametric models			
Gumbel	0.147	0.136	0.174	0.109	0.129
Survival Gumbel	0.158	0.177	0.227	0.202	0.232
Clayton	0.169	0.237	0.476	0.663	0.644
Frank	0.157	0.108	0.135	0.146	0.197
Gauss	0.305	0.310	0.741	1.116	1.461
Student t	**0.027**	**0.022**	**0.022**	**0.007**	**0.014**

observations falling into the square $[0; 2^{-t}] \times [0; 2^{-t}]$, for $t = 1, 2, \ldots, 10$. Then the simple regression $y = a + bx + \varepsilon$ can be used with $y = n_t/(n+1)$ and $x = 2^{-t}$. For every pair of indexes the intercept a in this regression turns out to be insignificantly different from zero, therefore the slope coefficient $b = \lambda^{(emp)}$ can be used as a rough sample characteristic of the tail dependence.

Let us give an example of the sample estimation of tail parameter for the pair of CAC (France, sample distribution function FempIND[,7]) and HIS (Hong Kong, sample distribution function FempIND[,14]) indexes.

```
y<-tail(FempIND[,7],FempIND[,14])
summary(lm(y~x))
Coefficient: Estimate  Std. Error  t value  Pr(>|t|)
(Intercept)  -0.012     0.006       -1.897    0.094
x             0.619     0.033       18.740    6.79e-08
Multiple R-squared: 0.9777,
Adjusted R-squared: 0.9749
F-statistic: 351.2 on 1 and 8 DF,  p-value: 6.79e-08
```

In this example we obtain sample tail parameter $\lambda^{(emp)} = 0.619$.

The model values of the tail parameter were calculated from the Bayes estimates of copula parameters using formulas (7.22), (7.24), and (7.25). The

Table 8.3 Comparison of copula models

Copulas	RSS_τ	RSS_F	RSS_λ
	Semiparametric models		
Gumbel	0.324	28.812	*
Survival Gumbel	0.324	14.090	**33.352**
Clayton	1.203	41.757	37.352
Frank	**0.002**	19.735	*
Gauss	4.418	123.679	*
Student t	0.015	**10.269**	116.211
	Parametric models		
Gumbel	0.180	55.452	*
Survival Gumbel	0.180	69.176	**37.314**
Clayton	1.250	116.149	44.503
Frank	**0.021**	62.004	*
Gauss	4.213	179.191	*
Student t	0.031	**10.939**	113.192

empirical approximation $\lambda^{(emp)}$ and the model values of the tail parameter for every pair of indexes and every copula are summarized in **results1.xlsx**.

A brief summary of three measures RSS_τ, RSS_{DF}, and RSS_λ for all 378 pairs is put together in Table 8.3.

In terms of general description of the strength of dependence, which is reflected in Kendall's concordance, Frank copula is the best overall for both parametric and semiparametric models confirming results of [15]. It is closely followed by t-copula, Gumbel–Hougaard copula, survival Gumbel–Hougaard copula, and then Clayton copula and Gaussian copula. The use of entire distribution functions provides the largest amount of data for the comparison of the models. This indicator gives a more complex picture than the scalar value of concordance. For both semiparametric and parametric estimates Student's copula is the best with Archimedean copulas lagging behind.

As far as risk management is concerned, forecasting a simultaneous extreme drop of indexes and evaluating the probability of such a drop are of the utmost interest. From this perspective, tail dependence plays the most important role. According to this criterion, the best model is survival Gumbel copula, Clayton copula is slightly worse. Student copula is characterized by two symmetric tails, which makes it less appropriate for modeling the lower tail behavior. The other three copulas do not allow for the tail dependence, which makes comparison impossible. It is also worth mentioning that even the best models significantly underestimate the tail dependence, therefore, copulas with heavier tails might be considered.

8.6 Goodness-of-Fit Testing

In Section 8.5 we compare multiple prior copula models and assess their relative performance without ever addressing the issue of their overall feasibility. Maybe, all these models are bad and should be rejected? Maybe the true model does not belong to the particular classes of copulas we consider?

These are not the most natural questions in Bayesian setting. From Bayesian standpoint, there is no such thing as a "true model," and no useful model should be rejected. We have discussed this difference between classical and Bayesian statistics in Section 7.2. Nevertheless, we can take a formal frequentist approach and carry out some goodness-of-fit testing, which is supposed to tell us if all or some of our models should be rejected from classical point of view. For instance, we may want to use Cramer–von Mises or Kolmogorov–Smirnov test defined for univariate case in Section 7.2. A comprehensive review to goodness-of-fit testing for copulas can be found in Berg [4] or Genest et al. [13]. For most recent development, see Huang and Prokhorov [20].

On this road we may face some serious issues. First of all, there is a problem of multiple comparisons. The standard goodness-of-fit testing procedure with significance level, say, 0.05, when applied multiple times to massive datasets, will randomly reject approximately 5% of perfectly good models, which in case of all 378 pairs we test with 12 different pair copula models amounts to more than forty five hundred models and more than 200 legitimate rejections. To verify these rejections, one has to use a special multiple comparison technique, for example, one developed by Benjamini and Hochberg [3].

Also, there is a technical issue of extending goodness-of-fit tests to multivariate distributions. Even in the bivariate case this extension is not trivial. This problem includes two interrelated aspects. First, as a rule, the distribution of the test statistic in two or more dimensions is not easy to find. It will depend on the hypothetical null distribution and the method of estimation and might require verification of rather complex conditions. That is why critical values of the test statistic often have to be calculated via Monte Carlo method generating multiple samples from the null distribution. This necessity creates considerable computational difficulties. Therefore, if we wanted to use Cramer–von Mises or χ^2 type of tests, we would have to resolve this issue. For the work in this direction we will refer to Fermanian [11] or Kojadinovic and Yan [29].

This issue still stands for Kolmogorov–Smirnov test, based on the maximum distance with the univariate version defined in (7.2). This is the test we will use for our study. Its wide use in one-dimensional setting is explained by a relatively easy derivation of the distribution of test statistic, which does not depend on null distribution. In two dimensions it is no longer distribution free. The distribution of the test statistic cannot be easily found, and its critical points have to be defined by Monte Carlo method. In this case we will also face another problem. The fundamental distinction of the multivariate case from the univariate case is an infinite number of break points of the empirical distribution function. To address both issues further we will follow the approach of Justel et al. [25]. We will confine ourselves to the discussion of the bivariate case, though a multivariate generalization is also possible as presented in [25].

Let the joint distribution of random variables X and Y be defined by copula $C(u, v)$ and marginal functions $F(x)$ and $G(y)$. Let $H(x, y) = C(u, v) = P(X \leq x, Y \leq y)$ denote the joint distribution function. Let us consider a sample of paired observations $(x_i, y_i) \sim H(x, y)$, $i = 1, 2, \ldots, n$, and let $H^*(x, y)$ be their empirical distribution function. The critical statistic $D = \sup_{x,y \in R} |H^*(x, y) - H(x, y)|$ is not distribution free. However, we can introduce Rosenblatt's transform [33], which in bivariate case is defined as

$$U = F(X), V = G(Y|X). \tag{8.10}$$

According to the theorem by Rosenblatt [33], for any choice of $F(x)$ and $G(y)$, transformed random variables U and V are independent and uniformly distributed on [0,1].

Joe [21] introduced a representation of the conditional distribution in the bivariate case as

$$G(y|X = x_0) = \left. \frac{\partial C(u, v)}{\partial u} \right|_{u_0 = F(x_0), v = G(y)}. \tag{8.11}$$

Let $Q^*(u, v)$ denote the empirical distribution function of the transformed sample. Then statistic

$$D^* = \sup_{u,v \in [0,1]} |Q^*(u, v) - uv| \tag{8.12}$$

will be used to test check whether random variables U and V are independent and uniformly distributed on the interval [0,1].

Calculation of the test statistic D^* causes considerable technical difficulties because of an infinite number of break points of the function $Q^*(u, v)$. Apart from that, this statistic is not invariant with respect to the choice of a conditional distribution. Transform $U = F(X|Y)$, $V = G(Y)$ would bring about a different distribution. Nevertheless, the authors of [25] suggested and validated an algorithm which bypasses these difficulties. We will describe this algorithm in general omitting some details.

Let us call a pair (x_j, y_i) an intersection point, if $x_i < x_j$, $y_i > y_j$. Let us also introduce the following notation:

$$D^+(u, v) = Q^*(u, v) - uv, u, v \in [0; 1]. \tag{8.13}$$

Then the following variables are to be calculated consecutively

- $D^1 = \max_{i=1,\ldots,n} D^+(u_i, v_i)$,
- $D^2 = \max_{i,j=1,\ldots,n} \{D^+(u_j, v_i) | u_j > u_i, v_j < v_i\}$,
- $D^3 = \frac{2}{n} - \min_{i,j=1,\ldots,n} \{D^+(u_j, v_i) | u_j > u_i, v_j < v_i\}$,

- $D^4 = \frac{1}{n} - min_{i=1,\dots,n}D^+(1, v_i),$
- $D^5 = \frac{1}{n} - min_{i=1,\dots,n}D^+(u_i, 1).$

The value of the test statistic is obtained as the maximum of these five numbers:

$$D^* = max\{D^1, D^2, D^3, D^4, D^5\}.$$

An implementation of this algorithm in R is provided in **Appendix 8.10**. The function implementing it is **bivar.ks**. Critical values of this statistic are provided in [25] for some significance levels and for some sample sizes up to 300. However, the length of z-series generated by indices is 780. That is why we need to calculate the critical points. To do that, we will generate 100 pairs of independent samples from $Unif[0, 1]$. For each pair we calculate the value of test statistic D^*. Sorting the obtained values in ascending order and choosing the 95th and the 99th values, we get the critical points for the 0.05 and 0.01 levels of significance, respectively. In our experiment these turned out to be 0.061 and 0.067. These critical points correspond to the case of known null distribution $H(x, y)$. If the distribution parameters were estimated additionally, then the critical values will be slightly smaller. To calculate sample test statistic values for the pairs of indexes, we need to execute Rosenblatt's transform. To do that we need to calculate conditional distributions. For the five copulas in our study it is not hard. In the following formulas $u = F(x)$ and $v_0 = G(y_0)$.

Frank copula:

$$F(x|Y = y_0) = \frac{(e^{-\alpha u} - 1)e^{-\alpha v_0}}{(e^{-\alpha u} - 1)(e^{-\alpha v_0} - 1) + e^{-\alpha} - 1}.$$

Clayton copula:

$$F(x|Y = y_0) = v_0^{-\alpha-1} \cdot \left(u^{-\alpha} + v_0^{-\alpha} - 1\right)^{-1/(\alpha-1)}.$$

Gumbel copula:

$$F(x|Y = y_0) = q_0^{\alpha-1}\left(p^\alpha + q_0^\alpha\right)^{1/(\alpha-1)} \cdot exp\left(q_0 - \left(p^\alpha + q_0^\alpha\right)^{1/(\alpha)}\right),$$

where $p = -\ln(u)$, $q_0 = -\ln(v_0)$.

Gumbel survival copula:

$$F(x|Y = y_0) = 1 - q^{\alpha-1}\left(p^\alpha + q_0^\alpha\right)^{1/(\alpha-1)} \cdot exp\left(q_0 - \left(p^\alpha + q_0^\alpha\right)^{1/(\alpha)}\right),$$

where $p = -\ln(1 - u)$, $q_0 = -\ln(1 - v_0)$.

Gaussian copula:

$$F(x|y_0) = \Phi\left(\frac{u - \rho v}{\sqrt{1 - \rho^2}}\right), \tag{8.14}$$

Table 8.4 Results of goodness-of-fit testing

Significance level	0.01		0.05	
Copula	The number rejected	%	The number rejected	%
Gumbel	11	3	33	9
Survival Gumbel	20	5	40	11
Clayton	140	37	213	56
Frank	16	4	21	6
Gauss	124	33	169	45

where $\Phi(x)$ is the function of standard normal distribution. Then we need to substitute the marginal distribution functions into these formulas instead of u and v_0. For t-copula, evaluation of conditional distributions requires numerical integration. We will leave this problem beyond the scope of this application. After performing Rosenblatt's transform, we can implement R procedure **bivar.ks**. Table 8.4 gives the counts and percentages (out of the total of 378) of pairs for which the sample test statistic (8.12) exceeded the critical values.

As we can see, bivariate distribution models built using Frank and Gumbel–Hougaard copulas (both regular and survival versions), prove to be adequate for most of the pairs. At the same time, the models constructed with Clayton and Gaussian copulas are more questionable. Notice that the test statistic (8.12) assesses the overall fit and does not specifically address the strength of dependence or the tail behavior.

8.7 Simulation and Forecasting

In this section we will discuss simulation from joint distributions defined by copulas and marginal distributions. Our main goal in this section is to illustrate the sampling methods, discussed in Section 6.6. Simulation is important as a forecasting tool and often becomes indispensable in predictive analytics. Simulation may also be used for model validation. In this case back-forecasting or **backcasting** may be used to confirm good model performance.

A standard way to assess the forecasting accuracy or predictive quality of a model is to use the mean squared error of the forecast and percentage or relative error. Let $P_1, \ldots, P_m$ denote the values of an index forecast for m periods, and let $Q_1, \ldots, Q_m$ denote the actual values of the index. The mean squared error of the forecast is defined as

$$s = \sqrt{\frac{1}{m} \sum_{k=1}^{m} (P_k - Q_k)^2}. \tag{8.15}$$

The mean squared forecast error divided by the mean index value during the forecast period is called the relative forecast error: $\hat{s} = s/\bar{Q}$. The relative error allows us to compare predictive quality of the forecast for different indexes scale free.

Another simple measure of the predictive quality of a pair copula model is its ability to capture the strength and character of dependence between the variables characterized by marginal distributions. We can define

$$d_\tau = \sum_{1 \le i < j \le 28} \left(\tau_{ij}^F - \tau_{ij}^0 \right)^2, \tag{8.16}$$

where τ_{ij}^F is Kendall's concordance of the values of z-series generated from a given copula model for the indexes numbered i, j; τ_{ij}^0 is the sample concordance calculated for the standardized residuals of the actual data.

The algorithms discussed in Section 6.6 enable generating samples from joint distributions defined by copulas. They all start with simulating copula models to obtain marginals $u_i = F(x_i)$ and $v_i = G(y_i)$. The first steps of these algorithms are copula specific and we will discuss them later. However, all of these algorithms share the common part: suppose u and v are obtained, the final step of each algorithm requires the inverse transformation of marginal distribution functions to obtain $x_i = F^{-1}(u_i)$ and $y_i = G^{-1}(v_i)$. Let us recall that in this study we use asymmetric t-distribution described in Chapter 1 to model the marginals. Calculating the inverse of the t-distribution function presents considerable computational difficulties, since this function does not have an analytical representation. It is possible though to calculate the values of this function using approximate numerical methods. We will use our **qat** function in R environment to do this. This function is defined in **Appendix 8.3**.

Thus we obtain the dependent pairs of (x_i, y_i), where each of the components x_i and y_i forms an z-series for a certain index from our list. Let us assume that values z_k; $k = 1, \ldots, m$ are generated to be used for the forecasting purposes. To forecast the index values we have to revert the order of operations we used to obtain standardized residuals. First, on each step k the value of conditional variance h_k is calculated in accordance with GARCH(1,1) model. Then the residual of the autoregressive model AR(1) is calculated as $e_k = z_k \sqrt{h_k}$. After that, the values of logarithmic returns are calculated recurrently using the equation of AR(1) process: $r_k = a_0 + a_1 r_{k-1} + e_k$. Finally, the index forecast value is obtained: $s_k^F = s_{k-1}^F e^{r_k}$.

Algorithms implementing sampling from distributions defined by elliptical copulas in R environment are quite simple. The algorithm for Gaussian copula is contained in **Appendix 8.11**. It also includes the inverse transformation of z-series to a series of index values.

Two algorithms suggested in Section 6.6 can be used to create samples from distributions defined by Archimedean copulas. The first algorithm uses

Kendall's distribution function $K_C(t)$. General formula (6.31) defines $K_{C_\alpha}(t)$ using the generator of the Archimedean copula C_α. Formulas for Clayton copula (7.34) and Gumbel–Hougaard copula (7.31) were used in minimum distance estimation and model comparison in Chapter 7, $K_{C_\alpha}(t)$ for Frank copula may also be expressed as

$$K_{C_\alpha}(t) = t + \frac{1}{\alpha}(1 - e^{\alpha t})ln\left(\frac{e^{-\alpha t} - 1}{e^{-\alpha} - 1}\right), \tag{8.17}$$

where α is the copula parameter. The second step of this algorithm requires solving nonlinear equations $K_{C_\alpha}(t) = u$ for t. For all the three copulas above it is not possible analytically, but can be done numerically. The algorithm for Clayton copula in R environment is described in **Appendix 8.12**.

An alternative simulation algorithm discussed in Section 6.6 uses Marshall–Olkin construction. This construction requires sampling variable W from a distribution associated with the Archimedean copula through its generator. The inverse of the generator is the Laplace transform of variable W. According to [18], W should be chosen from:

gamma distribution $\Gamma\left(\frac{1}{\alpha}, 1\right)$ for Clayton;
logarithmic distribution with the parameter $\theta = 1 - e^{-\alpha}$ for Frank;
stable distribution $S\left(\frac{1}{\alpha}, 1, \left(cos\left(\frac{\pi}{2\alpha}\right)\right)^\alpha, 0, 1\right)$ for Gumbel–Hougaard.

In all the three cases α is the association parameter.

Random number generators for these distributions are built into R package **copula** [38], or an alternative described in [18]. Tools from these packages can be used to obtain samples from Archimedean copulas. However, most of our readers will not find it hard to write their own forecasting code in R or a different software environment. An algorithm for the Frank copula is suggested as an example in **Appendix 8.13**. This algorithm also includes the inverse transforms of the z series.

Let us discuss the results. Forecasts for 22 days ahead are generated for each pair of indexes and for five copula models. The forecasts for Archimedean copulas are generated using both algorithms. The forecasts are compared to the actual index values for the trading days in January 2012. This data is stored in the **new_indexes.xls** file. Relative forecast error is calculated for each forecast. Then these errors are averaged for all the pairs of indexes to provide a comparison between the copula models and the simulation techniques. The distance between concordance matrices (8.16) is also calculated for each copula as an additional tool of comparison. A brief review of the results is given in Table 8.5.

As we can see, the forecasts made with the Marshall–Olkin method lead to smaller relative errors. This can be due to many factors including the implementation of the steps of the algorithm. Among the copulas the best structural forecast is obtained by Frank copula. Most relative errors stay within 12%, which

Table 8.5 Results of the simulation

	Relative error	d_τ
Gaussian copula	0.119	11.46
Student's copula	0.121	8.89
Kendall's function method		
Clayton copula	0.500	9.73
Frank copula	0.439	11.43
Gumbel–Hougaard	0.394	10.45
Marshall–Olkin's method		
Clayton copula	0.121	12.04
Frank copula	0.125	6.69
Gumbel–Hougaard	0.125	9.71

can be considered acceptable in prediction for 22 trading days. Figures 8.14, 8.15, and 8.16 represent scatterplots of the relative forecast error for pairs of indexes depending on the strength of association.

In Figures 8.14, 8.15, and 8.16 we observe that the relative error differs considerably depending on association. It should be noted that for Archimedean copulas the quality of the forecast tends to increase with the growth of dependence. However, different copulas give the best forecasts for different pairs of indexes. To illustrate this fact we will look into the forecast of AEX (the Netherlands) which was paired with ASX (Australia) for different copulas. The forecast was made using Marshall–Olkin method for the Archimedean copulas.

In Figure 8.17 we can see that Student's t and Frank's copulas give reasonable forecasts, while Clayton's does not. However, this is just an example. The

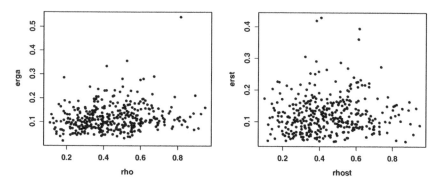

Figure 8.14 Scatterplots of relative forecast error for Gaussian and t- copulas.

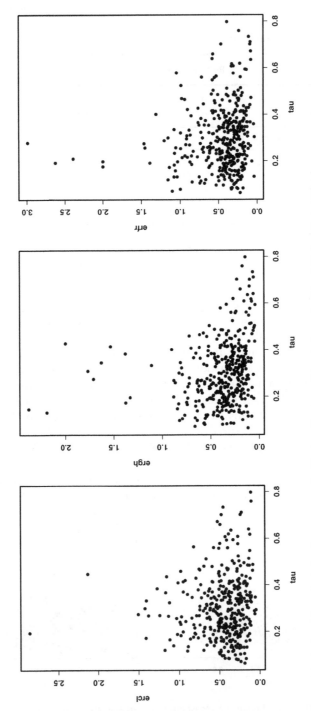

Figure 8.15 Scatterplots of relative forecast error for Clayton, Frank, and Gumbel copulas. Kendall's function method.

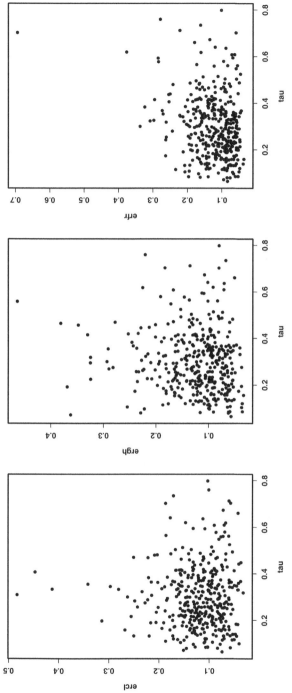

Figure 8.16 Scatterplots of the relative forecast error for Clayton, Frank, and Gumbel copulas. Marshall–Olkin's method.

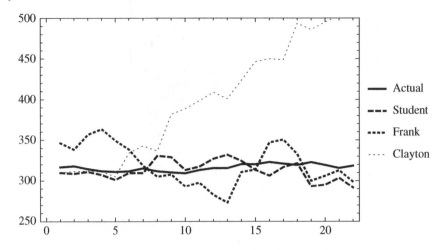

Figure 8.17 Forecasting AEX with Frank, Clayton, and Student *t*-copulas

situation may change dramatically for other pairs of indexes. Definitely, one-dimensional forecasts can help to compare pair copula models, but do not give a good idea of the advantages provided by copulas comparatively to other models of dependence. A better illustration of the copula approach is probably provided by modeling extreme co-movements of two indexes involving estimation of the tails of joint distributions as was done in [26] and [27].

It would be also more informative to depict the joint behavior of two related indexes in one picture. However there is much more going on in such pictures than the patterns of association between two indexes: there are autocorrelations, heteroskedasticity, and other effects, which do not allow us to clearly see the patterns of inter-market association. Besides, analysis of pairwise dependence is definitely not the end of the road. It should be just a bridge to more complex multivariate models. Pair copulas should be seen as building blocks for high-dimensional copula constructions. Nevertheless, without understanding of these blocks entire construction is not possible.

Recent developments in copula studies and further increasing power of modern computers make it possible to build copula models of higher and higher dimensions [8]. However, from the authors' experience, most of practical applications tend to avoid overly complicated models. Therefore pair copulas deserve to remain in the center of attention.

References

1 Aas, K., Czado, C., Frigessi, A., and Bakken, H. (2009). Pair-copula constructions of multiple dependence. *Insurance: Mathematics and Economics*, 44(2), 182–198.

2 Ane, T., Ureche-Rangau, L., and Labidi-Makni, C. (2008). Time-varying conditional dependence in Chinese stock markets. *Applied Financial Economics*, 18, 895–916.

3 Benjamini, Y., and Hochberg, Y. (1995). Controlling the false discovery rate: a practical and powerful approach to multiple testing, *Journal of the Royal Statistical Society, Series B*, 57(1), 289–300.

4 Berg, D. (2007). Copula goodness-of-fit testing: an overview and power comparison. *Statistical Research Report No. 5*, University of Oslo, Oslo.

5 Brechmann, E.C. (2014). Hierarchical Kendall copulas: properties and inference. *Canadian Journal of Statistics*, 42(1), 78–108.

6 Czado, C., Schepsmeier, U., and Min, A. (2012). Maximum likelihood estimation of mixed C-vines with application to exchange rates. *Statistical Modelling*, 12(3), 229–255.

7 Czado, E., Brechmann, C., and Gruber, L. (2013). Selection of vine copulas. In: P. Jaworski, F. Durante, and W. Haerdle, (editors). *Copulae in Mathematical and Quantitative Finance*, Berlin: Springer.

8 Embrechts, P., and Hofert, M. (2014). Statistics and quantitative risk management for banking and insurance. *Annual Review of Statistics and its Application*, 1, 492–514.

9 Embrechts, P., McNeil, A., and Straumann, D. (2003). Correlation and dependency in risk management: properties and pitfalls. In: *Risk Management: Value at Risk and Beyond*, 176–223. Cambridge University Press.

10 Evans, T., and McMillan, D. G. (2009). Financial co-movement and correlation: evidence from 33 international stock market indices. *International Journal of Banking, Accounting and Finance*, 1(3), 215–241.

11 Fermanian, J-D. (2005). Goodness of fit tests for copulas. *Journal of Multivariate Analysis*, 95, 119–152.

12 Fortin, I., and Kuzmics, C. (2002). Tail-dependence in stock-return pairs. *Economics Series 126*, Institute for Advanced Studies.

13 Genest, C., Remillard, B., and Beaudoin, D. (2009). Goodness-of-fit tests for copulas: a review and a power study. *Insurance: Mathematics and Economics*, 44, 199–213.

14 Genest, C., and Rivest, L.-P. (1993). Statistical inference procedures for bivariate Archimedean copulas. *Journal of the American Statistical Association*, 88, 1034–1043.

15 Gordeev, V. A., Knyazev, A. G., and Shemyakin, A. (2012). Selection of copula model for inter-market dependence. *Model Assisted Statistics and Applications*, 7, 315–325.

16 Hansen, B. E. (1994). Autoregressive conditional density estimation. *International Economic Review*, 35, 3, 705–730.

17 Hodrick, R., and Xiaoyan, Z. (2014). International Diversification Revisited. Columbia Business School.

18 Hofert, M., and Maechler, M. (2011). Nested Archimedean copulas meet R: the nacopula package. *Journal of Statistical Software*, 39, 1–20.

19 Hofert, M., and Scherer, M. (2011). CDO pricing with nested Archimedean copulas. *Quantitative Finance*, 11(5), 775–787.

20 Huang, W., and Prokhorov, A. (2014). A goodness-of-fit test for copulas. *Econometric Reviews*, 98, 533–543.

21 Joe, H. (1997). *Multivariate Models and Dependence Concepts*. London: Chapman & Hall.

22 Joe, H. (2005). Asymptotic efficiency of the two-stage estimation method for copula based models. *Journal of Multivariate Analysis*, 94, 401–419.

23 Joe, H. (2014). *Dependence Modeling with Copulas*. London: Chapman & Hall/CRC.

24 Jondeau, D., and Rockinger, M. (2006). The Copula-GARCH model of conditional dependencies: an international stock market application. *Journal of International Money and Finance*, 25, 827–853.

25 Justel, A., Pena, D., and Zamar, R. (1997). A multivariate Kolmogorov–Smirnov test of goodness of fit. *Statistics & Probability Letters*, 35(3), 251–259.

26 Kangina, N., Knyazev, A., Lepekhin, O., and Shemyakin, A. (2015). Bayesian copula models for statistical dependence of national stock indices. In: Proceedings of XV April International Academic Conference on Economic and Social Development, HSE, Moscow, Russia, v. I, pp. 403–413.

27 Kangina, N., Knyazev, A., Lepekhin, O., and Shemyakin, A. (2016). Modeling joint distribution of national stock indices. *Model Assisted Statistics and Applications*, 11, 15–26.

28 Kniazev, A., Lepekhin, O., and Shemyakin, A. (2016). Joint Distribution of Stock Indices: Methodological aspects of construction and selection of copula models. *Applied Econometrics*, 42, 30–53.

29 Kojadinovic, I., and Yan, J. (2010). Comparison of three semiparametric methods for estimating dependence parameters in copula models. *Insurance: Mathematics and Economics*, 47, 52–63.

30 Longin, F., and Solnik, B. (2001). Extreme correlation of international equity markets. *The Journal of Finance*, 56(2), 649–676.

31 Patton, A. J. (2001). Modelling time-varying exchange rate dependence using the conditional copula. *Discussion Paper 2001-09*, San Diego, CA: University of California.

32 Puzanova, N. (2011). A hierarchical Archimedean copula for portfolio credit risk modelling. *Discussion Paper Series 2: Banking and Financial Studies 2011*, 14, Deutsche Bundesbank, Research Centre.

33 Rosenblatt, M. (1952). Remarks on a Multivariate Transformation, *The Annals Mathmatical Statistics*, 23(3), 470–472.

34 Sharma, A., and Seth, N. (2012). Literature review of stock market integration: a global perspective. *Qualitative Research in Financial Markets*, 4(1), 84–122.

35 Shemyakin, A., Kniazev, A., and Lepekhin, O. (2016). Bayesian Model Selection for Hierarchical Copulas and Vines, In: *JSM Proceedings, Section on*

Bayesian Statistical Science. Alexandria, VA: American Statistical Association, 1371–1385.

36 Silva, R. S., and Lopes, H. F. (2008). Copula, marginal distributions and model selection: a Bayesian note. *Statistical Computing*, 18, 313–320.

37 Su, E. D. (2013). Measuring and Testing Tail Dependence and Contagion Risk between Major Stock Markets. *MPRA Paper 48444*, University Library of Munich, Germany.

38 Yan, J. (2007). Enjoy the joy of copulas: with a package *copula. Journal of Statistical Software*, 21(4), 1–21.

Exercises

8.1 Plot the graphs of daily index prices for CAC, HIS, and JSE.

8.2 Plot the graphs of logarithmic returns for CAC, HIS, and JSE.

8.3 Estimate the parameters of AR(1) and GARCH(1,1) for the logarithmic returns of CAC, HIS, and JSE. Plot the graphs of these series.

8.4 Calculate the standardized residuals (z-series) for CAC, HIS, and JSE. Plot the graphs of z-series.

8.5 Obtain MLE of the parameters of asymmetric t-distribution for z-series of CAC, HIS, and JSE. Use function **mlat** from **Appendix 8.3**.

8.6 Obtain Bayes estimates of the parameters of asymmetric t-distribution for z-series of CAC, HIS, and JSE indices. Use the algorithm from **Appendix 8.4**.

8.7 Evaluate Kendall's sample concordance $\hat{\tau}$ for z-series of CAC and SPX.

8.8 Using the method of moments obtain estimates of Clayton, Gumbel–Hougaard, Frank, and Gaussian copula parameters for z-series of CAC and SPX.

8.9 Obtain Bayes estimates of Clayton, Gumbel–Hougaard, Frank, and Gaussian copula parameters for z-series of CAC and SPX using RWMA. Use both semiparametric and parametric approaches. Select such a value of scaling parameter σ that the acceptance rate stays within the range from 25% to 50%. Use the estimates obtained in the previous exercise as the initial values.

8.10 Calculate the model-induced values of Kendall's concordance $\tau(\alpha)$ based on the estimates obtained in Exercise 8.9. Which copula gives the best approximation of τ for the given pair of indexes? Does your conclusion agree with the conclusion for the entire set of indices?

8.11 Calculate the sum of squared deviations RSS_{DF} from the empirical distribution for the model distributions obtained in Exercise 8.9. Which copula gives the smallest RSS_{DF} value for the given pair of indexes? Does your conclusion agree with the conclusion for the whole set of indexes?

8.12 Perform Kolmogorov–Smirnov test for Clayton, Gumbel–Hougaard (regular and survival versions), Frank, and Gaussian copula for z-series of CAC and SPX. Use semiparametric and parametric estimates. Which models can be rejected for this pair of indices with significance levels 0.01 and 0.05?

8.13 Suggest a forecast for the pair of CAC and SPX using Gaussian copula (use the algorithm from **Appendix 8.11**).

8.14 Write an algorithm in R to make a forecast based on t-copula (use the algorithm from Section 6.6). Make a forecast for the pair of CAC and SPX.

8.15 Make a forecast for the pair of CAC and SPX indices using Clayton copula and Kendall's distribution (use the algorithm from **Appendix 8.12**).

8.16 Make a forecast for the pair of CAC and SPX indices using Frank copula and Marshall–Olkins construction (use the algorithm from **Appendix 8.13**).

8.17 Write an algorithm in R to make a forecast based on Gumbel–Hougaard copula and Marshall–Olkin's construction (use the algorithm from Section 6.6). Make a forecast for the pair of CAC and SPX.

8.18 Compare the forecasts for the pair of CAC and SPX obtained using different methods. Which of the forecasts is more accurate? Does your conclusion agree with the one made for the entire set of indexes?

Index

Introduction to Bayesian Estimation and Copula Models of Dependence, First Edition.
Arkady Shemyakin and Alexander Kniazev.
© 2017 John Wiley & Sons, Inc. Published 2017 by John Wiley & Sons, Inc.
Companion Website: http://www.wiley.com/go/shemyakin/bayesian_estimation

Printed and bound by CPI Group (UK) Ltd, Croydon, CR0 4YY

16/04/2025

14658372-0001